Lecture Notes in Computer Science 4534

Commenced Publication in 1973
Founding and Former Series Editors:
Gerhard Goos, Juris Hartmanis, and Jan van Leeuwen

Ioannis Tomkos   Fabio Neri
Josep Solé Pareta   Xavier Masip Bruin
Sergi Sánchez Lopez (Eds.)

Optical Network Design and Modeling

11th International IFIP TC6 Conference, ONDM 2007
Athens, Greece, May 29-31, 2007
Proceedings

 Springer

Volume Editors

Ioannis Tomkos
Athens Information Technology (AIT) Center
Peania, Athens, Greece
E-mail: itom@ait.gr

Fabio Neri
Politecnico di Torino (PoliTO)
Turin, Italy
E-mail: neri@polito.it

Josep Solé Pareta
Xavier Masip Bruin
Sergi Sánchez Lopez
Universitat Politecnica de Catalunya (UPC)
Barcelona, Spain
E-mail: {pareta,xmasip,sergio}@ac.upc.edu

Library of Congress Control Number: 2007927098

CR Subject Classification (1998): C.2, C.4, B.4.3, D.2.8, D.4.8

LNCS Sublibrary: SL 5 – Computer Communication Networks and
Telecommunications

ISSN   0302-9743
ISBN-10   3-540-72729-9 Springer Berlin Heidelberg New York
ISBN-13   978-3-540-72729-3 Springer Berlin Heidelberg New York

Springer is a part of Springer Science+Business Media

springer.com

© IFIP International Federation for Information Processing, Hofstrasse 3, A-2361 Laxenburg, Austria 2007
Printed in Germany

Typesetting: Camera-ready by author, data conversion by Scientific Publishing Services, Chennai, India
Printed on acid-free paper   SPIN: 12068876   06/3180   5 4 3 2 1 0

Preface

The optical networking field is seen to be rapidly emerging with a strong and sustained technological and business growth. Research has led to massive development and deployment in the optical networking space leading to significant advancements in high speed network core and access networks.

The 11th International Conference on Optical Network Design and Modeling brought together scientists and researchers to meet and exchange ideas and recent work in this emerging area of networking.

The conference was sponsored by IFIP and supported by the e-Photon/ONe and COST 291 projects. The conference proceedings have been published by Springer and are also available through the Springer digital library. The conference program featured 14 invited presentations and 41 contributed papers selected from over 90 submissions. A series of sessions focusing on recent developments in optical networking and the related technology issues constituted the main conference program.

The international workshop "Optical network perspectives vs. optical technologies reality" was collocated with ONDM 2007 and took place on May 29th. It was organized by the EU COST 291 action. The objective of the workshop was to focus on cross layer issues and address various challenges with respect to the implementation of optical networking concepts based on available optical technology capabilities.

ONDM 2007 was located in the beautiful and historic city of Athens, Greece, May 29–31, 2007. The conference venue was the well equipped facilities of the Athens Information Technology Center – AIT (www.ait.gr), where participants enjoyed a warm hospitality.

Finally, we would like to thank all those who helped in the organization of this event and especially the distinguished members of the Technical Program Committee and the Local Organizing Committee.

April 2007

Ioannis Tomkos, General Chair
Fabio Neri, General Co-chair
Josep Sole-Pareta, TCP Co-chairs
Sergio Sanchez, TCP Co-chairs
Xavier Masip, TCP Co-chairs
Dimitrios Klonidis, Organizing Committee Chair

Table of Contents

Performance comparison of multi-wavelength conversion using SOA-MZI and DSF for optical wavelength multicast

Jorge del Val Puente[1,2], Ni Yan[1], Eduward Tangdiongga[1], Ton Koonen[1]

[1] Electro-optical communication group, Eindhoven University of Technology, Eindhoven, Netherlands.
[2] Grupo de Comunicaciones Ópticas, Universidad de Valladolid, Valladolid, Spain.
j.del.val.puente@student.tue.nl

Abstract. The electronic layer multicast is going to face the speed and capacity bottleneck of the future optical data networks. Transparent optical wavelength multicast by multi-wavelength conversion is an effective way of achieving data multicast in the optical domain without any optical-electronic-optical conversion. In this paper, two multiple wavelength conversion technologies for 10 Gb/s data rate are investigating and discussed. The first technology is based on cross-phase modulation in a semiconductor optical amplifier – Mach-Zehnder interferometer, and the second is based on four-wave mixing in a dispersion-shifted fiber. We present the simulated performance comparison of two approaches obtained using *VPItransmissionMaker^{TM}WDM* simulator. Afterwards, we analyze these results in comparison with our previous experimental results of the same schemes.

Keywords: Multi-wavelength conversion, multicast, SOA-MZI, FWM, DSF.

1 Introduction

In the last few years, optical layer multicasting development has been continuous. Data multicast at the optical layer avoids the needs for Optical-Electronic-Optic (OEO) conversion, and thus can improve the transparency, efficiency and effectiveness of the optical networks [1]. Moreover, it provides other benefits such as simplification of network layer protocols and optical network designs. Multi-wavelength conversion (MWC) is a simple way to realize optical wavelength multicast [2]. MWC is very attractive because it can potentially reduce the number of converters in a routing node without adding more complexity in the switch design. Thus switches can be easily adapted for optical layer multicasting [3].

Recently, several methods for MWC have been reported and demonstrated. The MWC and multicasting approaches that are studied in this paper are based in semiconductor optical amplifier – Mach-Zehnder interferometer (SOA-MZI) [4], and in four-wave mixing (FWM) in a dispersion-shifted fiber (DSF). Other methods are based in FWM in a SOA [5], in XGM in a SOA [6], electroabsorption modulator (EAM), etc. Each MWC technique has advantages and disadvantages, and all of them

I. Tomkos et al. (Eds.): ONDM 2007, LNCS 4534, pp 1-10, 2007.

have to be studied to know which are more suitable to be adopted in optical switches [7]. This study is normally carried out in laboratories by experimental performances. After that, it is necessary to simulate in a computer the same schemes that have been studied in the laboratory. By this way, we can know if our simulation models developed are correct or not and if the behavior of the devices are similar in the simulations and in the experiments.

In this paper, two MWM aproaches are discussed and compared. Furthermore, simulations of these MWC techniques using 10 Gb/s data rate are presented. *VPItransmissionMaker*[TM]*WDM* simulator was used to create the setups of each MWC technique and to make the required simulations. To summarize, this simulation results are compared with the experimental results reported in [7], where experimental performance validation of four MWC techniques was demonstrated.

2   MWC simulation model

Simultaneous single-to-multiple-channel wavelength conversions have been simulated with two methods at 10 Gb/s bit rate, to evaluate if they are suitable for multicasting. These two techniques are FWM in a DSF and cross phase modulation (XPM) in an SOA-MZI. These two techniques were chosen because, according to our experimental experience, these two schemes were the most promising MWC approaches we have worked with. We also have worked with FWM in a SOA and cross gain modulation in a SOA but the results were not as satisfactory.

The general schematic for these two MWC simulations are shown in Fig.1 [7]. To compare the simulation results with the experiments, we adopted the same general schemes. In all the simulations, four continuous wave (CW) lasers were applied the wavelength probes. Another laser was encoded with 2^{23}-1 pseudorandom bit sequence by an intensity modulator to obtain the 10 Gb/s NRZ data signal. After EDFA amplification, out-of-band ASE filtering and polarization control, the data signal was injected into the MWC medium. This medium was a SOA-MZI or a DSF. Couplers and a multiplexer were employed to combine data signal and CWs. At the output of the MWC medium, the signal spectrum is monitored and each converted wavelength channel was selected through an optical filter, amplified by another EDFA and the bit-error rate (BER) was measured [7].

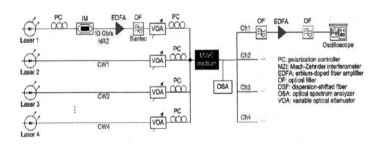

Fig. 1. General schematic for the two MWC simulations.

2.1 MWC by SOA-MZI

SOA-MZI-based MWCs has been reported with good performance [1], [4]. SOA-MZI for MWC has the advantages of compactness and integration probability [4]. High data rate wavelength conversion of 10 Gb/s or more generally requires SOA-MZIs which can operate at the same speed, or more complicated setups such as differential scheme have to be employed. In this section, simulation results of one-to-four MWC with 200 GHz and 100 GHz channel spacing (distance between the wavelength multicast channels) are presented.

In the simulation, the signal channel wavelength was set to be 1541.35 nm. 400 GHz detuning was applied to the closest CW channel. Data signal power before the SOA-MZI was -1 dBm and the total CW probe power was 0 dBm. Other parameters of the SOA-MZI are summarized in Table 1. The schematic for the SOA-MZI MWC is shown in Fig 2.

Table 1. Parameters for the SOA-MZI.

Parameter name	Value	Units
Injection Current SOA 1		
100 GHz Channel Spacing	0.35	A
200 GHz Channel Spacing	0.313	A
Injection Current SOA 2		
100 GHz Channel Spacing	0.32	A
200 GHz Channel Spacing	0.281	A
SOA Length	1	mm
SOA Width	1	μm
SOA Height	0.2	μm
Optical Confinement	0.3	
Internal Losses	3000	1/m
Differential Gain	2.8e-20	m^2
Carrier Density at Transparency	1.4e24	$1/m^3$
IndexToGainCoupling	3	
Linear Recombination Constant (A)	1.43e8	1/s
Bimolecular Recombination Constant (B)	1.0e-16	m^3/s
Auger Recombination Constant (C)	1.3e-41	m^6/s
Initial Carrier Density	2.0e24	$1/m^3$

200 GHz channel spacing MWC. The output spectrum is illustrated in Fig. 3. The eye diagrams of each wavelength-converted channel are represented in Fig. 4. All output channels had clear open eyes. BER measurements of the back-to-back signal (b2b) and converted channels are shown in Fig.5, from which we observe around 0.4 dB power penalty for the copies at BER equal to 10^{-9}. Besides, eye extinction ratio (ER) of the converted signals measured is around 29,5 dB in each channel.

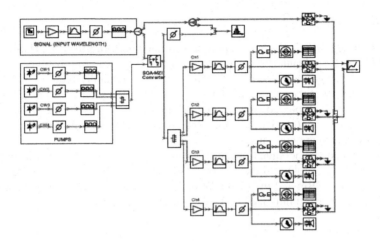

Fig. 2. Schematic for the SOA-MZI MWC

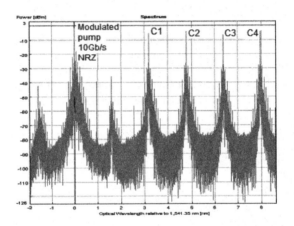

Fig. 3. SOA-MZI MWC: 200 GHz channel spacing Spectrum

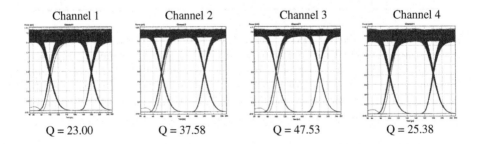

| Channel 1 | Channel 2 | Channel 3 | Channel 4 |
| Q = 23.00 | Q = 37.58 | Q = 47.53 | Q = 25.38 |

Fig. 4. SOA-MZI MWC: 200 GHz channel spacing converted eye diagrams

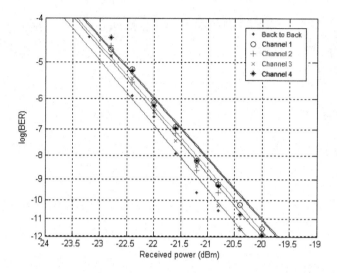

Fig. 5. SOA-MZI MWC: 200 GHz channel spacing BER measurements

100 GHz channel spacing MWC. The results are shown in Fig. 6, Fig. 7 and Fig. 8. In this case, the eyes were less opened than in previous one. This was because channels were nearer now and the crosstalk between them was bigger. Thus, the simulated Q factors were also lower (from 10.5 in the first channel, to 14.4 in the second one) comparing to that of the 200 GHz channel spacing MWC. From the spectrum we also observe more satellite products generated. Also from the BER curve we observe an error floor for channel 1 and 4. The power penalty at 10^{-9} increased to about 3 dB. Finally, eye ER is reduced to around 12.6 dB in this simulation.

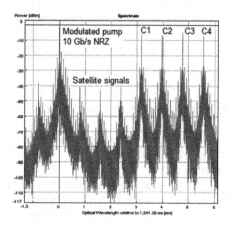

Fig. 6. SOA-MZI MWC: 100 GHz channel spacing Spectrum

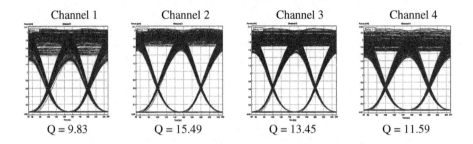

Q = 9.83   Q = 15.49   Q = 13.45   Q = 11.59

Fig.7. SOA-MZI MWC: 100 GHz channel spacing converted eye diagrams

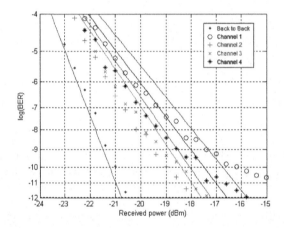

Fig. 8. SOA-MZI Wavelength Converter: 100 GHz channel spacing BER measurements

The much worse results that we got from bring the channels closer were due to the increased inter-channel crosstalk, though at 100 GHz spacing we still got open eyes and reasonably good performance. In latter case, channel 2 had the shortest power penalty, while with the former channel spacing channel 3 had the shortest power penalty.

2.2 MWC by FWM in dispersion-shifted fiber (DSF)

In principle, FWM for wavelength conversion offers strict transparency in both bit rate and modulation format. Moreover, in theory FWM in DSF has no limitations in relation to bit rate or operation speed.

In this MWC schematic, data signal was placed at 1547.72 nm, the zero-dispersion wavelength of the DSF, and four other tunable lasers were used as CWs (continuous waves). The laser was encoded with $2^{23}-1$ bits by a MZI in order to form the 10 Gb/s pseudorandom pump NRZ data. 200 GHz detuning and 100 GHz channel spacing were used. The schematic for the FWM in DSF MWC is shown in Fig 9.

The output signal power was set to 12 dBm, pump average powers were 0 dBm each. Just before the SOA, signal power was 9.61 dBm and the total CW power was 0 dBm. Pump signal here needed much higher power than that in the SOA-MZI MWC setup.

In this case, polarization controllers were very important because FWM in DSF was polarization sensible.

The simulated output spectrum is illustrated in Fig. 10. The measured average MWC conversion efficiency was around -18 dB. Eye diagrams are shown in Fig. 11. It is noticeable that all the eyes had a quite good opening, but they had more noise that in SOA-MZI case. Q factor values were good (from 9.2 in the first channel, to 14.3 in the last one). BER measurements of the back-to-back signal (b2b) and converted channels are shown in Fig.12. The output ER of our scheme is near 23 dB in each converted channel. In this case, at BER of 10^{-9} the power penalty was nearly 0 dB for the third and forth channels, about 0.3 dB for the second one, and about 0.8 dB for the first channel.

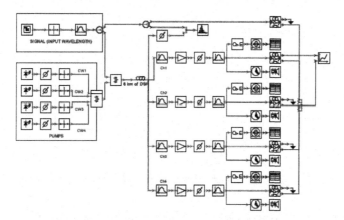

Fig. 9. Schematic for the FWM in DSF MWC

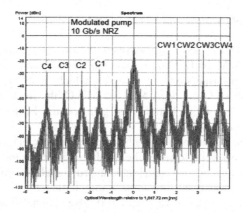

Fig. 10. FWM Wavelength Converter: Spectrum at DSF output

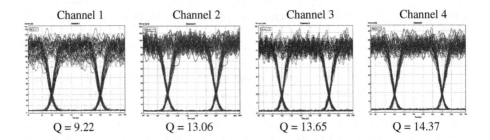

Fig. 11. FWM Wavelength Converter in DSF: converted eye diagrams

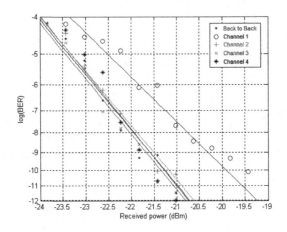

Fig. 12. FWM Wavelength Converter in DSF: BER measurements

2.3 Comparison of SOA-MZI and FWM in DSF MWCs

Both SOA-MZI and DSF-based multi-wavelength converters have their advantages and disadvantages. FWM in DSF needs the modulated pump signal to be placed around the fiber zero-dispersion wavelength, which restricted the wavelength flexibility of the scheme. Using SOA-MZI for MWC, the modulated signal can be placed on any wavelength within both SOAs' gain spectra. Besides, SOA-MZI setup needs less input power than the other.

All the eye diagrams obtained in both schemes have clear opening and noise suppression at the zero level. However the quadratic effect to the signal power also amplified the noise at logical 'one' level and caused signal distortion at this level. Q factor from the simulation result are quite good: higher than 9 in FWM in DSF scheme, higher than 10.5 in SOA-MZI with 100 GHz channel spacing and higher in SOA-MZI with 200 GHz spacing. Because of the simulation environment, these results cannot give absolute credit for comparing the MWC scheme in real, but it helps us to a certain extend to understand the important parameters that matters for each scheme stand alone.

FWM in DSF has the least power penalties and excellent eye ER measurements. These results are better than SOA-MZI results with the same channel spacing (100 GHz). Besides, FWM in DSF detuning is half of the other scheme detuning and crosstalk between channels should be higher.

3 Comparison of simulation and experimental results

Simulation results shown in the previous section are largely coherent to experimental results shown in [7]. However simulation provides more convenience in changing the parameters and assesses their importance to the scheme.

Comparing to the SOA-MZI MWC experimental results, with 200 GHz channel spacing, obtained power penalty was around 0.5 dB, while in the simulations it was around 0.7 dB. In both the simulations and experiments, every converted channel had clear open eyes, but in experiments, the eye 'zero' level has more noise than what was shown in simulations. It might due to the fact that the simulator neglected some effects that added noises at 'zero' level. In experiments, converted signals Q factor values were also quite good (around 10-11), which was confirmed by the simulation results.

With 100 GHz channel spacing SOA-MZI MWC, experiments achieved much less power penalty than in the simulations (less than 0.5 dB in experiments, and around 3dB in simulations). Both simulation and experimental results had similar eye opening diagrams (more noisy in the experiments) and Q factor (near 10).

Experimental results in FWM in DSF MWC scheme were as good as simulation ones: similar eye diagrams with a noisy 'one' level, conversion efficiency around -15 dB (-13.5 dB in experiments and -18dB in simulations), small power penalties and Q factor values around 11.

4 Conclusions

For MWC at bit rate below 10-40 Gb/s, SOA-MZI-based wavelength converter can be a very good choice. This technique has an excellent compactness and conversion flexibility in wavelengths. It also requires very low optical input power, limiting a possible high crosstalk between channels. Both simulations and our previous experiments show that this scheme produces MWC with low power penalty between the converted channels and the back-to-back signal at 200 GHz channel spacing. Besides, converted channels have quite good eye opening and Q factor values.

On the other hand, FWM in DSF offers transparency in data rate and modulation format, which can be suitable for very high bit rate operation towards Tb/s range. In this case, the number of converted channels has little impact on conversion efficiency and converted Q factor as long as they are placed within the effective conversion detuning range from the modulated signal pump. However, this setup needs high optical input power and is very sensitive to the polarization of all the channels.

Finally, our simulation results and previous experimental results showed similar behavior, which confirmed the correctness of our simulation models developed. Some slight difference is due to the fact that it was impossible to emulate the real experimental environment, and the component models provided in the simulator cannot be completely adjust to present the exact devices we used.

Acknowledgments. This work has been supported by the European Commission through the Network of Excellence e-Photon/ONe+ project. The author would like to thank the IST-LASAGNE project for funding these research activities, and Dr. Javier Herrera Llorente for his help with the simulator.

References

1. N. Yan, et al: Simultaneous one-to-four multi-wavelength conversion using a single SOA-MZI, 11th European Conference on Networks and Optical Communications (NOC), 2006.
2. F. Ramos et al., IST-LASAGNE: Towards All-Optical Label Swapping Employing Optical Logic Gates and Optical Flip-flops, IEEE Journal of Lightwave Technology, October 2005.
3. N. Yan, A. Teixeira et al.: Simultaneous multi-wavelength signal conversion for transparent optical multicast. European Conference on Networks Optical communications (NOC), 2005
4. H. S. Chung et al.: All-optical multi-wavelength conversion of 10 Gbit/s NRZ/RZ signals based on SOA-MZI for WDM multicasting. Electron. Lett., Vol. 41, no.7, Mar. 2005.
5. S. Diez, C. Schmidt et al.: Four-Wave Mixing in Semiconductor Optical Amplifiers for Frequency Conversion and Fast Optical Switching, IEEE Journal of selected topics in Quantum Electronics, Vol. 3, No. 5, October 1997.
6. T.G. Silveira, et al.: Cross-gain modulation bandwidth enhacement in semiconductor optical amplifiers by means of detuned optical filter, IEEE Electronics Letters, Vol.41, no. 13, 23rd June 2005.
7. N. Yan, A. Teixeira, T. Silveira, I. Tafur Monroy, H.-D. Jung, T. Koonen, Optical multicast technologies by multi-wavelength conversion for optical routers, Proc. IEEE 2006 10th International Conf. on Communication Technology (ICCT'2006), 27-30 Nov 2006, Guilin, China, paper SRGN (#2006072901).

80Gb/s Multi-wavelength Optical Packet Switching using PLZT Switch

Katsuya Watabe[1], Tetsuya Saito[2], Nobutaka Matsumoto[1],
Takuo Tanemura[3], Hideaki Imaizumi[3], Abdullah Al Amin[4],
Mitsuru Takenaka[4], Yoshiaki Nakano[4], and Hiroyuki Morikawa[3]

[1]Graduate School of Frontier Sciences, The University of Tokyo
Kashiwanoha 5-1-5, Kashiwa-shi, Chiba, 277–8561 JAPAN
[2]Graduate School of Science and Engineering, Chuo University
Kasuga 1–13–27, Bunkyo-ku, Tokyo, 112–8551, JAPAN
[3]School of Engineering, The University of Tokyo
Hongo 7–3–1, Bunkyo-ku, Tokyo, 113–8656, JAPAN
[4]Research Center for Advanced Science and Technology, The University of Tokyo
Komaba 4–6–1, Meguro-ku, Tokyo, 153–8904, JAPAN
k_watabe@mlab.k.u-tokyo.ac.jp
http://www.hotaru-project.net/

Abstract. This paper proposes 80Gb/s multi-wavelength optical packet switching(OPS) using a PLZT switch. The Multi-wavelength OPS Network can achieve low implementation costs compared to existing OPS networks in which the number of wavelengths is large. In this network, the header is processed separately from the payload. The payload is divided into multiple segments. Each segment is then encoded into different wavelengths. After assignment of wavelength to each payload segment, payload segments are multiplexed into an optical signal which is then transmitted through an optical fiber. Therefore, the number of components required for intermediate node functionality stays the same for the number of wavelengths used for payload transmission. This paper shows a fundamental experiment of the Multi-wavelength OPS network using a PLZT switch. PLZT switch provide a fast optical switching with low noise, independent polarization, and low drive voltage. In this paper, we describe the detail of the experiments and results.

Keywords: multi-wavelength, WDM, optical packet, PLZT switch

1 Introduction

Wide deployment of broadband Internet access and new multimedia services such as real-time streaming have caused enormous traffic growth in the Internet. It is expected that this traffic will keep increasing with the emergence of new online services such as on-demand service for high-definition movies.

For transmission of large traffic volumes, the WDM technology has been widely researched and a number of test-bed systems have been developed recently[1]. According to studies in [2], the capacity of transmission within a single fiber can exceed 10 Tb/s.

I. Tomkos et al. (Eds.): ONDM 2007, LNCS 4534, pp 11-20, 2007.
© IFIP International Federation for Information Processing 2007

In addition, the optical packet switching (OPS) network technology has been researched in order to achieve high bandwidth utilization and low power consumption as well as to minimize the number of devices in the WDM network[3, 4]. However, the existing OPS network technologies transmit the whole packet that consists of the header and the payload through a single wavelength. This functionality requires intermediate nodes to have necessary components to support it. The node must have label processing units and contention resolution units for each wavelength. This will cause the number of devices for construction of WDM networks to grow with each additional wavelength [5].

In order to solve the problem, the Multi-wavelength OPS Network Technology has been researched for OPS networks to achieve scalability with the number of wavelengths. In contrast with the existing OPS technologies, the header of the packet is processed separately from the payload. The header is encoded and transmitter at dedicated wavelengths. The payload itself is divided into multiple segments which are encoded and transmitted at different wavelengths. Therefore, this requires core nodes to contain necessary components for each port, not for each wavelength.

The feasibility of the multi-wavelength OPS technology has been confirmed with Semiconductor Optical Amplifier (SOA) switch and LiNbO$_3$(LN) switch [6, 8]. However, it has never been confirmed with a PLZT switch which has the capabilities of fast switching, low noise, polarization independency, and low drive voltage.

In this paper, we experiment the multi-wavelength OPS technology using a PLZT switch in order to evaluate its feasibility. In this expetiment, we demonstrate 80Gb/s (10Gb/s × 8 wavelengths) multi-wavelength OPS using a PLZT switch and measure Bit Error Rate (BER) and the eye pattern.

This paper is organized as following. Section 2 describes details of multi-wavelength OPS technology. Section 3 explains the experimental setup and its detail, and Section 4 shows the experimental result. Finally, we conclude this paper in Section 5.

2   Multi-wavelength OPS Network

This section describes the characteristics of the multi-wavelength OPS network with comparison to the existing OPS networks.

In the existing OPS network, every packet consisting of the header and the payload is encoded into individual wavelength. Therefore, after input optical signal is demuxed into multiple wavelengths, independent processing modules are required for each wavelength in order to perform necessary functions for packet switching such as packet switching, wavelength conversion, and contention resolution as illustrated in Fig.1. Therefore, this property will not provide good scalability for the larger number of devices because of the cost, the physical size of the system and the complexity for controlling the node in the future WDM network with the large number of wavelengths.

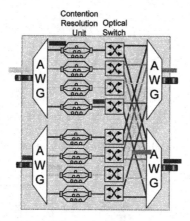

Fig. 1. Node Architecture Diagram of Traditional OPS Networks

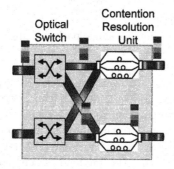

Fig. 2. Node Architecture Diagram of Multi-wavelength OPS Networks

In contrast to that, in the multi-wavelength OPS network, each packet is encoded into multiple wavelengths as illustrated in Fig.2. The header is encoded into one wavelength. The payload is divided into multiple segments. Each segment is encoded into different wavelengths. At the intermediate node in this network, the payload is forwarded in accordance with the header information. Therefore, when the number of wavelengths for the payload increases, the number of devices in the intermediate node does not change.

In the multi-wavelength OPS network, the payload can contain any data in the traditional OPS network. However, due to the characteristics that the payload in the multi-wavelength OPS network is encoded into multiple wavelengths, the composition of the payload influences the implementation cost and the characteristics of the network.

Currently two methods of the composition have been proposed: 1) the payload contains multiple packets of upper layer protocols[6], and 2) the payload contains one packet of upper layer protocols[7] as illustrated in Fig.3 and Fig.4 respectively.

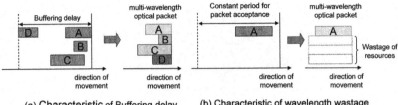

(a) Characteristic of Buffering delay   (b) Characteristic of wavelength wastage

Fig. 3. Payload containing multiple packets of upper layer protocols

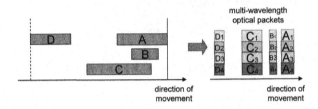

Fig. 4. Payload containing a single packet of upper layer protocols

With the former method, incoming packets of upper layer protocols are buffered at an ingress edge node for defined time period. The buffered packets which have the same destination form a payload. The maximum number of the packets in the buffer is equal to the number of available wavelengths. Hence, when the number of the packets with the same destination in the buffer reaches the number of the wavelengths, the payload is sent so that each upper layer packet from the payload is transmitted on separate wavelength. This method does change the structure of the upper layer packet so fragmentation and reassembly of those packets are unnecessary. Therefore, it makes the composition of the edge node simple. However, the time period for buffering affects the jitter. In addition to that, the utilization of wavelengths is low when the number of packets towards the same edge node does not reach the number of the wavelengths within defined time period. The feasibility of the multi-wavelength OPS network with this method has been confirmed by Onaka et al.[6]. Their node architecture was based on broadcast-and-select switch which consists of SOA switches with multistage configuration.

In contrast with described method, the second method uses only one single upper layer packet for the payload. Namely, a single upper layer packet is divided into multiple segments. The advantage of this method is that the ingress edge node transmits immediately upon packet reception. In the previous method, a node must wait until the buffer will receive enough packets for the payload or go by defined time period. In addition, the utilization of wavelengths is higher than one of the former method. This is because in the method with packet fragmentation, the available wavelengths for the payload are always utilized. However, the implementation of this method is difficult because there must be a function which divides and reassembles packets. Furukawa et al. constructs the multi-wavelength OPS network with this method using LN switches[8]. They also constructed the implementation of the edge node with the function which converts Ethernet frames into multi-wavelength optical packets. In this paper, we realize multi-wavelength OPS by this method on PLZT switches.

3   Experiment Setup and Detail

This section describes the setup and its detail of the experiment for confirming the feasibility of the multi-wavelength OPS network using PLZT switches.

The characteristics of PLZT switches are the most suitable for the multi-wavelength OPS network. The switching speed of SOA, LN, and PLZT switches is fast enough to perform optical packet switching. However, Optical Signal-Noise Ratio(OSNR) of SOA switches is low due to its high Amplified Spotaneous Emission(ASE) noise. On the other hand, LN switches provides even faster switching speed than one of SOA switches but it has polarization dependency. High drive voltage lowers the dependency but it becomes difficult to drive electric circuits of the switch. In addtion, the LN switches causes DC drift phenomena that its refraction index becomes unstable with direct current. Unlike LN and SOA switches, PLZT switch provides noise tolerance with high OSNR, polarization independence, and operation with low voltage[9].

OPS node mainly consists of four components: 1) the label processing unit which analyzes the destination of packets and determines the output port for the packet, 2) the routing control unit which exchanges routing information with other nodes and calculates the optimized routes, 3) the switching unit which transmits incoming optical packets into the particular port indicated by the label processing unit, and 4) the contention resolution unit which schedules and buffers out-going packets to the same port in order to avoid the packet collisions.

In order to evaluate the feasibility and the performance of the multi-wavelength OPS network, we develop 1) the label processing unit and 3) the switching unit.

3.1   Detail of Subsystems for Evaluation of The Multi-wavelength Optical Packet Switching Node Performance

The system for experiments mainly consists of three sub-systems: 1) the payload generation sub-system, 2) the label generation sub-system, and 3) the monitoring sub-system as illustrated in Fig.6.

The payload generation sub-system consists of six main devices generating the signal of the payload and five supplementary devices coordinating the signal. The main devices include the light source (LS), the arrayed waveguide grating (AWG), the LiNbO$_3$ modulator (LNM), the pulse pattern generator (PPG), the AO modulator (AOM), and the Field Programmable Gate Arrays(FPGA). The supplementary devices includes the polarization controller (PC), the polarization maintainer fiber (PMF), the polarization beam splitter (PBS), the EDFA, and the band pass filter(BPF).

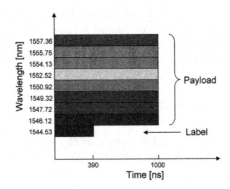

Fig. 5. Packet format: the payload length is 1000ns

The payload generation sub-system works as follows. First, the LS generates eight continuous waves (CWs), which are independent wavelengths ranging between 1546.12 nm and 1557.36 nm at 200GHz intervals. Then, after the polarization, the signal of each wavelength is coordinated by the PC and the PMF. The AWG multiplexes eight signals into a single optical signal. The polarization of the signal is adjusted to suitable one for the LNM by the PBS. After the PPG and the LNM modulate the signal into the 10 Gb/s NRZ signal(2^7-1 PRBS), the power of the modulated signal gains by the EDFA and the noise of the signal is lowered by the BPF. Finally, the AOM cut the signal into impulses of 1000 ns or 1500 ns. These impulses are treated as the payload.

The label generation sub-system consists of the LS, the AOM, and the FPGA. The LS generates CW of 1544.53 nm and the AOM controlled by the FPGA modulates the CW into 7.69 Mb/s NRZ as a label. Through this experiment, the label is one of two labels: "111" and "100".

The monitoring sub-system consists of the two AWGs and the oscilloscope. The oscilloscope shows the input and output signals for each wavelength.

The arrival interval of the packet is 2500 ns. The packet format with 1000 ns length is illustrated in Fig.5 and the parameters for the experiment are shown in Table.1.

3.2   Details of Main Components for Multi-wavelength Optical Packet Switching

The diagram of this experiment is illustrated in Fig.6.

The label processing sub-system consists of the Photo Detector (PD) and the FPGA. The PD converts an optical signal of the label into an electric signal. The optical signal of the label is received from the label generation sub-system. The FPGA analyzes the electric signal and recognizes the encoded label. To analyze the encoded label we used perfect matching method. Then, FPGA determines the output port for transmission and notifies switching sub-system about the determined output port.

Table 1. Parameters of multi-wavelength optical packets

	Label	Payload
Bit rate	7.69Mbps NRZ	10Gbps NRZ
Data length	390ns	1000ns or 1500ns
Data signal	"111" or "100"	$2^7 - 1$ PRBS
Number of wavelength	1	8
Range of wavelength	1544.53nm	1546.12nm～1557.36nm

The switching sub-system, which consists of the 1×2 PLZT switch, transmits the multiplexed payload to the port indicated in the control signal from the label processing sub-system. In this experiment, the switching sub-system transmits the payload with the label "111" to the port 1 and one with the label "100" to the port 2.

4   Experimental Result

Fig.7 shows the results received with the monitoring sub-system. Fig.7(a), (b), and (c) represent the input signals, the output signals from the port 1, and the output signals from the port 2, respectively. The signal on the top of the figure represents a signal of the wavelength assigned to the label. The rest of the signals represent the wavelengths of payload fragments. This results show that the packet encoded into multiple wavelengths is correctly switched in accordance with the label information. Therefore, this experiment has confirmed that the multi-wavelength OPS network can be achieved using a PLZT switch.

In addition to the fundamental experiment for the multi-wavelength OPS network, this experiment evaluates the signal degradation coming out of the PLZT switch. The eye patterns of the input and output signals at 1557.36 nm are shown in Fig.8. The eye patterns of the rest of the wavelengths were similar to patterns presented in Fig.8.

In addition, this experiment measures Bit Error Rate (BER) for all the wavelengths. Fig.9 shows the results of the signal at 1557.36 nm. This results shows that the BER of the output ports was less than 10^-10 when the input power was -27dBm. This proposes that this experimental system can perform error-free transmission with high input power.

5   Conclusions

In this paper, we have achieved 80Gb/s multi-wavelength OPS using a PLZT switch for evaluation of its feasibility. Moreover, we have confirmed that error-free transmission is possible in this multi-wavelength OPS node model.

As a future work, we plan to solve the problem caused by group velocity dispersions. Also, we plan to construct 640Gb/s(40Gb/s $\times$ 16 wavelengths) multi-wavelength OPS system.

Acknowledgments. We would like to thank Dr. M. Ohta, Dr. H. Harai, and Dr. N. Wada for their encouragements, discussions, and advices. This reseach is supported by NICT(National Institute of Information and Communicat ions Technology).

References

1. Mukherjee, B.: WDM Optical Communication Networks: Progress and Challenges. IEEE JOURNAL ON SELECTED AREAS IN COMMUNICATIONS, Vol. 18, No. 10, (2000) 1810–1824 Springer-Verlag, Berlin Heidelberg New York (1997) 415–438
2. Fukuchi, K., Kasamatsu. T., Morie, M., Ohhira, R., Ito, R., Sekiya, K., Ogasahara, D., and Ono, R.: 10.92-Tb/s (273 × 40-Gb/s) triple-band/ultra-dense WDM optical-repeatered transmission experiment. Optical Fiber Communication Conference and Exhibit 2001, Vol. 4, PD24, (2001) 1–3
3. Hunter, D.K., Andonovic, I.: Approaches to optical internet packet switching. IEEE Communications Magazine, Vol. 38, No. 2, (2000) 116–122
4. Carena, A., Feo, V.D., Finochietto, J.M., Gaudino, R., Neri, F., Piglione, C., and Poggiolini, P.: RingO: an experimental WDM optical packet network for metro applications. IEEE JOURNAL ON SELECTED AREAS IN COMMUNICATIONS, Vol. 22, No. 8, (2004) 1561–1571
5. Takara, H., Ohara, T. Yamamoto, T. Masuda, H. Abe, M. Takahashi, H. Morioka, T.: Field demonstration of over 1000-channel DWDM transmission with supercontinuum multi-carrier source. Electronics Letters, Vol. 41, (2005) 270–271
6. Onaka, H., Aoki, Y., Sone, K., Nakagawa, G., Kai, Y., Yoshida, S., Takita, Y., Morito, K., Tanaka, S., and Kinoshita, S.: WDM Optical Packet Interconnection using Muti-Gate SOA Switch Architecture for Peta-Flops Ultra-High-Performance Computing Systems. European Conference on Optical Communication 2006, Vol. **, (2006) **-**
7. Ohta, M..: Components for All Optical Data Path Routers. TECHNICAL REPORT OF IEICE, PN11, (2005)
8. Furukawa, H., Wada, N., Harai, H., Naruse, M., Otsuki, H., Katsumoto, M., Miyazaki, T., Ikezawa, K., Toyama, A., Itou, N., Shimizu, H., Fujinuma, H., Iiduka, H. Cincotti, G., Kitayama, K.: All-Optical Multiple-Label-Processing Based Optical Packet Switch Prototype and Novel 10Gb Ethernet / 80 (8 Lambda x 10) Gbps-Wide Colored Optical Packet Converter with 8-Channel Array Burst-Mode Packet Transceiver. to be presented at Optical Fiber Communication Conference and Exhibit 2007, (2007)
9. Nashimoto, K., Tanaka, N., LaBuda, M., Ritums, D., Dawley, J., Raj, M., Kudzuma, D., and Vo, T.: High-speed PLZT optical switches for burst and packet switching. 2nd International Conference on Broadband Networks, Vol. 2, (2005) 1118–1123

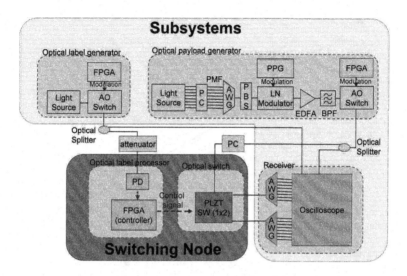

Fig. 6. Experimental Block Diagram

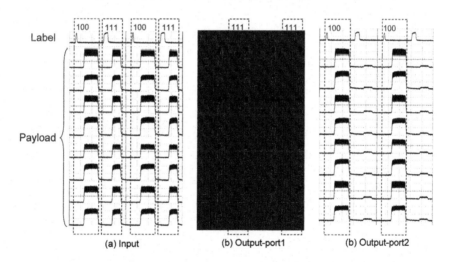

Fig. 7. Input signal & Output signal

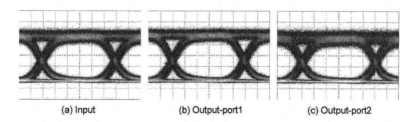

Fig. 8. Eye Pattern diagram @ 1557.36nm

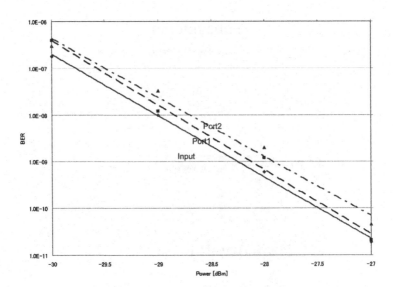

Fig. 9. BER @ 1557.36nm

2x2 Bismuth-Oxide-Fiber Based Crossbar Switch for All-Optical Switching Architectures

O. Zouraraki[1], P. Bakopoulos[1], K. Vyrsokinos[1], and H. Avramopoulos[1]

[1] Photonics Communications Research Laboratory, National Technical University of Athens, Zographou, GR 15773, Athens, Greece
ozour@mail.ntua.gr

Abstract. We demonstrate an optically controlled 2x2 Crossbar switch suitable for all-optical switching architectures. The switch is based on an Ultrafast Nonlinear Interferometer (UNI) configuration and uses 80 centimeters of highly nonlinear Bismuth Oxide fiber. Error-free operation is demonstrated with 10 and 40 Gb/s input signals, with power penalties in its BAR and CROSS states of less than 0.3 dB and 1 dB respectively. The short length of the highly nonlinear Bismuth Oxide fiber ensures compact size, small latency and good operational stability of the switch.

Keywords: Optically controlled 2x2 switch, Bismuth Oxide Nonlinear fiber, Ultrafast Nonlinear Interferometer (UNI), all-optical signal processing.

1 Introduction

The increasing demand in ultrahigh capacity, global data networks has pushed intelligent switching functionalities to the optical layer. [1]. All-optical switching is believed to play a vital role in future broad-band communications systems as it uses capacity more efficiently, increases throughput, provides greater flexibility than pure wavelength-division multiplexing [2] and offers format and bitrate transparency of data [3]. Towards this end, optical switching has been investigated in a plethora of architectures that incorporate optical cross-connects for optical packet routing, time-slot interchangers for packet re-arranging and collision avoiding, optical add/drop multiplexers, optical buffers and optical signal monitoring schemes [2]. In most optical switch-based approaches, 2x2 and 1x2 switches are used as the core switching elements [3] since their appropriate cascading enables the implementation of switching subsystems that can accommodate more than two input signals. For example switches of size larger than 2x2 can be implemented using 2x2 switches as the basic building block. In the past, various efficient architectures for building large switches using fundamental switching elements have been presented [2][4]. Another application for cascaded 2x2 switches is also optical buffering schemes [5][6]. Such applications pose significant performance requirements on the 2x2 switching elements, such as switch bitrate scalability in order to increase the maximum link utilization, cascadability and high extinction ratio [1][7] whereas signal-to-noise ratio deterioration is especially crucial in feedback applications.

I. Tomkos et al. (Eds.): ONDM 2007, LNCS 4534, pp 21-28, 2007.
© IFIP International Federation for Information Processing 2007

So far, 2x2 all-optical switches have been associated with semiconductor-based devices [11][12]. Semiconductors have been considered more suitable for all-optical switching applications as they offer significant advantages including high nonlinearity, compact size and capability for integration. Nonlinear fibers, on the other hand, provide the unique advantages of fs nonlinearity response, nearly penalty-free switching and operation without electrical power supply, but have traditionally required long, bulky fiber spans. However recent advances in the fabrication of highly-nonlinear fibers (HNLF) led to the construction of high quality nonlinear fibers which allow for extensive size reduction of fiber-based implementations. One of the most attractive candidate for fiber-based switching applications is Bismuth Oxide (Bi_2O_3) based nonlinear fiber (Bi-NLF) that exhibits an extremely high nonlinearity exceeding 1000 $W^{-1}km^{-1}$. This means that a meter or less [9][10] of this kind of fiber would be enough for the implementation of ultra-compact, all-optical fiber switches. Besides compact size, short nonlinear fiber lengths enhance operational stability and most significantly guarantee low switch latency. This could be very crucial when cascading a large number of switches, as well as in feedback applications such as feed-back buffering schemes.

In this letter we demonstrate an optically controlled 2x2 Crossbar switch using 80 centimeters of highly nonlinear Bi-NLF. The switch is based on a single-arm, Ultrafast Nonlinear Interferometer (UNI) gate [11], which uses the 80-cm long Bi-NLF as the nonlinear interaction medium in order to obtain a complete π phase shift. The below-200 fs nonlinearity response time of the highly nonlinear fiber [13] ensures that the switching window of the switch is defined by the width of the control pulses. This has enabled bitwise operation at input data rates of 40 Gb/s, whereas in principle switch operation can be extended at even higher rates. Error-free operation has been achieved with power penalties in its BAR and CROSS states of less than 0.3 dB and 1 dB respectively. The proposed 2x2 Crossbar switch can comprise the key unit for the implementation of a full range of 2x2 switch-based circuits including switching matrices [2][4], time-slot interchangers [14], header reinsertion circuits [15] and optical buffers [1] [16].

2   Experimental Setup

The 2x2 crossbar switch was tested to operate with both 10 Gb/s and 40 Gb/s input data signals as follows. A single 10 Gb/s data channel along with a 40 Gb/s time-division-multiplexed (TDM) signal enter the fiber UNI, from inputs IN 1 and IN 2 respectively. In the absence of the control signal, the input signals exit the switch from OUT 1 and OUT 2 unchanged. Injection of the control signal causes the Cross function, namely one 10 Gb/s channel of IN 2 is switched to OUT 1, whereas the input signal of IN 1 is switched to OUT2 to form the new 40 Gb/s TDM output signal.

Fig. 1 shows the experimental setup used for the evaluation of the 2x2 crossbar switch. A 1553 nm DFB laser diode was gain switched at 10 GHz to generate 10-ps pulses. These pulses were compressed to 3 ps in a nonlinear fiber pulse compressor and were then modulated with a Ti:LiNbO$_3$ modulator (MOD1) to form a pseudorandom data pattern. The signal was subsequently split in a 3 dB coupler into two parts: The first part entered an all-optical wavelength converter consisting of an integrated, SOA-based Mach Zehnder interferometer (MZI 1) to generate a 10 Gb/s input signal (IN 1) to the 2x2 switch at 1558 nm. The second part was rate quadrupled into a fiber bit interleaver, to produce a 40 Gb/s 2^7-1 PRBS data pattern (IN 2). The control signal was obtained by another 10-GHz gain switched laser (DFB 2) operating at 1534 nm and producing 8.2 ps clock pulses. These pulses were modulated in a Ti:LiNbO$_3$ modulator (MOD2) to produce bursts of clock pulses with 2.3 nsec duration and 40.5 nsec period and were amplified in a high power EDFA.

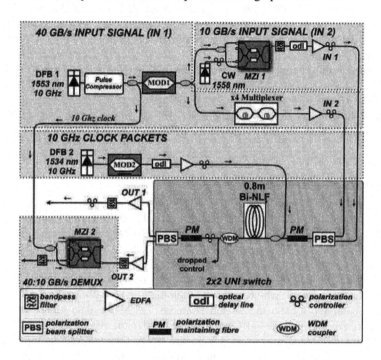

Fig. 1. Experimental setup

These three signals were injected into the 2x2 Crossbar switch which was implemented with a 2-input, 2-output UNI gate as shown in Fig. 1. The nonlinear element in the UNI was a 80-cm long, highly nonlinear (γ = 950 W^{-1}Km^{-1}) Bi-NLF which was pigtailed with single mode fiber (SMF). Core diameter of the Bi-NLF was 2 μm. The Bi-NLF exhibited 1.3 dB/m propagation losses whereas the input pigtail splice introduced an additional 2.1 dB loss. The time-synchronized input signals entered the switch in orthogonal polarizations from the ordinary and extraordinary ports of a Polarization Beam Splitter (PBS) respectively. A 45° splice analyzed each signal in two equal orthogonal components which are temporally separated by the polarization maintaining (PM) fiber section providing 10 ps of birefringent delay. The control signal was launched into the switch through a 90:10 fiber coupler so that it co-travels with the input signals. Fine bitwise synchronization between the three optical signals was achieved with two optical delay lines, so that the control pulse coincided with the trailing component of both input signals. Re-alignment of each input signal component was achieved by another PM section of equal length where the fast/slow signal components were aligned to the slow/fast PM fiber axis respectively by means of a polarization controller. A 45° splice followed by a PBS perform spatial separation of the resultant signals at the corresponding PBS outputs according to their polarization state. At the output of the switch and in the absence of the control signal, data signal IN 1 passes through to OUT 1 and data signal IN 2 passes through to OUT 2. In the presence of the control signal, the polarization state of each input data component which temporally coincides with the control is rotated by 90°. As a result, the 10 Gb/s channel of IN 1 and the synchronized 10 Gb/s data channel of IN 2 are interchanged at the outputs of the switch. After interacting with the two input data signals in the nonlinear fiber, the control signal was removed with a 1550/1530 nm wavelength selective coupler. Finally, bit error rate (BER) measurements were performed at both outputs of the switch. The BER performance of the 40 Gb/s signal of OUT 2 was assessed with a 40-to-10-Gb/s data demultiplexer, so as to evaluate each of the 10 Gb/s tributary channels. Demultiplexing was performed with a second integrated Mach Zehnder interferometer (MZI 2) operating in push-pull control configuration, which was controlled with 3 ps clock pulses at 1553 nm.

3   Results and Discussion

The experimental results of the 2x2 Crossbar switch are illustrated in Figures 2, 3 and 4. Fig. 2 depicts the oscilloscope traces for the BAR state operation of the switch whereas Fig. 3 shows the corresponding traces for the CROSS state along with the respective control signal. The arrows shown in Fig. 3 indicate the time slots which coincide with the control pulses. 2x2 switching occurs throughout the entire time window shown in Fig. 3, which is shorter than the burst length. In the BAR state (Fig. 2), data signal IN 1 exits the switch from OUT 1 while data signal IN 2 passes through to OUT 2. In the CROSS state (Fig. 3), the 10 Gb/s channel of IN 2 coinciding with the control pulses is switched to OUT 1, whereas the input signal of IN 1 is switched to OUT2 to form a new 40 Gb/s output signal. Fig. 4 illustrates the input and output eye diagrams for both states of the switch as obtained at both output ports simultaneously. The eye diagrams demonstrate an extinction ratio of approximately 10.1 dB for OUT1 and 11 dB for OUT2 in BAR state operation. For the CROSS state, the extinction ratio was 9.5 dB for OUT1 and 10 dB for OUT2.

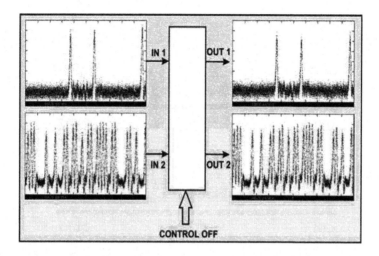

Fig. 2. Oscilloscope traces in the BAR state: The time base is 100 ps/div

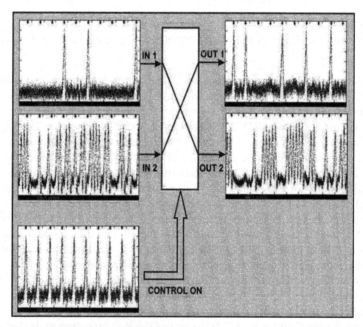

Fig. 3. Oscilloscope traces in the CROSS state: The time base is 100 ps/div. Arrows indicate the time slots which coincide with the control pulses.

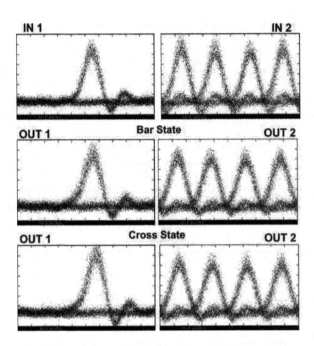

Fig. 4. Eye diagrams for the 2x2 Crossbar operation of the switch: First row: input signals, second row: BAR state operation, third row: CROSS state operation. (time scale: 10 ps/div).

The Bit-Error-Rate (BER) performance of the switch was investigated with a 10 Gb/s Bit Error Rate Tester and is illustrated in Fig. 5. BER curves are presented for the 10 Gb/s signals of IN 1 and OUT 1 along with the corresponding 10 Gb/s tributary data channels of IN 2 and OUT 2 for both BAR and CROSS states. Error-free 2x2 operation was achieved in both BAR and CROSS states with low power penalties. In BAR state operation, a power penalty of 0.3 dB and 0.28 dB was obtained for OUT 1 and OUT 2 ports respectively. CROSS state operation introduced a power penalty of 0.8 dB and 1.06 dB for OUT1 and OUT2 ports. For this operation the peak power of the control pulses was 8.5 W, corresponding to 69.7 pJ pulse energy and a complete π nonlinear phase change. These power levels refer to the input of the Bi-NLF fiber and exclude the splice losses between the SMF pigtail and the Bi-NLF fiber. The polarization of both input signals was adjusted so as to maximize power transmitted through the input PBS, whereas control signal polarization was set with view to optimum switching. Once all polarizations were adjusted, the switch provided stable operation without the need for further polarization control, owing to the shortness of the Bi-NLF. The additional 2 dB power penalty at the input 10 Gb/s signal (IN 1) compared to the input 40 Gb/s (IN 2) is due to OSNR degradation in the wavelength conversion process in MZI 1.

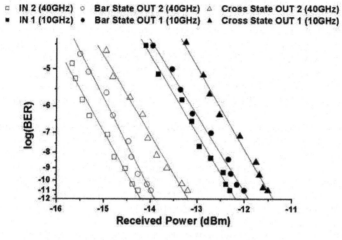

Fig. 5. BER measurements in BAR and CROSS states.

4 Conclusion

We have demonstrated an optically controlled 2x2 Crossbar switch using only 80 centimeters of highly nonlinear Bismuth Oxide fiber in a UNI configuration. The short length of the Bi-NLF fiber ensures to the switch compactness, enhances polarization stability and allows for small latency of the device. Error-free 2x2 operation was demonstrated with low power penalties for both 10 Gb/s and 40 Gb/s signals, indicating the switch may be cascaded or used in feedback applications, without requiring regeneration. The proposed 2x2 Crossbar switch can comprise the basic building block for the implementation of a full range of 2x2 switch-based circuits including large-scale switching subsystems, time-slot interchangers and optical buffers.

Acknowledgments. This work has been supported by the European Commission through project IST-ePhoton/One + (027497). The authors would like to thank Dr N. Sugimoto and ASAHI GLASS Company Ltd for providing the highly nonlinear Bi-NLF.

References

1. Emily F. Burmeister, John E. Bowers, "Integrated Gate Matrix Switch for Optical Packet Buffering", IEEE Photon. Technol. Lett., vol. 18, issue 1, pp. 103-105, Jan. 2006.
2. G.I. Papadimitriou, C. Papazoglou, A.S. Pomportsis, "Optical switching: switch fabrics, techniques, and architectures", IEEE/OSA J. Lightwave Technol., vol. 21, issue 2, pp. 384-405, Feb. 2003.

3. D.K. Hunter, D.G. Smith, "New architectures for optical TDM switching", IEEE/OSA J. Lightwave Technol., vol. 11, issue 3, pp. 495 - 511, Mar. 1993.
4. 7. Rajiv Ramaswami, Kumar N. Sivarajan, "Optical Networks. A Practical Perspective", Sec-ond Edition, Academic Press, 2002.
5. D.K. Hunter, M.C. Chia, I. Andonovic, "Buffering in optical packet switches", IEEE/OSA J. Lightwave Technol., vol. 16, issue 12, pp. 2081 - 2094, Dec. 1998.
6. D.K. Hunter et al, "2×2 buffered switch fabrics for traffic routing, merging, and shaping in photonic cell networks", IEEE/OSA J. Lightwave Technol., vol. 15, issue 1, pp. 86-101, Jan. 1997.
7. C.P. Larsen, M. Gustavsson, "Linear crosstalk in 4×4 semiconductor optical amplifier gate switch matrix", IEEE/OSA J. Lightwave Technol., vol. 15, issue 10, pp. 1865 - 1870, Oct. 1997.
8. T. Hasegawa et al, "Bismuth-based extra-high nonlinear optical fiber", in Proc. Conference on Laser and Electro-Optics (CLEO), Baltimore, 2005, pp. 2094-2096, CFC2.
9. G. Meloni et al, "Ultrafast All-Optical ADD–DROP Multiplexer Based on 1-m-Long Bismuth Oxide-Based Highly Nonlinear Fiber", IEEE Photon. Technol. Lett., vol. 17, issue 12, pp. 2661-2663, Dec. 2005.
10. J. H. Lee et al, "Wide-Band Tunable Wavelength Conversion of 10-Gb/s Nonreturn-to-Zero Signal Using Cross-Phase-Modulation-Induced Polarization Rotation in 1-m Bismuth Oxide-Based Nonlinear Optical Fiber", IEEE Photon. Technol. Lett., vol. 18, issue 1, pp. 298-300, Jan. 2006.
11. G. Theophilopoulos, et al, "Optically Addressable 2x2 Exchange Bypass Packet Switch", IEEE Photon. Technol. Lett., vol. 14, issue 7, pp. 998-1000, Jul. 2002.
12. C.K. Yow, Y. J. Chai, C. W. Tee, R. McDougall, R. V. Penty, and I. H. White, "All-Optical Multiwavelength Bypass-Exchange Switching Using a Hybrid-Integrated Mach-Zehnder Switch", in Proc. Eur. Conf. Optical Communication, vol. 3, 2004, pp. 704-705, We4.P.118
13. N. Sugimoto, H. Kanbara, S. Fujiwara, K. Tanaka, Y. Shimizugawa, and K. Hirao, "Third-order optical nonlinearities and their ultrafast response in Bi2O3 - B2O3 - SiO2 glasses ," J. Opt. Soc. Am. B 16, 1904-1908 (1999)
14. Varvarigos, E.M., "The "packing" and the "scheduling packet" switch architectures for almost all-optical lossless networks", IEEE/OSA J. Lightwave Technol., vol. 16, issue 10, pp. 1757-1767, Oct. 1998.
15. D. Tsiokos et al, "All-Optical 10 Gb/s Header Replacement for Variable Length Data Packets", in Proc. Eur. Conf. Optical Communication (ECOC), Rimini, Italy 2003, We4.P83.
16. S.Yao et al, "All-optical packet switching for metropolitan area networks: opportunities and challenges", IEEE Communications Magazine, vol. 39, issue 3, pp. 142-148, Mar. 2001.

Impact of Transient Response of Erbium-Doped Fiber Amplifier for OPS/WDM and its Mitigation

Yoshinari Awaji[1], Hideaki Furukawa[1], Naoya Wada[1], Eddie Kong[2], Peter Chan[2], Ray Man[2]

[1]National Institute of Information and Communications Technology (NICT), 4-2-1, Nukuikita, Koganei, Tokyo 184-8795, Japan
{yossy, furukawa, wada}@nict.go.jp
[2]Amonics Ltd., Unit 101, 1/F, Winning Centre, 29 Tai Yau Street, San Po Kong, Kowloon, Hong Kong
{eddie, peter, rayman}@amonics.com

Abstract. We investigated impairments of optical packets when these packets amplified by cascaded EDFAs in several distinctive cases in WDM environment. The origin of impairments were mainly gain excursion and additional noise come from inter-channel cross talk between WDM channels, and the degree of impairment changed according to link utilization. Moreover, the impairment is cumulative by increase of the number of EDFAs cascaded. It means that the tolerance for error of optical packet receiver substantially reduces according to the traffic intensity and the number of hops each packet experiences. We proposed the mitigation by adopting special class of EDF which have larger active erbium core. Our methodology can supplement conventional electrical controlling measures against transient responses which have insufficient response speed in the application for short optical packet of sub-micro seconds long.

Keywords: EDFA, Gain excursion, Cross gain saturation, Optical packet, WDM

1   Introduction

Development of optical packet transporting stratum is a fundamental issue to assure transparency, granularity, high efficiency of bandwidth utilization, agility of forwarding, and of course, high wire rate in optical telecommunications. Indeed, continuous bit stream or flame train with fixed repetition are easy to be transmitted on optical domain and to be operated by simple architecture in control plane, however, it can promise low efficiency on physical layer. Such kind waste of given resource is inherently irretrievable by effort in network architecture. Sophisticated techniques to handle optical packets are indispensable.

Optical packet carries not only messages of users but also functional signals upon requests of control plane. Therefore, the length and frequency of optical packet are various. As well known, such bursty intensity-modulated signals cause dynamic

I. Tomkos et al. (Eds.): ONDM 2007, LNCS 4534, pp 29-37, 2007.
© IFIP International Federation for Information Processing 2007

transient of envelope of packet when it is optically amplified by erbium-doped fiber amplifiers (EDFAs) [1].

As a matter of fact, similar distortion of envelope matters in electronic amplifiers mainly because of capacitance and drift of DC balance. Therefore, several artifices, such as finely tuned additional capacitance, electronic gain controlling, coding, and so on, were necessary to realize workable physical layer to handle electric packets.

The dynamic transient in EDFA has another origin. It is caused by consumption of excited population of Er^{3+} ions in gain media as a result of amplification itself and cross gain saturation in wavelength division multiplexing (WDM) environment [2]. Therefore, the transient in EDFA is inherently unavoidable as far as using intensity-modulated optical packets.

The measures previously reported are roughly classified to two categories, one is electronic approaches [3] including the automatic gain controlling (AGC) and another is optical approaches [4] including optical feedback loop. The electronic approaches suppose that the natural lifetime of Er^{3+} ions is around 10ms which define the gain dynamics of EDFA. On the other hand, the AGC can respond within sub-micro second [3]. It seems to be enough, but if we assume 10Gbps and 500bytes (typical average size on the Internet backbone) packet, the duration of the packet is only 400ns and the AGC cannot treat such short packet. Perhaps the transient in such relatively short optical packet had ignored comparing the natural lifetime. However, it is not negligible [5].

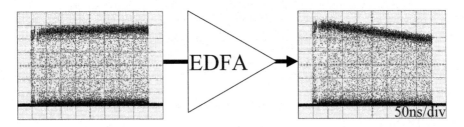

Fig. 1. An example of transient in optical packet envelope cause by EDFA. A packet consists of 128bits preamble and 3814bits payload at 9.95328Gbps, then the duration of packet was about 400ns.

We have observed the dynamic transient of EDFA for 400ns packet in single wavelength [5]. Because of short duration, it did not seem as optical surge but gradual gain excursion shown in Fig. 1. The degree of transient might be cumulative [1], and it is clear that such transient can seriously affect clock and data recovery (CDR) on receiver. Moreover, it is very difficult to be compensated except 2-R methodology using optical-electrical-optical (OEO) conversion which makes the optical forwarding opaque and limits available bandwidth.

On the other hand, optical controlling approaches resulted as insufficient suppression of transient or needed unfavorable complexity.

As one of effective mitigation, we have suggested the adoption of new class of EDF which can be applied in parallel with existing method to control EDFA [5].

In this paper, we observed the properties of transient response of cascaded EDFAs with optical packet over WDM. We confirmed that our proposed mitigation presented the best result within available devices under test (DUTs) even in WDM environment, hence the validity of our approach was proved. Finally, we discuss the impact of transient for optical packet switch (OPS) with WDM. We respectfully call attention about the gain excursion issue in designing of EDFA dedicated to OPS with WDM.

2   Concept of Examination

2.1   Formula of gain excursion

The gain excursion of EDFA can be approximated by the following formula [2].

$$G'(0) = \frac{[G(\infty) - G(0)]}{\tau_0} [1 + \sum_j \frac{P^{out}(\lambda_j)}{P^{IS}(\lambda_j)}] \tag{1}$$

$$P^{IS}(\lambda_j) = \frac{h\nu S}{[\sigma_a(\lambda_j) + \sigma_e(\lambda_j)]\Gamma_j \tau_0} \tag{2}$$

$G(0)$ is the gain before the transient and $G(\infty)$ is the steady gain after the transient, respectively. τ_0 is the intrinsic lifetime of the upper level of the erbium ions (We assumed two level systems of Er^{3+} even for three level pumping [5]). P^{IS} is the intrinsic saturation power at the wavelength channels. S is the active erbium area of EDF, σ_a and σ_e are the absorption and stimulated emission cross section at the wavelength channels, Γ is the confinement factor which is the overlap integral between the Er^{3+} ions and the mode field of light, respectively. In this formula, signal and pump light can be treated identically [6].

The $G'(0)$ is the initial slope of gain excursion and is inverse proportional to S. An improvement on gain excursion should be made by using EDF with larger active erbium area. We compared a two types of commercially available EDFAs (called as Type A and B) with our prototype of EDFA (called as Type C) optimized for OPS adopting an EDF with enhanced active area whose diameter was 4.3μm.

2.2   Form of optical packet

We have developed optical packet transmitter which generated WDM packet stream segmented to eight wavelengths from 10G-Ether flame input [7,8]. Fig.2 shows packet format and traffic intensity in demonstration. Bit rate of payload conformed to OC-192 (9.95328Gbps) and a packet consists of 128bits preamble and 3814bits payload, then the duration of a packet was about 400ns. We compare two different link utilization (LU [%] = summarized total duration of packets within a second) of 33% and 0.4%.

Comparing the actual experimental setup, the formula (1-2) present only approximative understanding because the extinction ratio of NRZ modulated signal was not ideal and typically 10dB. Hence, there is residual cw light and amplified spontaneous emission (ASE). For example, the summation of optical energies forming mark ("1") bits are only 1/50 of total optical energy in the case of LU=0.4%, taking into account that average mark ratio of packet was 50:50. The rest of optical energy belongs to DC component on space ("0") bits or blank time between packets. The energy for effective optical pulse is not dominant. On the other hands, the formula assumed fast switched WDM networks in which the bit stream is continuous, and pumping power for EDF was successfully consumed by effective optical pulses and explicit ASE component. However, our experience qualitatively agreed with the formula up to now. The precise and quantitative examination of formula should be future issue.

Packet format

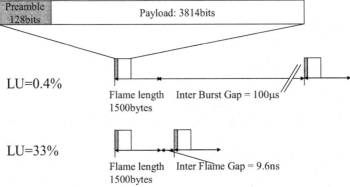

Fig. 2. Packet format and two different traffic intensities. LU: Link Utilization (summarized total duration of packets within a second)

Fig. 3 shows wavelength and timing allocations of packets in several distinctive cases in WDM environment. The packet transmitter originally generates 100GHz spaced eight wavelengths of packet at same timing (Pattern I). We also observed the de-correlated WDM packet stream (Pattern II) for comparison. The choice of Pattern I and Pattern II should depend on the network architecture, and these are typical case from the point of view of effect of cross gain saturation in EDFAs resulting in transient response.

Reminding that the formula also indicates that optical power on other wavelength affect the gain excursion (cross gain saturation), it should be more serious in WDM than our previous report in single wavelength [5]. To emphasize it we also observed Pattern III and IV.

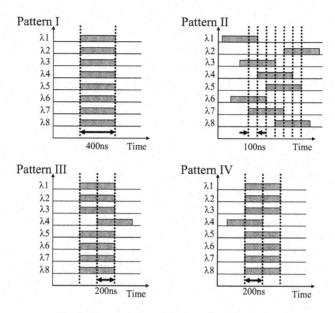

Fig. 3. Wavelength and timing allocations of packets

2.3   Experiment

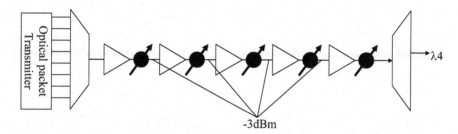

Fig. 4. Schematic drawing of experimental setup. Black circle with arrow indicates variable attenuator assuming optical losses derived from optical components.

Fig.4 shows schematic drawing of experimental setup. We assumed the model of setup as a typical configuration of OPS and cascaded up to five EDFAs. It can also be applicable to the model of optical packet network with several hops. The initial average input power into the first EDFA was -8 ~ -12dBm depend on the LU and it was amplified up to +13dBm. Then it was filtered with 7.6nm band pass filter and attenuated to -3dBm and injected into latter EDFA. Each gain of EDFAs was adjusted to 16dB. We observed λ4 (1550.12nm) only.

Table 1. Combination of condition

Type of EDFA	LU(Link Utilization)[%]	Wavelength and timing allocation
A (Commercial)	33%	I
B (Commercial)	0.4%	II
C (Proposed)		III
		IV

3   Results

At first, we summarized the result of observation.

As a comparison of the type of EDFA, Type A is the worst and Type C is the best. Type B is similar to Type C but slghtly worse.

As a comparison of LU, LU=33% present no significant impairment in almost case except Type A. We observed slight additional noise for Type A. To the contrary, LU=0.4% condition was severe and many impairment were observed.

It was also confirmed that the the degree of impairment was cumulative depending on the number of EDFAs cascaded by comparing with single stage of EDFA.

And we show several distinctive results according to wavelength and timing allocations in Fig. 5.

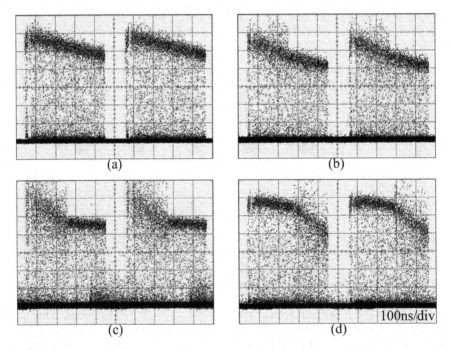

Fig. 5. Impairment derived from gain excursion and cross talk (caused by cross gain saturation). LU=0.4%. (a) Pattern I, Type B, (b) Pattern II, Type A, (c) Pattern III, Type A, (d) Pattern IV, Type A

Fig. 5 (a) shows that optical packet suffer gain excursion as a result of five cascades of EDFAs. Such degree of impairment was unavoidable in present. Type B tends to relatively low excursion but accumulation of excursions occerred even all wavelength have same timing.

Fig.(b)~(d) show the worst case with Type A. (b) There is additional bursty noise around the time when density of de-correlated packets on other wavelengths is high. It is clearly shown in (c) and (d) that seven channels other than observed one seriously affect on λ4 when they are dropped or added. To emphasize the impairment, the timing of seven channels are same. It points out that usage of unoptimized EDFA in WDM environment eventually result in such destruction of optical packet.

We observed the best result in all cases with Type C adopting proposed mitigation which is larger erbium core. Fig.6 shows the observed result with Type C corresponding to Fig.5.

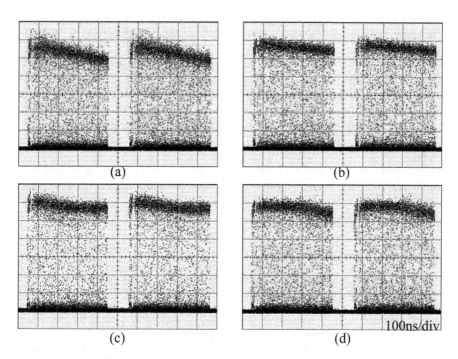

Fig. 6. The best result with Type C adopting proposed mitigation corresponding Fig.5. LU=0.4%. (a) Pattern I, (b) Pattern II, (c) Pattern III, (d) Pattern IV

Fig.6 (a) is almost same with Fig.5 (a) but 30% improved. Moreover, (b)~(d) show slight excursion but no significant additional noise come from cross talk while Type A shows heavy impairement. It is clear that our approach certainly mitigate gain excursion and cross talk.

4   Discussion and Concluding remark

We observed several impairments on WDM packet amplification caused by cross gain saturation. It appears not only as gain excursion but also additional noise come from inter-channel crosstalk. We have proposed to adopt EDF with larger erbium core for mitigation, previously. And we proved in this paper that our proposal is also effective in WDM enviroment.

There are three major features in our experimental result.

At first, reminding that there is no significant impairment with Type B and C in the condition of LU=33% while we observed serious gain excursion occured in LU=0.4%, the change of link utilization according to the request of control plane induces the change of packet envelope which affect receiver condition. This dependency on link utilization is more serious than the gain excursion itself.

Secondly, the degree of impairment was cululative. It is not only the problem inside of independent OPS, but also the problem of optical packet network because the packets from different paths and hops should suffer different impairment.

Therefore, mitigation of excursion is important as same as enhancing the tolerance of receiver.

Thirdly, our observed excrusion occured within several hundred nano seconds and existing AGC cannot respond during packet. Fortunately, our mitigation inherently reduces the transient and it can be easily applied to conventional relatively low speed controlling method to supplement it. As far as the application of optical packet with short duration, it is desirable to adopt a EDF with larger active erbium core.

Further issues of our proposal are quantitatively investigated with modified formula.

Nonetheless, mitigation of gain transient of EDFA is one of most important issues not only for OPS but also for optical links, for whole optical packet transporting stratum.

References

1. Schulze, E., Malach, M., Raub, F.: All-Raman amplified links in comparison to EDFA links in case of switched traffic. 28th European Conference on Optical Communication (2002) Symposium 3.8.
2. Sun, Y., Zyskind, J. L., Srivastava, A. K., Zhang, L.: Analytical formula for the transient response of erbium-doped fiber amplifiers. Applied optics 38, 9 (1999) 1682-1685
3. Nakaji, H., Nakai, Y., Shigematsu, M., Nishimura, M.: Superior high-speed automatic gain controlled erbium-doped fiber amplifiers. Optical fiber technology 9 (2003) 25-35
4. Zhao, C-L., Tam, H-Y., Guan, B-O., Dong, X., Wai, P. K. A., Dong, X.: Optical automatic gain control of EDFA using two oscillating lasers in a single feedback loop. Optics communications 225 (2003) 157-162
5. Awaji, Y., Furukawa, H., Wada, N., Chan, P., Man, R.: Mitigation of transient response of erbium-doped fiber amplifier for burst traffic of high speed optical packets. Submitted to Conference on Lasers and Electro-Optics (2007)
6. Saleh, A. A. M., Jopson, R. M., Evankow, J. D., Aspell, J.: Modeling of gain in erbium-doped fiber amplifiers. IEEE Photon. Technol. Lett. 2, 10 (1990) 714-717

7. Furukawa, H., Wada, N., Harai, H., Naruse, M., Otsuki, H., Katsumoto, M., Miyazaki, T., Ikezawa, K., Toyama, A., Itou, N., Shimizu, H., Fujinuma, H., Iiduka, H., Cincotti, G., Kitayama, K.: All-optical multiple-label-processing based optical packet swtich prototype and novel 1-Gb ethernet / 80 (8λ x 10) Gbps-wide colored optical packet converter with 8-channel array burst-mode packet transceiver. Optical Fiber Communication Conference (OFC), OWC5 (2007)
8. Harai, H., Wada, N.: More than 1-Gbps photonic packet-switched networks using WDM-based packet compression. 8th OptoElectronics and Communications Conference (2003) 15E3-3

Mutual Impact of Physical Impairments and Grooming in Multilayer Networks

Szilárd Zsigmond[1], Gábor Németh[1], Tibor Cinkler[1],

[1] Budapest University of Technology and Economics,
Department of Telecommunications and Media Informatics
2nd Magyar Tudósok Krt., Budapest, H-1117
{zsigmond, nemeth, cinkler}@tmit.bme.hu

Abstract. In both, metropolitan optical networks (MON) and long haul optical networks (LHON) the signal quality is often influenced by the physical impairments, therefore a proper impairment based routing decision is needed. In the absence of all-optical 3R regenerators, the quality of transmission has a strong impact on the feasibility of all-optical transmission. It is assumed that signal regeneration can be done only in electrical layer. Once the signal is in electrical layer there are some features supported e.g. the traffic grooming. We show that by taking into account both, the physical impairments characterized by the Q-factor, as we propose, and the features of the electrical layer, will have a strong impact onto the routing that is based on impairment constraints.

Keywords: ASE, PMD, ICBR, Grooming, OSNR, Q-factor

1 Introduction

The tremendous growth in broadband communication services, brought for the phenomenal expansion of the internet, has triggered an unprecedented demand for bandwidth in telecommunication networks. Wavelength division multiplexing (WDM) has been introduced to increase the transmission capacity of existing optical links. Multi-wavelength technology appeared as the solution for the bandwidth hungry applications. WDM has been introduced to increase the transmission capacity of existing optical links. It has been soon recognized that the switching decision can be made according to the incoming wavelength without any processing of the data stream. In single hop WDM based All Optical Networks (AON) a wavelength is assigned to a connection in such a way that each connection wavelength is handled in the optical domain without any electrical conversion during the transmission [1], [2]. Routing and Wavelength Assignment (RWA) takes a central role in the control and management of an optical network. Many excellent papers deal with design, configuration and optimization of WDM networks. See e.g. [3]-[5]. The majority of these RWA algorithms assume that once the path and wavelengths have been identified, connection establishment is feasible. This is true when we consider that in each node the signal is regenerated but may not be true in transparent networks, where the signal quality degrades as it is transmitted through optical fiber and nodes.

I. Tomkos et al. (Eds.): ONDM 2007, LNCS 4534, pp 38-47, 2007.

Impairment constraint-based routing (ICBR) may be used in transparent networks as a tool for performance engineering with the goal of choosing feasible paths while obtaining the optimal routes regarding the RWA problem. Many excellent papers have been written about constraint based routing which obeys physical effects [6-9].

There is no doubt, that the near future info-communications will be based on optical networks. In general for networks of practical size, the number of available wavelengths is lower by a few orders of magnitude than the number of connections to be established. The only solution here is to join some of the connections to fit into the available wavelength-links. This is referred to as traffic grooming. The main idea of our optimization was that in optical layer we do not make signal regeneration. We assume that in the optical layer, there is no signal regeneration, and the noise and signal distortion accumulate along a lightpath. Actually, re-amplification, re-shaping, and re-timing, which are collectively known as 3R regeneration, are necessary to overcome these impairments. Although, 3R optical regeneration has been demonstrated in laboratories, only electrical 3R regeneration is economically viable in current networks.

We have already mentioned that in the electric layer it is possible to do traffic grooming. If we investigate the physical limitations in the optical domain, and take them into consideration, we will have to include new optical-electrical-optical conversion just to ensure the quality prescriptions. These new optical-electrical-optical conversions will have influence onto the RWA process.

The rest of the paper is organized as follows. Section 2 describes the investigation of the physical layer impairments. In Section 3 we describe the routing and wavelength assignment process. In Section 4 the results are presented and finally in Section 5 we conclude our work.

2   Modeling the Physical Layer Impairments

The signal quality of a connection is characterized by Bit Error Ratio (BER). Experimental characterization of such systems is not easy since the direct measurement of BER takes considerable time. Another way of estimating the BER is to degrade the system performance by moving the receiver decision threshold value, as proposed in [10]. This technique has the additional advantage of giving an easy way of estimating the signal quality (Q) of the system, which can be more easily modeled than the BER. [11] explains well and gives a definition to it. The Q-factor is the signal-to-noise ratio of the decision circuit in voltage or current units, and can be expressed by:

$$Q = \frac{\langle I_1 \rangle - \langle I_0 \rangle}{\sigma_1 + \sigma_0} \quad (1)$$

where: $I_{1,0}$, is the mean value of the marks/spaces voltages or currents, and $\sigma_{1,0}$ is the standard deviation

In our model we consider a chain of amplifiers and optical cross-connects (OXC). The calculation of the Q is based on [12] where fully transparent optical cross connection architecture is presented. In this study the OXC architecture is based on wavelength selective architecture, as can be seen in Figure 1. The switching is done for each wavelength by an $(N+1) \times (N+1)$ switch that is included between the demultiplexer and the multiplexer.

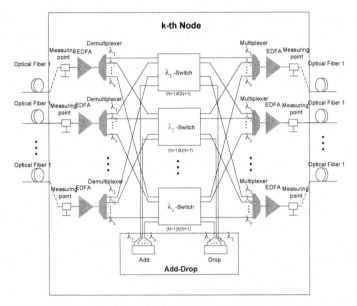

Figure 1: Architecture of an OXC

In this approach the noise, power and distribution for ones and zeros are calculated recursively. Assuming we know the ASE and crosstalk parameters at node k-1 and the parameters of node k, then we can calculate the ASE and crosstalk parameters at node k. In this approach the crosstalk is introduced only in the OXC nodes and ASE is introduced by the erbium-doped fiber amplifiers (EDFA), which the signal passes through. We assume that in every 80 km there is an inline amplifier.

The impact of PMD onto the signal quality can be calculated based on [13], where the PMD-induced degradation is assessed by an eye-opening penalty (EOP) along the lines. This EOP is subsequently translated to a Q-factor penalty. [14]

The main physical parameters of the network can be seen in Table1.

N_f	4,8 dB	Noise figure of the EDFA
X_{sw}	40dB	Crosstalk of the switch
D_{pmd}	0,1ps/nm*km	PMD coefficient of the fiber
Alpha	0.2 dB/km	Fiber Attenuation
L_{tap}	1dB	Attenuation of the measuring point
L_{mx}	4dB	Attenuation of the multiplexer
L_{dmx}	4dB	Attenuation of the demultiplexer
L_{sw}	8dB	Attenuation of the switch
OP	10^{-3}	Outage probability
P_{out}	8dBm	Total Output of an EDFA

Table 1: The main physical parameters of the network

3   The Routing Model

The routing algorithm is a highly complex algorithm which can handle optical nodes, electrical nodes and optical nodes with electrical regenerations. We consider two layer architecture, an electrical layer and an optical layer. The electrical layer supports some features such as traffic grooming and λ-conversion. The routing is realized by a shortest path algorithm. Each link and node has its own cost. In this way we can choose the lowest cost path by implementing Dijkstra's algorithm. This algorithm can route demands dynamically. The input of the optimization is the network topology and the demands. The output of the algorithm is the set of optimal routes and statistical data on the blocking in the network. The routing parameters contain information about the blocking ratio and the reason why the route has been blocked. A route can be blocked due to the RWA problem, or because of the physical impairments. A route is blocked due to RWA problem if there is not enough resource to route the demand between the source and destination node. This happens when all the wavelengths are used or in case of grooming there is not enough free capacity to groom the demand We consider a route blocked due to physical impairments if Q value of the route is lower than 3.5 which is still acceptable if using coherent detection schemes.

3.1   Routing Algorithm

The setup of the algorithm can be split in two main parts. The first one is the routing part and the second one is the calculation of physical impairments (CPI), which can be switched on or off, (Figure 2). The communications between these two parts are as follow: The routing algorithm chose an optimal route, between the source and destination node and if the CPI is switched on, it sends the description of the route to CPI. The description of the route contains the lengths of the optical fibers between the nodes. The CPI calculates the signal quality and if it is adequate it sends a message back to the routing part, that the connection can be established. If the signal quality is not adequate the CPI determines the maximum reachable node (MRN) along the path and sends this information back to the routing model. The routing model establishes the connection between the source and the MRN, than chooses another route between the MRN and the destination node. If the MRN is the source node e.g. there is no possible connection due to the physical layer, the route is blocked.

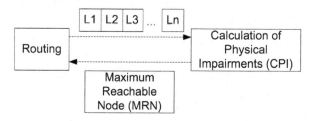

Figure 2: Set-up of the algorithm

To perform the effects of grooming onto the ICBR we made fore *simulation* types.

- The first one when there is no grooming in the RWA and the physical effects are negligible
- The second one when there is *grooming* and the physical effects are negligible
- The third one when there is no grooming in the RWA and we take into consideration the *physical effects*
- The fourth one when there is *grooming* and the *physical effects* are taken into consideration

As it was mentioned before the routing is done by a shortest path algorithm, when each link has its own cost. By using different cost values for the links of the network we can optimize an RWA oriented, or a physical impairments oriented routing. For this purpose we use four *metrics*.

- The first one where the cost of each link is the same. Will be referred as *hop routing.*
- The second one when the cost of each link is equal to the length of the link. Will be referred as *length routing.*
- The third one when the cost of the link is equal to the $1/Q$ where the Q is the Q-factor of the link
- The fourth one when the cost of the link is equal to the $1/Q^2$ where the Q is the Q-factor of the link

In the case of the third and of the fourth metrics we calculate the Q-factor of each link as a point-to-point connection between the two end nodes of the link. The Q-factor based routings are not obviously the best routings for the point of view of the physical layer. This is due to the nonlinear behavior of the Q-factor. If we have two lightpats, each lightpath has its representative Q, for example Q_1 and Q_2. Consider a route which contains these two lightpats in chain. The overall Q can not be calculated from these two Q-factors, if we take both the PMD and ASE effects into calculation. The only assumption which we can make, is that, if Q_1 and Q_2 have a high values than the overall Q will be high as well.

The exact flow of the algorithm can be seen in Figure 3.

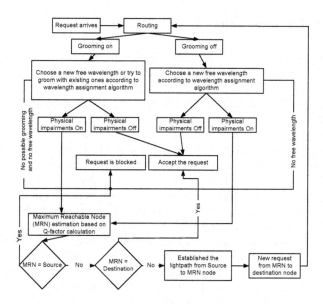

Figure 3. Flow chart of the algorithm

3.2 Network and traffic generation

The used network scenario is one of the COST266 [15] reference networks, Figure 4. Each link contains 24 wavelengths. The used bit rate is 10Gbit/s. The generation of the demands is based on the traffic matrices for year 2006 of the COST266 European Reference Network. More than 9000 demands were generated and routed in each simulation. The arrival of the demands occurred according to a Poisson process with the intensity of 0.005.

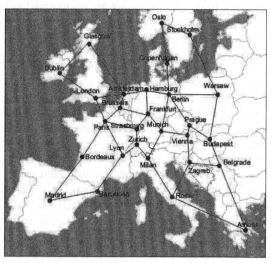

Figure 4: Used network scenario

4   Simulation Results

We compared the four metrics used for representing the cost values of the links, in Figure 5-6. In Figure 5 the calculation of the physical impairments was switched off and the grooming capability was switched on, and in figure 6 both modules were switched on. In the X axis the scale of the network can be seen. The meaning of it is that we changed the used network link lengths by multiplying the original lengths with the scale parameter. This resulted in increase of impairments. On the Y axis the blocking ratio is plotted. In Figure 5 it is to be seen that the best metric from the point of view of the blocking ratio is the hop-metric followed by $1/Q$ and $1/Q^2$ metrics while length metric yields the worst results. We expected that in case when the physical impairments are switched off the scale of the network has no influence onto the blocking ratio. This is true when the grooming is switched off. In case of grooming there are several routing decisions which have the same ratio so it is done randomly. These random decisions lead to the non-deterministic behavior.

In figure 6 when the physical effects are taken into consideration the differences between the four metrics decrease. To understand this behavior we investigated the blocking ratio dependency on to the physical effects, see figure 7. In the X axis the network scale and in the Y axis the blocking ratio due to physical effects is plotted. This blocking ratio contains only the blockings due to physical effects without rerouting. This means that the routing module chooses an optimal lightpath and the CPI module calculates its Q-factor. If the Q is lower than 3,5 then the request is blocked. This is a more simplified scenario than the one presented in Section 3. In Figure 7 it is to be seen that the characteristics of the curves are what we expected. In case of low network scales, where the lengths of the links are very small, where the physical effects have no influence, the blocking ratio is very low. While increasing the link lengths, we increase the influence of the physical effects, the blocking ratio is increasing. We compared the four metrics from the point of view of blocking ratio due to physical impairments i.e. grooming and rerouting capabilities were not used at all. Length routing has the best performance while hop routing has the worst. Between these two are the Q-based routings. Of course it is possible to find a metric which is the function of Q, f(Q), that gives better results than the length based metric, however this is not the scope of this paper.

Returning to Figure 6 the blocking ratio subsidence between the four metrics is due to the constraints on the physical effects. In the aspect of physical effects the best metric is the length followed by the $1/Q^2$, and the $1/Q$ while the worst is the hop metric. From the point of view of RWA the order of these four metrics is reverse. Taking into account both the physical effects and the RWA problem, as we did, will leads to the behavior.

The other interesting property is that while increasing the scale of the network the blocking ratio decreases. This is due to the fact that increasing the lengths of the network increases the influence of the physical effects. The effect of this influence is that we have to do more optical-electrical-optical regenerations (OEO). If there are more points where the signal goes to the electrical layer, and we are capable to groom in these nodes, the network will be more optimally used. This leads to decreased blocking ratio.

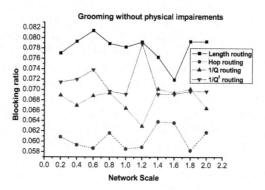

Figure 5: Blocking ratio dependency from the scale of the network in case of grooming without physical effects

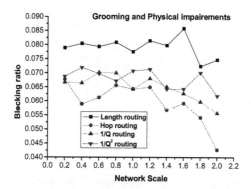

Figure 6: Blocking ratio dependency from the scale of the network in case of grooming and physical effects

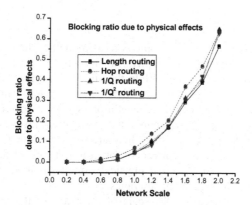

Figure 7: The dependence of the blocking ratio on physical effects from the as the network scales.

In Figure 8 we plotted the blocking ratio dependency on the scale of the network for the four routing scenarios using the length routing metric. The characteristics of the curves were the same for each metric. As it was expected there is a huge difference in the blocking ratio when the grooming capability is switched on or off. The other interesting property is that in case when the grooming is switched off and the physical impairments constraints are taken into consideration while increasing the scale of the network the blocking ratio is increasing. This is because increasing the lengths of the network the physical effects become dominating so we have to do more often OEO regeneration which increase the overall load of the network. In case when the nodes are capable to groom this trend of blocking growth can not be observed. As we mentioned before, see Figure 6, the blocking ratio is even decreasing while increasing the scale of the network.

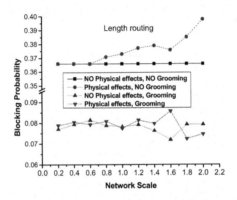

Figure 8: The dependency of the blocking ratio from the scale of the network for the four routing scenarios using length routing metric.

5   Conclusion

In this paper we presented a simple way of modeling the physical effects of the network. We demonstrated that the physical limitation must be taken into account while routing decisions are taken. While the signals pass through the network their quality deteriorates. Considering that signal regeneration can be done only in electrical layer, since all optical 3R regeneration is not commercially available, there are some nodes where the signal is converted to electrical layer. Once the signal is in the electrical layer except regeneration there are some additional features that can be performed such as wavelength conversion or grooming. If all this effects are taken into account the results express that physical limitations have serious impact onto routing and wavelength assignment strategies.

We show that by increasing the scale of the network the influence of physical impairments grows. The effect of this influence is that we have to do more optical-electrical-optical regenerations (OEO). If there are more points where the signal is in electrical layer and we are capable to groom in these nodes the blocking ratio is even decreasing while increasing the scale of the network.

Acknowledgments. This work has been supported by COST291 (http://palantir.ait.edu.gr/cost291), and e-Photon/ONe+ (http://www.e-photon-one.org) projects. The authors would like to thank to Dr. Ioannis Tomkos and Dr. Anna Tzanakaki for their advices regarding to physical impairments calculations.

References

1 P. E. Green "Optical Networking Update" , IEEE Journal on Selected Areas in Communications , vol 14 no. 5 pp 764-779, June 1996

2 I. Chlamtac, A. Ganz, G. Karmi, "Lightpath Communications: An Approach to High Bandwidth Optical WANs", IEEE Transactions on Communications, vol. 40, no. 7, pp.1171-1182, July 1992

3 N. Wauters, P. Demister, "Design of the Optical Path Layer in Multiwavelength Cross-Connected Networks", IEEE Journal on Selected Areas in Communications, vol. 14, no. 5 pp. 881-892, June 1996

4 R. Ramaswami, K.N. Sivarajan, "Routing and Wavelength Assignment in All-Optical Networks", IEEE Transaction on Networking, vol. 3 no. 5 pp. 489-500, Oct. 1995

5 D. Banerjee, B. Mukherjee, "A practical Aproach for Routing and Wavelength Assignment in Large Wavelength-Routed Optical Networks", IEEE Journal on Selected Areas in Communications, vol. 14 no. 5 pp. 903-908, June. 1996

6 M. Ali, D. Elie-Dit-Cosaque, and L. Tancevski, "Enhancements to Multi-Protocol Lambda Switching (MPlS) to Accommodate Transmission Impairments," *GLOBECOM '01*, vol. 1, 2001, pp. 70–75.

7 M. Ali, D. Elie-Dit-Cosaque and L. Tancevski "Network Optimization with Transmission Impairments-Based Routing" *European Conf. Opt. Commun.2001*, Mo.L.2.4, Amsterdam, The Netherlands, 2001, pp. 42–43.

8 B. Ramamurthy Debasish Datta, Helena Feng, Jonathan P. Heritage, Biswanath Mukherjee, "Impact of Transmission Impairments on the Teletraffic Performance of Wavelength-Routed Optical Networks," *IEEE/OSA J. Lightwave Tech.*, vol. 17, no. 10, Oct. 1999, pp. 1713–23.

9 I. Tomkos *et al.*, "Performance Engineering of Metropolitan Area Optical Networks through Impairment Constraint Routing" *OptiComm*, August 2004

10 N. S. Bergano, F.W. Kerfoot and C.R. Davidson "Margin measurements in optical amplifier systems," IEEE Photon. Technol. Lett. Vol. 5, pp. 304-306, Mar. 1993

11 G. P. Agrawal, Fiber-Optic Communication Systems. New York: Wiley, 1997.

12 B. Ramamurthy, D. Datta, H. Feng, J. P. Heritage, B. Mukherjee, "Impact of Transmission Impairments on the Teletraffic Performance of Wavelength-Routed Optical Networks," IEEE/OSA J. Lightwave Tech., vol. 17, no. 10, Oct. 1999, pp. 1713–23.

13 C.-J. Chen. System impairment due to polarization mode dispersion. Proc. Optical Fiber Conference and Exhibit (OFC), 77–79, paper WE2-1. 1999.

14 J. Kissing, T. Gravemann, E. Voges. Analytical probability density function for the Q factor due to pmd and noise. IEEE Photon. Technol. Lett., vol. 15(4):611-613. 2003.

15 J. Kissing, T. http://www.ure.cas.cz/dpt240/cost266/index.html

Impairment Aware based Routing and Wavelength Assignment in Transparent Long Haul Networks

George Markidis[1], Stelios Sygletos[1], Anna Tzanakaki[1], and Ioannis Tomkos[1]

[1] Athens Information Technolgy (AIT),19,5 Markopoulou Avenue 68,
19002 Athens, Greece
{gmar, ssyg, atza, itom}@ait.edu.gr

Abstract. We investigate new routing and wavelength assignment algorithms considering as constraints physical impairments that arise in transparent networks. Accounting both linear and nonlinear impairments we propose a scheme that integrates the routing and wavelength assignment to achieve an optimal combination of physical and networking performance. Through simulations of a typical long haul network the improvement achieved using the proposed approach compared to conventional RWA algorithms (i.e shortest path and first fit) is demonstrated and the significance of impairment aware routing and wavelength assignment schemes is recognized.

Keywords: Routing and Wavelength Assignment (RWA), Impairment Constraint Based Routing (ICBR), transmission impairments, transparent networks.

1 Introduction

Transparent WDM optical networks have gained strong importance, over the last years, and are considered as a viable solution to meet the rapidly increasing bandwidth demands imposed by the explosive growth of the Internet. In these networks signals are transported end-to-end optically, without being converted to the electrical domain along their path and a control mechanism is required to establish and tear down all-optical connections, called lightpaths, between source and destination nodes. This is known as connection provisioning and is achieved utilizing specific routing and wavelength assignment algorithms (RWA). The development of an intelligent control plane able to provide efficient connection provisioning is an important traffic engineering problem for minimizing the cost and improve the efficiency of resource utilization. The traditional RWA schemes and formulations [1][2] make the routing decision based only the network level conditions such as connectivity, capacity availability etc. and they don't include the impact of physical layer in the overall network performance. However, in transparent networks the transmitted signal has an analogue nature and when it propagates through optical fibers and OADMs/OXC nodes it experience impairments that accumulate.
Recently, various Impairment Aware Routing and Wavelength Assignment (IA-RWA) algorithms have been proposed that take into consideration the impact of the

I. Tomkos et al. (Eds.): ONDM 2007, LNCS 4534, pp. 48–57, 2007.

physical network performance when assigning connections.[3][4][5][6] The challenges of an IA-RWA approach originate from the seemingly diverse nature of the networking and physical performance issues that have to be considered. From one hand there are the impairments that are existent in the physical layer and on the other hand there are the networking aspects (blocking probability, end-to-end delay, and throughput) that capture and describe the overall performance of the WDM optical network. It is evident, that all these heterogeneous issues have to be modeled and unified under a properly designed framework that will provide a solution for the RWA problem, which will be both feasible and efficient at the same time.

This paper extends the work presented in [7] by implementing an IA-RWA scheme for an all optical long haul network. This is achieved by also considering all the nonlinear degradations that are imposed on the signal when it transverses the fiber Therefore, the degradations due to cross-phase modulation (XPM), four-wave-mixing (FWM) and self phase modulation (SPM) have been analytically modeled and their combined influence has been integrated into the same figure of merit function, the Q-factor, along with the linear degradations. In addition, due to the wavelength dependence of nonlinear impairments, a novel wavelength assignment scheme is proposed to provide improve performance compared to conventional wavelength assignment schemes like first fit and random wavelength assignment.

2 Modeling of transmission induced impairments

2.1 Transmission link architecture

A challenging task in the engineering of transparent and dynamically reconfigurable optical networks, where regenerators are absent from the switching nodes is to deal with the accumulation of the impairments over the cascaded fiber sections. Different techniques to increase propagation distances and achieve transparency at least for a few links, have been identified by inserting specific devices, such as Raman amplification, dynamic gain equalizers [8], or dynamic dispersion compensators. This solution has the drawback of high cost, competing with the savings of optoelectronic interfaces. Alternatively, it was suggested to use dispersion management rules and EDFAs to achieve acceptable signal quality at the end of the path. In our simulation case the second approach is adopted by using a specific dispersion management scheme that eliminates the resonant built up of nonlinear penalties. Moreover it is simple and cost-effective to implement particularly in B&S OADM and OXC networks [9]. According to this scheme each amplifier span is under-compensated, whilst pre-compensators are required at the OADMs and OXC sites to reduce the total amount of the accumulated dispersion.

The schematic diagram of the transmission link that has been considered for this study is illustrated in fig. 1. At the end of each SMF span there is a double stage EDFA that is used to compensate the corresponding losses. Also, a DCF fibre module exists at the intermediate stage of the EDFAs to be used for the appropriate dispersion management. At the beginning of the link there is also a pre-compensating fiber module, whilst post-compensation may also be provided at the end of the link. The parameters P_{inSMF}, P_{inDCF}, $P_{inPRE,}$ and P_{inPOST} represent the peak power levels of the signal pulse-stream at the input of the SMF, DCF and PRE, POST fibers sections, respectively. The rest of their corresponding parameters are summarized in Table 1.

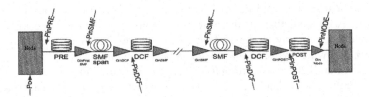

Fig. 1. The transmission link architecture.

Parameters	SMF	DCF
Attenuation a (dB/km)	0.25	0.5
Nonlinear index coefficient n (m^2/W)	2.6 10^{-20}	3.5 10^{-20}
Chromatic Dispersion Parameter D(s/m^2)	17 10^{-6}	-80 10^{-6}
Dispersion Slope dD/dλ(s/m)	0.085 10^3	-0.3 10^{-3}
Effective Area A$_{eff}$(m^2)	65 10^{-12}	22 10^{-12}

Table 1. Parameters of the fibers used on each link

2.2 Q factor formulation

As the signal propagates through a transparent network it experiences the impact of a variety of degrading phenomena that introduce different types of distortions. For example, there are distortions of almost "deterministic" type related only to the single channel's pulse stream, such as the interplay of SPM and GVD or the optical filtering introduced by the MUX/DEMUX elements at the OXC/OADMs. The other category includes degradations of pertubative nature, introduced by the ASE noise, WDM nonlinearities such as Four-Wave-Mixing (FWM) and Cross-Phase-Modulation (XPM) and finally crosstalk.

To incorporate the physical layer impact of a WDM network into the routing procedure there is a need the corresponding impairments to be evaluated in an accurate and time efficient way. Numerical modeling of a long haul WDM network is computationally a very heavy procedure that might take several hours to produce a single result. On the other hand, analytical or semi-analytical models can be also derived for each degradation to provide an instantaneous estimation of their relative influence. What should be noted in this case is the way all the different types of degradations can be integrated to a single physical layer performance metric.

In this work, Q-factor metric has been used to integrate all the different types of degradations and thus to reflect the overall signal quality. Several assumptions have been considered to achieve this. The first is that any interplay among the different types of degradations is ignored. Furthermore, the statistics of the distortions that have pertubative nature (such as ASE, Crosstalk, FWM and XPM) follow a Gaussian distribution. Finally, concerning the SPM/GVD and optical filtering effects, these are introduced through an eye closure penalty metric calculated on the most degraded bit-pattern. The corresponding Q-factor penalty on the k-link into the network is then given according to the following equation (1).

$$Q_k = \frac{pen_k \cdot P}{\sqrt{\sigma^2_{ASE,k} + \sigma^2_{crosstalk,k} + \sigma^2_{XPM,k} + \sigma^2_{FWM,k}}} \tag{1}$$

where pen_k is the relative eye closure attributed to optical filtering and SPM/group velocity dispersion (GVD) phenomena, calculated semi-analytically through single channel simulations according to the model proposed in [10]. Furthermore, $\sigma^2_{XPM,k}, \sigma^2_{FWM,k}$ represent the electrical variances of the XPM and FWM induced degradations, which are calculated according to [11], [12], and are added to the corresponding electrical variances of the ASE noise $\sigma^2_{ASE,k}$ and the generated crosstalk $\sigma^2_{crosstalk,k}$.

3   Proposed IA-RWA Scheme

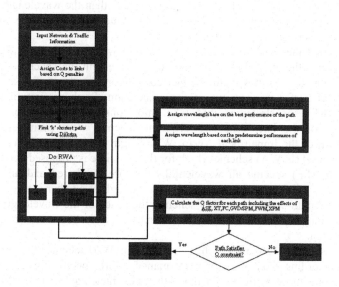

Fig. 2. Flow chart of proposed IA-RWA.

The flowchart of the proposed IA-RWA scheme is shown in Fig. 2. Initially the preprocessing phase collects all the information related to the network and the traffic demands. Information such as the topology of the network, the link capacities, the fiber characteristics, dispersion map applied (pre, post and inline dispersion compensation), span lengths, attenuation of each span, the launched powers at each fiber segment, the noise figure of the amplifiers, the nodes architecture, the channel spacing and the link capacities are required by the algorithm for the physical impairments evaluation. Moreover information concerning the number of requests, the bit rate and the source destination pairs are required to identify the traffic demands for the static routing that is considered here. Then, based on the collected information a Q-factor penalty is assigned (according to eq.1) as a cost parameter to each link reflecting the corresponding degradation on the signal quality.

The RWA phase is initiated once the link costs have been found. This phase assigns paths (incorporating the physical impairments as weight of the links or applying conventional shortest path by utilizing link lengths as link weights) and wavelengths to all the demands. The RWA problem is handled in two steps. In the first step it is treated as a single-joint optimization problem as this has been shown to be the optimum approach [3], [13]. In order to simplify the optimization task and reduce the computation complexity the following approach was taken: The "*Find k shortest paths*" part of the algorithm identifies the *k* shortest paths for each source-destination pair. These *k* paths are the input for the "*Do RWA*" module (Fig. 2), which gives to the algorithm the flexibility to select the optimum path among the *k* input paths in terms of load balancing and physical layer performance. We should note that the *k* parameter is set to 3 since this value proved to be a good compromise between efficiency and computational overhead. At the end of this step the algorithm identifies the minimum number of wavelengths required to carry the requests and specifies the paths that should be established. If this number is less than the wavelengths that are available in the network then an additional blocking percentage due to lack of network resources is calculated at this stage.

Having identified the minimum number of required wavelengths along with all the established paths, the algorithm may be used to implement the specific wavelength assignment scheme. This may include either conventional strategies that are unaware of the physical performance status of the connection, (such as first fit (FF), and random fit (RF)) or strategies that take into consideration the corresponding impairments. For the later case two different schemes were examined, a direct implementation of the Impairment Aware Wavelength Assignment-(IAWA), as well as, a Pre-Specified (IAWA) scheme which for the first time is proposed here.

In the first fit (FF) scheme all wavelengths that are numbered and among those available on every link of the path the one with the highest number is selected. In the Random wavelength assignment (RF) scheme, the space of the wavelengths that are available on the required path is firstly identified and then one wavelength is randomly chosen.

In the Impairment Aware Wavelength Assignment (IAWA) scheme, the lightpaths are established according their Q-factor performance. More specifically, each potential lightpath, among those available on the path, is characterized in terms of Q-factor taking into consideration the already established wavelength connections. The one having the optimum performance is finally selected.

A novel Pre-Specified Impairment Aware Wavelength Assignment scheme (PS-IAWA) that offers an advanced performance in terms of computational efficiency has been also considered. According to this scheme, and prior to the wavelength assignment process, all the wavelength locations on per link basis are characterized and ordered in terms of their Q-factor value. Then the algorithm will define the paths and for each one of them the space of the common wavelengths that are available across its links. The final selection will be made based on the pre-specified order created at the beginning.

After the wavelength assignment is completed the control is transferred to the Impairment Constraint Based Routing module which verifies the Q factor constraint considering all the physical impairments involved across the path. In this case the overall Q factor for each selected lightpath is calculated considering the accumulated amount of degradation A path is accepted when the Q-factor value at the destination node is higher than 11.6dB, which corresponds to a BER of 10^{-15} after forward error correction (FEC) is utilized at 10Gbit/s and the connection is established, in any other case the path is rejected and the connection is blocked.

4   Transparent Network Simulations

The performance of our proposed IA-RWA scheme was compared with other RWAs under a Long Haul network representing a core Pan-European network.

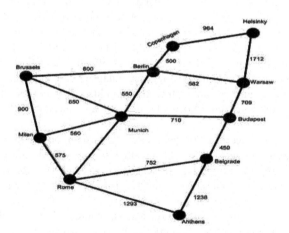

Fig. 3. The network used for the simulations

The network topology as illustrated in Fig. 3 consists of 11 nodes corresponding to 11 different European cities interconnected by 16 bidirectional links. Link lengths cover a range from 400 to 1750km, with the average link length being 813km. We assume that there are connection requests between every possible pair of nodes and therefore 55 end-to-end connections are requested to be established.

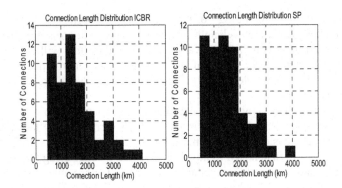

Fig. 4. The connection length distribution for ICBR and Shortest Path (SP).

The distributions of the lengths of these connections are depicted in Fig. 4 when Impairment Constraint Based Routing (ICBR) and Shortest Path (SP) algorithms are used to establish the connections. The average connection length when SP is used is 1480km and for the case where ICBR is used to satisfy the requests the average connection length is a bit higher at 1518km.

Based on the topology introduced above the network performance is evaluated in terms of blocking percentage which is defined as the ratio of the total number of connections that are unable to be established due to capacity and performance constraints divided by the total number of connections requests. In the following simulations we have considered that 40 wavelengths can be located at each link in a 50GHz channel spacing. Initially for a specific dispersion map where the residual dispersion after each 80 km SMF-DCF segment is 30ps/nm and the amount of pre dispersion is -400ps/nm, the overall blocking percentage is calculated as a function of the channel power levels at the input of the inline modules. The corresponding results are presented in figures 5a,c and 5b,d for both ICBR and shortest path (SP) routing schemes.

It can be noticed that by selecting a proper wavelength assignment scheme the overall blocking percentage of the network can be considerably improved. Our proposed PS-IAWA outperforms first fit and random fit schemes and exhibits similar performance behavior with the more computational intensive IAWA. This is because the latter scheme calculates the Q factor of all potential wavelengths that can be used to establish a lightpath each time a path is considered, whereas the PS-IAWA does not require any further calculation once the order of wavelengths has been defined at the beginning. For low power levels in the DCF segment (-4dB to 2dB) the observed improvement is around 5% between the IAWA schemes and the random WA scheme, and more than 20% compared with first-fit scheme. The significant advantage of IAWA schemes is viewed when the power levels of the DCF increase. In such cases, a considerable improvement of 20% to 40% is earned by introducing the IAWA schemes compared with the random fit-case. This is even more for the first fit wavelength assignment scheme. Similar conclusion can be derived as the power in the SMF fiber increases. For this case the benefit is more than 10% for the majority of the input powers between the IAWA schemes and the random WA. Therefore by applying the proposed IAWA schemes in the network a wider range of input powers can be tolerated in both SMF and DCF segments. Also introducing our ICBR algorithm for the path computation procedure proves to be beneficial against the conventional shortest path as demonstrated by comparing figures 5a and 5c with 5b and 5d respectively. An appreciable improvement, around 10% is indicated at least for cases where the blocking percentage is at an acceptable level as it appears when IAWA schemes are implemented. Conclusively, the combination of a proper WA assignment scheme with an ICBR algorithm provides significant performance improvement in the network.

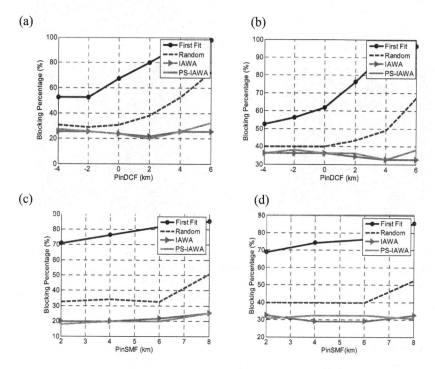

Fig. 5. Blocking percentage for ICBR (a,c) and SP (b,d) as a function of the power level at the DCF and the SMF segments for different Wavelength Assignment schemes.

In the next step of our analysis we investigate the blocking percentage of the network as a function of the pre and inline dispersion parameters. The results depicted in Fig. 6 represent the blocking percentage for each of the considered wavelength assignment schemes when the ICBR algorithm is involved. The advantage of the proposed PS-IAWA scheme comparing with the typical first fit and random-fit is quite considerable. The wide regions of optimum performance that are identified when impairment aware schemes are implemented designate flexible dispersion engineering. The whole spectrum of the implemented dispersion maps results in a blocking percentage that varies between 20% and 35% for both IAWA schemes. This range of blocking percentage values is impossible to be obtained using first-fit assignment, whilst and it can be observed only for a small range of dispersion parameters when random WA scheme is used.

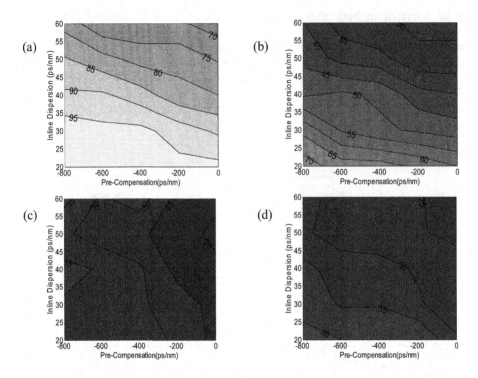

Fig. 6. Blocking percentage as a function of the dispersion mapping implementing (a) First Fit (b) Random Wavelength Assignment (c) Impairment Aware Wavelength Assignment and (d) Pre Specified IAWA scheme in combination with the ICBR algorithm

5 Conclusions

In this study a new Impairment Aware Routing and Wavelength Assignment algorithm was developed applicable for long haul applications. A number of linear and nonlinear impairments have been taken into account including ASE noise, crosstalk, filtering effects, SPM/GVD, cross phase modulation and four wave mixing. Through extensive simulations it is shown that the overall network performance can be significantly improved if ICBR in combination with Impairment Aware Wavelength Assignment is used to establish the connections.

Acknowledgments. This work was supported by the EU via the IST/NoE e-Photon/ONE+project and IST-PHOSPHORUS

References

1. R. Mewanou, S. Pierre, "Dynamic routing algorithms in all-optical networks", Electrical and Computer Engineering, 2003, IEEE CCECE 2003, Canadian Conference, vol. 2, no 4-7, pp.773-776, May 2000
2. D. Cavendish, A. Kolarov, B. Sengupta, "Routing and wavelength assignment in WDM mesh networks", Globecom 2004, pp. 1016-1022, 29 Nov. -3 Dec. 2004.
3. B. Ramamurthy, D. Datta, H. Feng, J.P. Heritagen and B. Mukherjee, " Impact of tranmsission impairments on the teletraffic performance of wavelength routed optical networks", Journal of Lightwave Technology, 17, no. 10, October 1999
4. A. Maher, D. Elie-Dit-Cosaque and L. Tancevski, " Enhancement to Multi-Protocol Lambda Switching (MPλS) to accommodate transmission impairments", Globecom 2001, 1, no 25-29, pp. 70-75, November 2001
5. M. Ali, D. Elie-Dit-Cosaque and L. Tancevski, " Network optimization with tranmsission impairments-based routing", ECOC 2001, Mo. L.2.4. pp: 42-43, Amsterdam, The Netherlands, 2001
6. J.F. Martins-Filho, C.J.A. Bastos-Filho, E.A.J. Arantes, S.C. Oliveira, F.D. Nunes, R.G. Dante, E. Fontana, "Impact of device characteristics on network performance from a physical- impairment- based routing algorithm", OFC 2004, 1, pp. 23-27, Feb. 2004
7. P. Kulkarni, A. Tzanakaki, C. Mas Machuka, and I. Tomkos, "Benefits of Q-factor based Routing in WDM Metro Networks", ECOC2005, September 2005.
8. I. Tomkos et al., "Ultra-long-haul DWDM network with 320 x320 wavelength-port broadcast & select OXCs," in Proc. ECOC'02, Copenhagen, Denmark, Sept. 2002
9. I. Tomkos, M. Vasilyev, J.-K. Rhee, A. Kobyakov, M. Ajgaonkar, and M. Sharma, "Dispersion map design for 10 Gb/s ultra-long-haul DWDM transparent optical networks," at the OECC July 2002, PD-1-2
10. N. Kikuchi, S. Sasaki, "Analytical Evaluation Technique of Self-Phase Modulation Effect on the Performance of Cascaded Optical Amplifier Systems", IEEE Journal of Lightwave Technology, vol. 13, no. 5, May 1995
11. A. Cartaxo, " Cross-Phase Modulation in Intensity Modulation Direct Detection WDM Systems with multiple optical amplifiers and dispersion compensators", J. Lightwave Technology, vol. 17, no. 2, pp. 178-190, February 1990
12. K. Inoue, K. Nakanishi, K. Oda, " Crosstalk and Power Penalty Due to Fiber Four-Wave Mixing in Multichannel Transmissions", J. Lightwave Technology vol. 12, no. 8, pp. 1423-1439, Aug. 1994
13. J.Strand, A.L. Chiu and R. Tkach, " Issues for routing in the optical layer", IEEE Communication Magazine, Feb. 2001, pp. 81-87

MatPlanWDM: An educational tool for network planning in wavelength-routing networks

P. Pavon-Mariño, R. Aparicio-Pardo, G. Moreno-Muñoz, J. Garcia-Haro,
J. Veiga-Gontan

Department of Information Technologies and Communications, Polytechnic University of
Cartagena, Plaza del Hospital 1, 30202 Cartagena, Spain
Pablo.Pavon@upct.es, Tlf: +34968325952, Fax: +34 968325973

Abstract. This paper presents the MatPlanWDM tool, an educational network planning tool for wavelength-routing WDM networks. It includes a set of heuristic algorithms for solving the virtual topology design, and the routing and grooming of traffic flows on top of it. In addition, an implementation of the linear programming problem to obtain the optimal solution of the complete design is included for comparison. The input parameters to the planning problem are the network physical topology, the traffic matrix, and technological constraints like the number of transmitters, receivers, optical converters and wavelengths available. The tool is implemented as a MATLAB toolbox. The set of heuristic algorithms can be easily extended. A graphical interface is provided to plot the results obtained from different heuristics and compare them with the optimal solution in small-scale topologies.

Keywords: Wavelength routing networks, network planning, educational tool.

1   Introduction

The tremendous increase of the transmission capacity in optical networks provided by the Wavelength Division Multiplexing (WDM) technology has created a gap between the amount of traffic we can transmit, and the amount of traffic that can be processed electronically in the switching nodes. This is called the "electronic switching bottleneck", and has a more evident impact on backbone networks, which carry the highest volume of traffic.

Several alternatives have been proposed in order to address this bottleneck. They intend to decrease the traffic that has to be switched electronically in the nodes, by allowing a more or less sophisticated switching at the optical layer. Optical Packet Switching (OPS) and Optical Burst Switching (OBS) paradigms are based on switching nodes capable of an optical processing of optical packets/bursts [1]. Both alternatives are still in a research or testing stage [2] because of the non-mature state of the photonic enabling technologies involved. Nowadays, the only commercial

I. Tomkos et al. (Eds.): ONDM 2007, LNCS 4534, pp 58-67, 2007.

alternative addressing the optical-electronic gap is given by the Wavelength-Routing (WR) switching paradigm. In WR networks, traffic is carried onto transparent lightpaths that may traverse a given number of nodes. This is performed by reconfigurable optical add/drop multiplexers (R-OADM) or reconfigurable wavelength crossconnects (R-WXC), which allow some wavelengths to be dropped or added in a link, while others are optically switched without electronic conversion [1], [3]. Figure 1 sketches a node of this type. Figure 1-(a) identifies the two types of lightpath configurations carrying traffic which is not processed electronically in the node, without (1) or with (2) wavelength conversion. In the first case, the lightpath is called to be the subject of the *wavelength continuity constraint*. Figure 1-(b) shows four lightpaths which carry traffic that is processed electronically in the G-fabric (grooming fabric) of the node. The 100% of the traffic in lightpath (3) is ingress traffic added by the node. The 100% of the traffic in lightpath (4) is egress traffic dropped in the node. Lightpaths (5) and (6) show a much more common situation. Some portion of the traffic in lightpath (5) is dropped in the node. The rest of the traffic is groomed. That is, it is allocated in other lightpaths like (6) which are initiated in this node, sharing the lightpath with other grooming and/or add traffic.

Figure 1 helps us to identify the main functional blocks in the switching nodes and the WR network itself, which impact the overall cost of the network: a) the number of input and output wavelengths, which may be different for each fiber link, b) the number of electro-optic transmitters (T) and opto-electronic receivers (R) in the node, c) the number of tunable wavelength converters (TWC) in the node and d) the electronic switching capacity required, given by the sum of the ingress, egress and grooming traffic.

The equipment required is determined by the network planning decisions taken:

(i)   *Virtual Topology (VT) Design*. The VT consists of the lightpaths to be configured in the network, their traversing fibers, and their transmission wavelengths in each hop. A lightpath consumes one transmitter in the initial node, one receiver in the ending node, and one TWC in any intermediate node where a wavelength conversion is required.

(ii) *Routing of the traffic flows on top of the VT*. This decision is related to the way in which the traffic (electronic traffic flows offered to the network) is routed on top of the established lightpaths. This determines the amount of traffic to be carried by each lightpath (which can be used to compose a congestion measure), and where and how the grooming is performed (which determines the electronic switching capacity required in each node).

The combined (i+ii) network planning problem to be solved can be summarized as: "for a given traffic demand and a given network topology (existing fiber links between nodes), determine the VT design and routing of traffic flows on top of it, which optimizes a given cost function". The VT design problem is also denoted as the RWA (Routing and Wavelength Assignment) problem. It is known to be a NP-hard problem ([4], [5], [6]). Obviously, the composed (i+ii) problem is also NP-hard.

In the last decade, several techniques have been presented to solve the aforementioned network planning problems. They can be classified into two categories: mathematical programming approaches, and heuristic methods. The former, are generally based on a mixed-integer linear programming (MILP) formulation of the planning problem, followed by a search of the optimum or a suboptimum solution by means of conventional optimization algorithms. Unfortunately, the search of the optimum solution is limited to small to medium size topologies because of the exponential complexity growth. The latter approach, is based on faster algorithms, designed specifically to solve problems (i), (i+ii), or some parts of these problems. These algorithms do not search for an optimum solution, but look for a good approximation which can be computed in polynomial time.

Table 1 summarizes some related works in this field.

Algorithms	(i)	(ii)	Wavelength Conversion	Objective Function	References
HLDA	YES	NO	NO	Maxim. Single-Hop Traffic, Minim. Congestion	[1] [7]
MLDA	YES	NO	NO	Maxim. Single-Hop Traffic, Minim. Message Delay	[1] [7]
TILDA	YES	NO	NO	Minim No. Used Physical Links	[7]
LPLDA	YES	NO	NO	Minim. Congestion	[7]
RLDA	YES	NO	NO	-- (Random Assignment)	[7]
SHLDA	YES	NO	NO	Minim. No. of Virtual Hops	[8]
LEMA	YES	NO	NO	Minim. Single-Hop Traffic, Minim. Multi-hop Traffic	[1] [9]
SHTMH	YES	NO	NO	Maxim. Single Hop Traffic	[1]
PDMH (Greedy)	YES	NO	NO	Minim. Propagation Delay	[1]
SABH (Sim. Ann.)	YES	NO	NO	Minim. Message Delay	[1] [6]
HRWA	YES	NO	NO	Minim. Required No. of Wavelengths	[10]
FAR-FF RWA with SWC	YES	NO	YES	Minim. Blocking Probability	[11]
LLR-FF RWA with SWC	YES	NO	YES	Minim. Blocking Probability	[11] [12]
MBPF under FAR-FF RWA	YES	NO	YES	Minim. Blocking Probability	[11]
MBPF under WLCR-FF RWA	YES	NO	YES	Minim. Blocking Probability	[12]
WMSL under LLR-FF RWA	YES	NO	YES	Minim. Sum of the maximum segment length	[11] [12]
WCA	YES	NO	YES	Minim. Blocking Probability	[13]
Flow Deviation Algorithm	NO	YES	NO	Minim. Network-Wide Average Packet Delay	[6]
--	NO	YES	NO	Minim. No. of Transceivers	[14]

Observations: the acronyms in the table can be found in the included references.

In this paper, an educational network planning tool for WR networks, with and without the wavelength continuity constraint, is presented. It is developed as a set of MATLAB [15] functions, along with a graphical interface. The toolbox can be publicly downloaded at the MATLAB Central site [15]. Its goal is to allow the testing of a set of heuristic algorithms, providing an integrated framework to observe, learn and study the VT planning concepts involved. The user can calculate and evaluate the results obtained when different algorithms are executed (in terms of the items from a) to d) mentioned above). A MILP optimum search of the (i+ii) problem [1] is also included in the tool, so that their results can be compared to the ones achieved by the heuristic algorithms (except for large topologies). The tool is fully extensible in terms of network topologies, traffic demands and optimization algorithms. These features make MatPlanWDM an adequate teaching tool in the field of network planning in WR networks. As shown in section 2, the focus on the algorithmic issues of the WR network planning is a novel approach not covered by other educational tools.

The rest of the paper is organized as follows. Section 2 summarizes related work. Section 3 describes the structure of the toolbox, and section 4 presents the graphical interface implemented. Finally, section 5 concludes the paper.

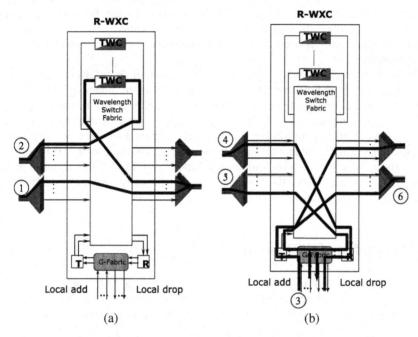

Figure 1. Functional blocks in a switching node in a WR network

2 Related work

This section briefly overviews some relevant educational planning tools, for WR networks. Naturally, a wide range of commercial utilities also exists, for the dimensioning of WDM networks [16-21]. Nevertheless, these tools are not designed for educational purposes (i.e. their underlying algorithmic details are not made public), and are not considered in this paper.

A notable educational tool is the *Optical WDM Network Simulator* (*OWns*) [22], designed as an extension to the well-known network simulator *ns*. The *OWns* facilitates the study of switching and routing schemes in WDM networks. It tries to incorporate the key characteristics of WDM networks in the simulator, such as optical switching nodes, multi-wavelength links, virtual topology constructions, related switching schemes and routing algorithms. In opposition to the dynamic event-driven *simulator* approach in *OWns*, the MatPlanWDM tool is specifically devoted for *static* or *off-line* network planning, concentrating the focus on the optimization algorithm design. The MATLAB language provides a simpler and more powerful framework than *OWns* for this purpose, which helps to clarify the planning concepts addressed.

Another educational network planning tool that can be employed for the analysis of WDM networks can be found in [23]. It employs the high level programming language Scilab for the user interface and C for computer intensive algorithms. This Scilab code analyses a given fiber plant and traffic requirements and returns the cost of different technological solutions. Unlike MatPlanWDM, the optimization engine in [23] provides little flexibility for testing different planning algorithms. It applies Dijkstra and Suurballe algorithms to obtain the shortest path between each pair of nodes, which further determines the equipment allocation. Therefore, again, this approach does not concentrate on the planning concepts addressed in the MatPlanWDM tool.

Delite [24] is a non-WDM specific educational tool. It follows a similar static philosophy of the MatPlanWDM tool, offering a set of planning algorithms for evaluation and comparison, under different traffic patterns and network topologies. Delite is implemented in C language, it is open source, extensible, and purely designed for academic purposes. Unfortunately, Delite tool does not consider the specifics of the RWA planning problem, or provide any optimum MILP search for comparison. In addition, when compared to MatPlanWDM tool, the advantages of the MATLAB language for the implementation of more complex mathematical operations is made evident.

3   Description of the toolbox

The MatPlanWDM tool has been implemented as a MATLAB [15] toolbox, consisting of a set of MATLAB functions. All the functions are documented, and open for free usage. Figure 2 helps us to describe the structure of the toolbox:

- *Input parameters.* The input parameters for the planning problem are: (1) the network topology (including the distances in kilometers between nodes), (2) the traffic matrix, (3) the maximum number of transmitters and receivers in each node, (4) the maximum number of wavelengths in each link, and (5) the maximum number of TWCs in each node. Zero TWCs in all nodes defines a problem restricted by the wavelength continuity constraint. MatPlanWDM provides some sample network topologies (i.e. the NSFNET), and sample traffic matrixes (i.e. sample matrix for NSFNET presented in [7]). In addition, a help function is provided to assist the generation of different types of traffic matrixes: uniform distribution between nodes, composition of heavy loaded and lightly loaded nodes with different proportions, and the method presented in [24] for creating traffic matrixes as a function of node population and inter-node distance.

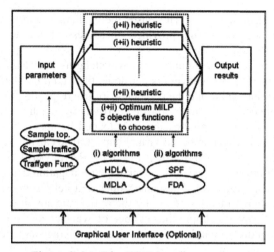

Figure 2. MatPlanWDM toolbox structure

- *Optimization algorithms.* The optimization algorithms are responsible for calculating a solution to the (i+ii) problem defined by the selected input parameters. These algorithms are implemented as MATLAB functions which follow a fixed signature, establishing the format of the input parameters, and the format of the output results. A broad range of heuristics has been implemented. Many of them result from the combination of one heuristic for the (i) problem, whose output is, in its turn, passed as the input of an algorithm solving the problem (ii). A library of functions of heuristics solving problem (i) is included in the toolbox. These heuristics are: HLDA, MLDA, TILDA and RLDA [7]. In addition, two algorithms for the routing of the flows are also included, an SPF routing (which minimizes the distance of each flow in terms of number of lightpaths traversed), and the Flow Deviation Algorithm (FDA) [6] routing (which minimizes the network-wide average packet delay). This scheme is fully and easily extended with further (i), (ii) or (i+ii) algorithms.

  The MatPlanWDM toolbox developed contains as well a function which implements a MILP programming of the (i+ii) problem as proposed in [1]. MatPlanWDM provides an optimum solution of this program by means of the TOMLAB/CPLEX solver [25]. Optimum solutions have been obtained for topologies of up to 12 nodes, in an Intel© Pentium© M730 processor with 1.6 GHz and 512 MB of RAM. Five different objective functions can be selected, which intend to optimize different metrics: (1) minimization of the average weighted number of hops of the flows on top of the VT, (2) minimization of the network congestion (defined as the utilization of the most loaded lightpath), (3) maximization of the single hop traffic, (4) minimization of the number of used wavelengths and (5) minimization of the maximum number of wavelength channels in any fiber link.

- *Output results.* The results of the (i+ii) optimization problem are the VT design and the routing of the flows on top of the VT. Some relevant measures of the WR network cost and performance can be calculated as a function of these results. This eases the evaluation and comparison among different solutions. The provided measures belong to the next three categories:

- Cost indicators: number of wavelengths per fiber, number of transmitters/receivers/converters per node, their total number in the network, their maximum value and the percentage of use respecting to the maximum.

- Performance indicators: traffic carried by each lightpath, network congestion (traffic carried by the most loaded lightpath), total traffic carried by the VT, average number of virtual hops, traffic carried by each fiber, total single-hop traffic, total traffic carried by the physical topology, average number of physical hops, average message propagation delay (µs).

- Impairment indicators. These are values that help to give an insight about the signal impairments suffered by the optical signals *in each lightpath*: the number of physical hops, the number of wavelength conversions and the propagation distance in kilometers. Also, the average and maximum of these numbers are provided.

- *Graphical User Interface (GUI)*. A user friendly GUI has been developed to ease the use of the MatPlanWDM tool for didactic purposes. Their main functionalities are described in the next section.

4   Graphical User Interface

The Graphical User Interface consists of a window workspace, whose configuration changes according to the function mode selected by the user. These modes are:

1) *Design Logical Lightpath Network.* This mode performs the virtual topology design by means of one of the heuristic algorithms or the MILP programming implemented. The user is requested for the design input parameters: the traffic pattern, the physical topology, the number of transmitters (T), receivers (R) and TWCs per node, and the number of wavelengths per fiber. Figure 3 shows the GUI in this operational mode.

2) *Evaluate Optimization Method.* This mode allows evaluating an optimization algorithm under different combinations of input parameters. A comparison metric is computed, such as the network congestion or the average number of virtual hops on the virtual topology, for several combinations of the next design parameters: the number of transceivers (transmitters (T) and receivers (R)) and the number of wavelengths per fiber. The tool displays a comparative graph with the results achieved.

3) *Compare Optimization Method.* This mode compares the results obtained from a set of selected optimization algorithms. The input variables are the number of transceivers (virtual topology maximum degree) for each algorithm. This GUI can be easily extended to compose comparisons modifying other parameters like the number of wavelengths per fiber, or the number of wavelength converters per node.

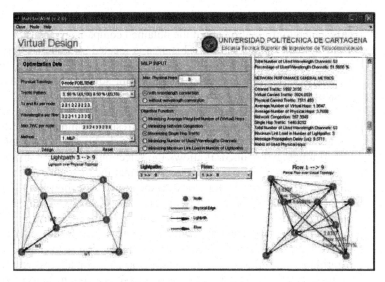

Figure 3. Graphical User Interface. *Design mode* view

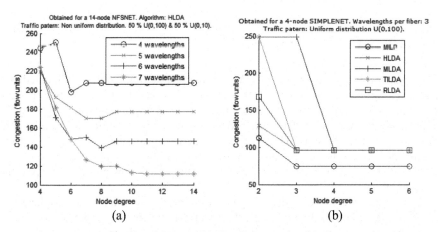

Figure 4. *Results Graphs Area.* (a) *Evaluate* mode, (b) *Compare* mode.

The workspace of the GUI is divided into four areas; each of them refers to a specific and different task according to the function mode selected by the user:

1. The *Optimization Data Area*. This area is located at the upper-left corner of the workspace. It is used to enter the design input parameters in the mode 1) *Design Logical Lightpath,* and the comparison parameters in the modes 2) *Evaluate Heuristic Algorithm* and 3) *Compare Heuristic Algorithms*.

2. The *Results Reports Area*. This area is located at the upper-right corner of the workspace. It reports the measures and the metrics calculated by the optimization. The measures provided are different in each operational mode. In the mode 1) an exhaustive report about the implemented virtual

topology design is shown. For the modes 2) and 3), the reports contain only general information about the obtained comparison values.

3. The *Physical Topology Area*. This area is located at the lower-left corner of the workspace. It presents the physical topology of the design. In mode 1), it allows to select a particular lightpath, so that the traversing fiber links of the lightpath are highlighted in the physical topology.

4. The *Results Graphs Area*. This area is located at the lower-right corner of the workspace. Its purpose is to draw the results obtained in the different modes in an intuitive graphical way. In mode 1), this area displays the virtual topology obtained. It allows the selection of a particular high level traffic flow from the input parameters, so that the traversing lightpaths are highlighted in the virtual topology. In the mode 2), it shows a graph where the network congestion is represented in the ordinate axis, and the number of transceivers in the abscissa axis. A curve is plotted for several values of the number of wavelengths per fiber. Figure 4-(a) illustrates an example of the *Results Graphs Area* with the congestion curves for a given algorithm with different design input parameters. Finally, when the mode 3) is selected, a similar graph is displayed with one curve for each compared algorithm (Figure 4-(b)).

5   Conclusions and further work

This paper presents an educational WDM network planning tool, implemented as a MATLAB toolbox with a graphical interface. Virtual design and flow routing algorithms can be combined and tested with the tool. An optimal linear programming implementation of the complete virtual design and traffic grooming problem is also provided for comparison. The graphical interface to the toolbox has been implemented to aid in the educational process of results evaluation and comparison, lightpath plotting on top of the virtual topology, and traffic flow plotting on top of the virtual topology.

The tool has been specifically designed to be easily extended with new heuristic algorithms. As a current line of work, the input parameters and output parameters to the tool are being completed, to facilitate the assessment of heuristics which also take into consideration impairment constraints in the optical signal, and protection and restoration issues.

Acknowledgments. This research has been funded by the Spanish MCyT grant TEC2004-05622-C04-02/TCM (ARPaq). The authors would like also to thank the EU e-photon/ONe+ network of excellence and the COST 291 project.

References

1. Murthy, C.S.R., Gurusamy, M.: *WDM Optical Networks (Concepts, Design and Algorithms)*. Prentice Hall PTR, Upper Sadle River (2002)

2. Ramaswami R.: *Optical Networking Technologies: What worked and what didn't.* IEEE Communications Magazine, vol. 44, no. 9 (2006)132-139
3. Sivalingam, K.M., Subramaniam, S.: *Optical WDM Networks (Principles and Practice).* Kluwer Academic Publishers, Norwell (2001)
4. Chlamtac, I., Ganz, A., Karmi, G.: *Lightpath communications: An approach to high bandwidth optical WANs.* IEEE/ACM Transactions on Communications, Vol. 40, No. 7 (1992) 1171-1182
5. Krishnaswamy, R., Sivarajan, K.: *Design of logical topologies: a linear formulation for wavelength routed optical networks with no wavelength changers.* Proc. IEEE INFOCOM (1998) 919-927
6. Mukherjee, B., Banerjee, D., Ramamurthy, S., Mukherjee, A.: *Some Principles for Designing a Wide-Area WDM Optical Network.* IEEE/ACM Transactions on Networking, Vol. 4, No. 5 (October 1996) 684-696
7. Ramaswami, R., Sivarajan, K.N.: *Design of Logical Topologies for Wavelength-routed Optical Networks.* IEEE Journal on Selected Areas in Communications, Vol. 14, No. 5 (June 1996) 840–851
8. Katou, J., Arakawa, S., Murata, M.: *A design method for logical topologies with stable packet routing in IP over WDM networks.* IEICE Transactions on Communications, E86-B (August 2003) 2350-2357
9. Banerjee, S., Yoo, J., Chen, C.: *Design of Wavelength Routed Optical Networks for Packet Switched Traffic.* IEEE/OSA Journal of Lightwave Technology, Vol. 15, No. 9 (September 1997) 1636-1646
10. Wauters, N., Demeester, P.: *Design of the Optical Path Layer in Multiwavelength Cross Connected Networks.* IEEE Journal on Selected Areas in Communications, Vol. 14, No. 5 (June 1996) 881–892
11. Chu, X., Li, B., Chlamtac, I.: *Wavelength converter placement under different RWA algorithms in wavelength-routed all-optical networks.* IEEE Transactions on Communications, Vol. 51, No. 4 (April 2003) 607-617
12. Li, B., Chu, X., Sohraby, K.: *Routing and wavelength assignment versus wavelength converters placement in all-optical networks.* IEEE Communications Magazine, Vol. 41, No. 8 (August 2003) S22-S28
13. Zhang, Y., Taira, K., Takagi, H., Das, S.K.: *An efficient Heuristic for Routing and Wavelength Assignment in Optical WDM Networks.* IEEE International Conference on Communications, Vol. 5 (2002) 2734- 2739
14. Konda, V.R., Chow, T.Y.: *Algorithm for Traffic Grooming in Optical Networks to Minimize the Number of Transceivers.* IEEE Workshop on High Performance Switching and Routing, (2001) 218-221
15. http://www.mathworks.com. MATLAB Central: http://www. matlabcentral.com.
16. OPNET SP Guru Transport Planner™, http://www.opnet.com
17. Cox Associates® NetAdvantage MeshPlanner™, http://www.cox-associates.com
18. Meriton Networks® Meriton 9500 (Network Planning Tool)™, http://www.meriton.com
19. Network Planning Systems NetMate™, http://www.netplansys.com/
20. Cisco ONS 15454 MSTP™ and MetroPlanner, http://www.cisco.com/
21. Atesio© DISCNET™ optimization engine, http://www.atesio.de/solutions/index.html
22. Wen B., *et al.*: *Optical Wavelength Division Multiplexing (WDM) Network Simulator (OWns): Architecture and Performance Studies.* SPIE Optical Networks Magazine Special Issue on "Simulation, CAD, and Measurement of Optical Networks", March 2001.
23. http://www.ee.iitm.ac.in/optics/np/
24. Cahn R. S.: *Wide Area Network design. Concepts and tools for optimization.* Morgan Kaufmann Publishers Inc. (1998)
25. TOMLAB Optimization. http://tomopt.com/

Centralized vs. Distributed Approaches for Encompassing Physical Impairments in Transparent Optical Networks

P. Castoldi[1], F. Cugini[2], L. Valcarenghi[1], N. Sambo[1], E. Le Rouzic[3], M. J. Poirrier[3], N. Andriolli[1], F. Paolucci[1], A. Giorgetti[1]

[1] Scuola Superiore Sant'Anna, Pisa, Italy, e-mail: castoldi@sssup.it
[2] Cnit, Pisa, Italy
[3] France Telecom, Lannion, France

Invited paper

Abstract. Transparent optical mesh networks are an appealing solution to provide cost-effective high bandwidth connections eliminating the need of expensive intermediate electronic regenerators. However, the implementation of transparent optical networks requires to take into account physical impairment information for effective lightpath set-up. In this paper, we present two distributed solutions to encompass physical impairments based on enhancements of the GMPLS protocol suite. Specifically, both GMPLS routing protocol and signaling protocol extensions are presented and discussed. An alternative centralized approach based on an impairment-aware Path Computation Element (PCE) is also proposed.

The distributed routing approach exhibits convergence limitations, while the distributed signaling approach is scalable and effective. The latter is then compared against the centralized PCE approach through simulations considering both a metro network and a more complex WDM network scenario. In addition, experimental implementations of the two approaches are presented. Results show the trade-off of the two approaches, demonstrating the general good performance in terms of lightpath set up time for both approaches.

Keywords: GMPLS, physical impairment, transparency, signaling protocol, Path Computation Element (PCE).

1. Introduction

Currently, the end-to-end lightpath provisioning over transparent optical mesh networks (i.e., networks without electronic regeneration at intermediate nodes) based on Generalized MultiProtocol Label Switching (GMPLS) protocols assumes that every route, eligible by the routing protocol, is characterized by a satisfactory signal quality. Indeed the GMPLS protocol suite does not include any information related to physical impairments [1]. Thus transparent optical network planning is based on worst case scenario, i.e. longest paths, and only limited size transparent networks are practically achievable.

I. Tomkos et al. (Eds.): ONDM 2007, LNCS 4534, pp 68-77, 2007.

A possible approach for the enhancement of the GMPLS protocol suite to encompass physical impairment parameters has been first proposed in [2]. This approach, namely Routing Approach (RA), is based on the extension of the routing protocol, e.g. the Open Shortest Path First with GMPLS extensions (OSPF-TE). RA requires additional extensions to the routing protocol to flood physical impairment parameters. In addition RA requires, besides the presence of a Traffic Engineering Database (TED) to store bandwidth information, the availability of an additional database in every network node, namely the Physical Parameter Database (PPD). The purpose of PPD is to maintain up-to-date information on physical parameters concerning each link of the transparent network. Local information (i.e., physical parameters of the local node and of the attached links) is included in the PPD resorting to automatic monitoring and/or management systems, while remote physical parameters are obtained by exploiting the routing protocol flooding. In this way, when a connection request arrives, the Constraint Shortest Path first (CSPF) algorithm, resorting to both TED and PPD, is able to compute a route satisfying both bandwidth and optical signal quality constraints [3]. The main advantage of RA is that it is fully distributed and only minor changes are needed to the current routing protocol version. However RA suffers from several potential drawbacks. For example RA is not capable of efficiently considering intra-node physical information such as node internal losses or cross-talks and the presence of shared regenerators. In addition, it may heavily suffer from PPD inconsistency, scalability and convergence problems particularly in case of frequent link parameter changes [4] or upon failure occurrence. Yet, if only few paths in a large network have to be avoided because of their high physical impairments, each node should store and manage a large amount of information coming from the whole transparent network. As a consequence, to avoid few impaired routes (e.g., longest paths), the routing protocol must continuously disseminate all the relevant parameters. Moreover, a multi-constrained path computation is required for achieving both optimal network performance and sufficient quality of the optical signal. This may heavily impact the load of the node's processing unit (CPU), determining large computation time and delaying the lightpath establishment. For these reasons, RA gives minor advantages as compared to the introduced complexity and alternative approaches have then been proposed.

In this paper, two different approaches alternative to RA are considered and evaluated.

The first considered approach, hereafter addressed as Signaling Approach (SA), has been proposed in [5] and elaborated more in [6]. SA is based on the enhancement of the signaling protocol, e.g. Resource Reservation Protocol with GMPLS extensions (RSVP-TE), to encompass physical impairments during lightpath establishment.

The second considered approach, hereafter addressed as Path Computation Element (PCE, [7]) based approach (PCE-A), has been first proposed in [8]. PCE-A is based on the computation of lightpath physical impairments performed in a centralized way by the PCE.

In this paper, for the first time, the SA and PCE-A performance are evaluated through simulations on realistic Metro network and WDM network scenarios considering an accurate physical impairment modeling. Moreover, the experimental implementations of the two different approaches are presented.

2. Signaling Approach

The Signaling Approach (SA) is based on the dynamic estimation of the optical signal quality during the signaling phase of the lightpath set up process. No modifications are introduced in the OSPF-TE routing protocol which elaborates routes ignoring physical impairments. The RSVP-TE signaling protocol is extended to collect the physical parameter values characterizing every traversed node (e.g., Photonic Cross-Connect (PXC)) from the source to the destination node. The source node generates an RSVP Path message extended with a novel *Cumulated Physical Parameter (CPP)* object, which contains the array of physical parameters of the transmitting interface and of its outgoing link. Every traversed node, before propagating the message, updates the CPP object by adding its own local parameter values (i.e., intra-node and outgoing link parameters). Admission control at the destination node compares the overall accumulated parameters with the required parameter values that characterize its receiver interface. If the accumulated parameters are within an acceptable range, the lightpath set up request is accepted and an RSVP Resv message is sent back to the source node. Otherwise the lightpath request is rejected and an RSVP Error is sent back to the source node. If the request is rejected, further set up attempts following possibly link-disjoint routes are triggered. However, the successive set up attempts may also fail, excessively delaying the lightpath establishment process.

To overcome these issues, a novel RSVP object called *Physical Parameter (PP) object* and a locally managed *Physical Parameter Database (PPD)* are introduced as optional extensions. PP object contains the array of physical parameter information that characterizes every traversed link and that contributes to the CPP object. PPD stores the physical parameters of all network links, retrieved resorting to every transit signaling message carrying PP objects. PPD is used to provide a prediction of the cumulated physical parameters along the computed path. If the predicted value is outside the acceptable range, then the computed path is not validated and a different path is considered. On the contrary, if the predicted value is within the acceptable range, the source node starts the signaling phase. PPD might be not up-to-date, so the set up attempt along validated paths might be rejected due to unacceptable physical impairments. In this case an alternative path, possibly link-disjoint, is then computed and evaluated. Proper expiration time of PPD parameters is also considered.

3. PCE-based Approach

Path Computation Element (PCE) is defined in [7] as an entity (component, application, or network node) that is capable of computing a network path or route based on a network graph and applying computational constraints during the computation. PCE has been introduced mainly to perform complex route computation on behalf of network nodes in particular in the cases of partial visibility of the network topology to the destination, i.e. multi-area, multi-domain and multi-layer networks. PCE typically takes into account bandwidth requirements, QoS parameters and survivability aspects.

The considered PCE-based Approach (PCE-A) resorts to the PCE utilization also in the case of transparent optical networks. For this purpose, the PCE has to be aware of additional information referring to the physical parameters. The complete set of physical parameters is stored by the PCE in a locally managed *Physical Parameter Database (PPD)* equal to the one used by SA. In this case, however, all the PPD parameters are obtained by the Management System or through a performance monitoring system. In this approach, no modifications are introduced into the Control Plane, thus neither the routing nor the signaling protocols are enhanced with further extensions. Whenever the PCE receives a path computation request from a Path Computation Client (PCC), it computes the required path taking into account the TE information and the physical impairment constraints. Then, the computed route is converted into an explicit route (i.e., a set of nodes and links) and returned to the PCC for establishing the lightpath by means of a distributed signaling protocol.

4. Network scenario and simulation results

The performance of SA and PCE-A approaches has been evaluated by means of a C++ event-driven network simulator considering two different network scenarios.

The first scenario aims at reproducing the behavior of a transparent Metro network made of just passive devices (e.g., PXC with no in-line optical amplification) between the source and destination transponders (e.g., optical Ethernet transponders. Fig. 1 shows the considered topology of 6 PXCs and 8 links. Network links are characterized by a physical parameter l_{ij} carried by the CPP object representing the link length between PXC i and j. Fig. 1 also reports the l_{ij} values in kilometers for the reference Metro network. The impairments introduced by every transit PXC i are also taken into account adding to CPP an equivalent distance l_i [2] equal to 3 km. A maximum acceptable cumulated value L_{MAX}=100 km is imposed.

The second scenario aims at reproducing the behavior of a transparent WDM backbone network. Fig. 2 shows the considered Pan-European topology with 32 links and 17 PXCs equipped with WDM transponders at 10 Gbps. In order to determine the optical signal quality, a mixed Q factor and Bit Error Rate (BER) criterion [9] with a direct relation to Optical Signal to Noise Ratio (OSNR) [10] is considered. OSNR value represents the optical signal quality and modeled physical impairments are expressed as penalties [9] on OSNR value. Beside noise, modeled impairments are: PMD (Polarization Mode Dispersion), CD (Chromatic Dispersion) and self-phase modulation (SPM) using the non linear phase shift parameter ϕ_{NL}. All modeled effects are represented by four parameters that cumulate linearly: 1/OSNR, PMD^2, CD and ϕ_{NL}. The four parameters are included in the PP and CPP objects to describe the impairments associated to links and paths respectively. The final OSNR value at the destination node is obtained from the cumulated OSNR value by considering the penalties induced by the other parameters [9]. The path is rejected if the final OSNR is below a threshold equal to 13 dB (this value includes also the margins for the non modeled impairments such as Polarization Dependant Loss, filtering, crosstalk, system ageing and other non linear effects). The path is also rejected if one parameter exceeds the model validity domain (e.g., 1 rad for ϕ_{NL}, 4 dB for penalty due to PMD or CD).

In both scenarios lightpath requests are dynamically generated with uniform distribution among all node pairs. Network load is kept limited in order to have lightpath blocking due exclusively to unacceptable optical signal quality.

The performance of SA is shown in Figs. 3-5. The curves show the percentage of established lightpaths within n set up attempts as a function of the generated unidirectional lightpath requests. The percentage of established lightpaths is computed for a 100 requests observation window and averaged over 100 repetitions. Fig. 3 refers to the Metro network scenario where just the equivalent length is considered. Fig. 4 refers to the WDM Pan-European network where the more accurate model is needed. Results show that in both Metro and WDM backbone network scenarios SA achieves constant performance above 96% of established lightpaths at the first set up attempt. Moreover Figs. 3 and 4 show that all the requests are successfully established within the second attempt in the Metro network and within the third attempt in the WDM backbone network. Fig. 5 shows the performance of SA in the WDM backbone network when also the optional PP extensions are utilized. Results show that the percentage of established lightpaths increases with the amount of generated requests up to 99% at the first set up attempt and 100% already at the second set up attempt, thus guaranteeing a limited amount of signaling messages and set up delay.

In PCE-A, PCE is aware of all network physical parameters. PCE then performs effective multi-constraint route computation [3] thus guaranteeing the establishment of all lightpaths at the first signaling attempt (if the PPD information is up-to-date).

5. Experimental evaluation: Signaling Approach (SA)

The testbed, in which the SA is implemented, is shown in Fig. 6. It reproduces a portion of a typical Metro network where two edge routers (A and B) are connected through a transparent network of four PXCs. Edge routers are equipped with optical Ethernet interfaces (1000BaseLX) whose maximum span reach is equal to 10 km. In this testbed again we assume the maximum distance as the only relevant physical parameter. PXC1 and PXC2 serve router A and B, respectively, to dynamically set up optical connections within the domain of transparency. Every link between PXCs is 3 km long except for PXC3-PXC2 link which is 6 km long. Linux Boxes (LB) are employed to manage PXCs via the parallel port. RSVP-TE messages extended with the previously described objects are exchanged among the LBs on the out-of-band Ethernet control plane. The proposed approach allows also to take into account intra-node information without causing, because of the absence of flooding, scalability problems (e.g., for PXC1 we estimate an equivalent distance of 1 km between port 1 and 2, 1.5 km between port 1 and 3, etc.). Every node also maintains a local PPD database containing the physical information, i.e. length, describing the attached links, e.g. LB3 maintains information for the internal PXC3 physical parameters and for links PXC1-PXC3 and PXC3-PXC2.

A connection request is generated at source router A to destination router B and it is handled by PXC1. Because the routing algorithm is unaware of physical impairments, the two possible paths are considered perfectly identical. As a worst case scenario, the route passing through PXC3 is selected. A first exchange of modified RSVP messages allows LB1 to know that excessive physical impairments do not allow the lightpath establishment and a second successful set up attempt is performed. The complete message exchange is presented in Fig. 7, showing that the overall RSVP packet exchange takes less than 40 milliseconds.

6. Experimental evaluation: PCE-based Approach (PCE-A)

The PCE architecture with the implemented activation modules is depicted in Fig. 8. The Label Switched Path (LSP) Request Collector (LSP-RC) module accepts both single and multiple LSP requests. LSP-RC produces an LSP XML file which contains, for every LSP request, the Ingress and Egress nodes, the bandwidth and the physical impairment requirements in case of transparent networks. The Traffic Engineering Database (TED) module collects information directly from the TED database stored in network nodes and maintained by the OSPF-TE protocol. The TED module downloads the complete TED from the node in the form of an XML file by connecting to the node proprietary User-to-Network Interface (UNI) through a TCP socket. TED download is triggered by the LSP-RC module upon a new LSP Request. The TED module requires a bidirectional communication between the PCE and the node management interface per routing Area. Indeed, TE information (i.e., Opaque Link State Advertisement (LSA)) is flooded only within the Area scope. EXtensible Stylesheet Language Transformation (XSLT) is then applied to the downloaded XML files in order to combine the information coming from different routing Areas and to prune redundant and useless information. XSLT elaboration produces a single TED XML file describing the network topology and the available bandwidth for every link of the network.

The Physical Parameter Database (PPD) maintains up-to-date information on physical parameters concerning any network link. Without loss of generality, we consider just the link length to model all the relevant physical impairments. An Optical Time Domain Reflectometer (OTDR) is used to populate the PPD database. One PPD XML file is then generated describing the physical parameter values of every network link. The three XML files obtained from the three aforementioned modules are then elaborated to compute all the required strict routes. In particular, the three files are first elaborated by a PCE Interface (PCE-I) module implemented in C code. The PCE-I generates an LP formulation of the lightpath provisioning problem and passes it to the PCE which is based on a Linear Programming (LP) solver. The implemented objective function minimizes the maximum link bandwidth utilization (least-fill policy). The output file generated by the PCE contains the strict routes of every LSP request which satisfy the required constraint in terms of bandwidth and optical signal quality. A further elaboration is performed by the Configuration File Builder (CFB) module to produce a set of XML files containing the computed routes formatted so that they can be directly uploaded as node configurations. One XML

Configuration file per Ingress node is generated. An LSP Activator (LSP-A) module, which exploits the node proprietary UNI, is then utilized to contemporarily upload and activate the proper XML Configuration File in each involved Ingress node.

The PCE-A implementation is applied to the general Metro scenario shown in Fig. 6. In this case, the PCE is able to select the feasible route traversing PXC3. The overall delay introduced by the PCE-A has been measured in only 0.320 s.

Discussion and conclusion

In this study two different approaches to encompass physical impairments in the Control Plane of transparent optical networks are evaluated. The Signaling-based (SA) and the PCE-based (PCE-A) approaches are analyzed through numerical simulations and experimental implementations.

Results show that under realistic Metro and WDM backbone network scenarios the two approaches perform well. SA is fast, requiring very few set up attempts before completing the lightpath establishment. Moreover it is fully distributed and it is potentially more effective in taking into account sudden physical parameter changes. However, it requires Control Plane extensions and it can provide less effective Traffic Engineering (TE) solutions. PCE-A guarantees effective TE solutions and it can also exploit ad-hoc routing strategies. Moreover, PCE-A does not require additional Control Plane extensions. However, it may suffer from scalability problems and may rely on outdated information.

Acknowledgments. This work has been partially supported by NoE e-Photon/One+ and by IST FP6 NOBEL Phase 2 project.

References

1. R. Martínez, C. Pinart, F. Cugini, N. Andriolli, L. Valcarenghi, P. Castoldi, L. Wosinska, "Challenges and requirements for introducing impairment-awareness into the management and control planes of ASON/GMPLS WDM networks", IEEE Comm. Mag., Dec. 06.
2. J. Strand, et al., "Issues for routing in the optical layer", IEEE Comm. Mag., Feb. 2001.
3. P. Kulkarni, A. Tzanakaki, C. M. Machuka, I. Tomkos, "Benefits of Q-factor based routing in WDM metro networks", ECOC 2005.
4. R. Chraplyvy, "Equalization in Amplified WDM Transmission Systems", PTL Aug 92.
5. F. Cugini, N. Andriolli, L. Valcarenghi, P. Castoldi, "Physical impairment aware signaling for dynamic lightpath set up" ECOC 2005.
6. N. Sambo, A. Giorgetti, N. Andriolli, F. Cugini, L. Valcarenghi, P. Castoldi, "GMPLS Signaling Feedback for Encompassing Physical Impairments in Transparent Optical Networks", GLOBECOM 2006.
7. A. Farrel, "A Path Computation Element (PCE)-Based Architecture", RFC 4655, Aug. 06.
8. F. Cugini, F. Paolucci, L. Valcarenghi, P. Castoldi, "Implementing a Path Computation Element (PCE) to encompass physical impairments in transparent networks", OFC 2007.
9. I. Kaminov, T. Li, "Optical Fiber Telecommunications", vol. IV, Elsevier, 2002.
10. Huang Y. et al., "Connection Provisioning with Transmission Impairment Consideration in Optical WDM Networks with High-Speed Channels", JLT Vol. 23, No. 3, Mar. 2005.

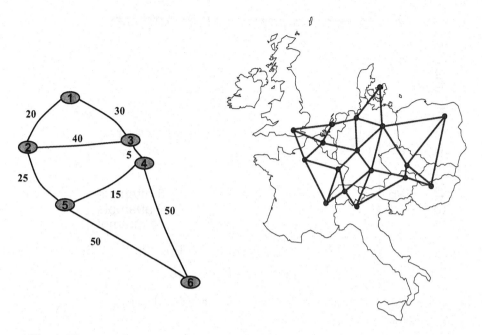

Fig. 1: Transparent Metro network   Fig. 2: WDM backbone network

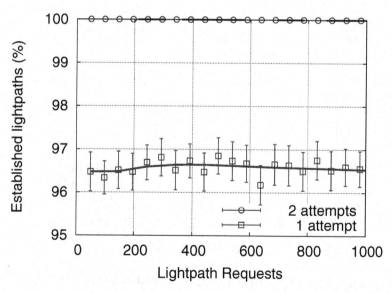

Fig. 3: Signaling Approach (SA), percentage of established lightpath (Metro network)

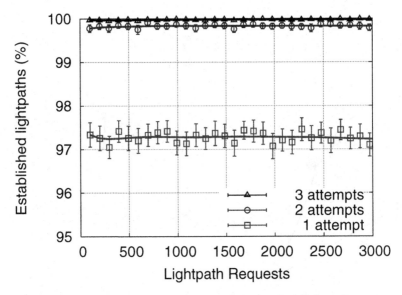

Fig. 4: Signaling Approach (SA), percentage of established lightpath (WDM backbone network).

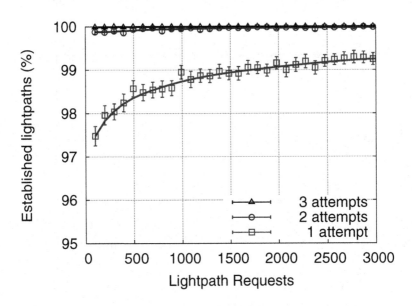

Fig. 5: Signaling Approach (SA), percentage of established lightpath resorting to PPD (WDM backbone network).

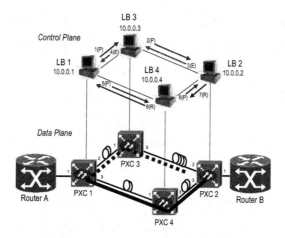

Fig. 6: Experimental implementation

No.	Time	Source	Destination	Protocol	Info
3	2.876711	10.0.0.1	10.0.0.3	RSVP	PATH Message. SESSION: IPv4
4	2.883147	10.0.0.3	10.0.0.2	RSVP	PATH Message. SESSION: IPv4
5	2.889287	10.0.0.2	10.0.0.3	RSVP	PATH ERROR Message. SESSION:
6	2.892644	10.0.0.3	10.0.0.1	RSVP	PATH ERROR Message. SESSION:
7	2.898569	10.0.0.1	10.0.0.4	RSVP	PATH Message. SESSION: IPv4
8	2.902485	10.0.0.4	10.0.0.2	RSVP	PATH Message. SESSION: IPv4
9	2.909444	10.0.0.2	10.0.0.4	RSVP	RESV Message. SESSION: IPv4
10	2.912848	10.0.0.4	10.0.0.1	RSVP	RESV Message. SESSION: IPv4

Fig. 7: GMPLS message exchange over the out-of-band control plane; time in seconds.

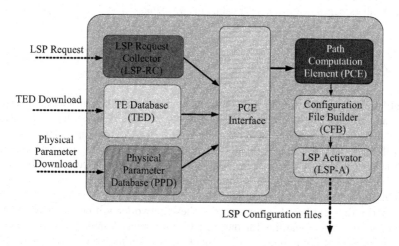

Fig. 8: PCE-based Approach implementation

All-optical signal processing subsystems based on highly non-linear fibers and their limitations for networking applications

Miroslav Karasek[1]
António Teixeira[2],
Giorgio Tosi Beleffi[3]
Ruben Luís[2,3],

[1] IREE, Prague - Czeck Republic
[2] Instituto de Telecomunicações, Universidade de Aveiro, 3810-193 Aveiro Portugal´
[3] ISCOM – Italian Communication Ministry Viale America n.201, Rome - Italy
[4] Siemens SA, Amadora, Portugal
teixeira@ua.pt

Abstract. All optical signal processing is a fundamental concept on which the next generation networks will be based. Implementation of key functionalities as multiwavelength regeneration, frequency conversion and high data rate demultiplexing performed all optically becomes a primary target in current research scenario. In this work, demultiplexing of high data rate signals is demonstrated. The base for the functionality is the cross phase modulation occurring in a span of highly non linear fiber. This fiber is designed to have low birefringence, therefore quite adapted for processing polarization dependent processes, which are the base for the achieved functionality. Demultiplexing from 40GHz and 80GHz to 10GHz is demonstrated.

Keywords: OTDM, Demultiplexing, Cross Phase Modulation, Highly non linear fibers, supercontinuum, polarization multiplexing.

1 Introduction

Next generation networks (NGN) will be bring high bandwidth multimedia services closer to the end user, maintaining a reasonable per bit cost. Such structures will be fundamentally based on novel network topologies like for example 10 and 100 Gigabit Ethernet (GE) Passive Optical Networks (PON), technologies convergence like Radio over Fiber (RoF), Free Space Optics (FSO) and fiber based applications on the possibility to manage a huge amount of data adopting high speed rates and wavelength division multiplexing technologies.

In this scenario, the next key developments of these networks will be the provision of all optical network functionalities for the high bit rate network nodes. The avoidance of OEO conversions and the integration of functionalities at the nodes, determines the adoption of all optical techniques, which can be achieved with standard high non linear fibers or in photonic crystal fibers (PCF) [1,2]. As referred, some of these

I. Tomkos et al. (Eds.): ONDM 2007, LNCS 4534, pp 78-85, 2007.

functionalities can be high bit rate wavelength conversion, for wavelength routing and collision avoiding, high bit rate time demultiplexing, for high speed WDM stream management, polarization multiplexing and signal reshaping.

All these functionalities can be achieved through all optical fast nonlinearities as cross-phase modulation (XPM), four wave mixing (FWM) and self phase modulation (SPM) in optical devices. Several approaches, in fact, have been proposed based on transferring the modulation to a carrier [3-5] or on broad spectrum generation with new carries [5], in semiconductor optical amplifiers [3] and optical fibers [3-6].

In this paper we present experimental results on high speed demultiplexing from 40 and 80 GHz to 10 GHz by means of XPM.

2   Cross phase in highly nonlinear fibers

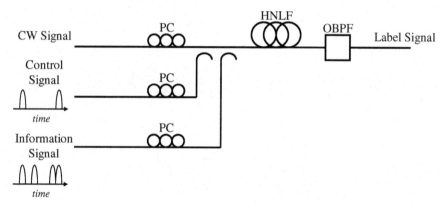

Fig. 1. All optical demultiplexing simulation set-up.

Consider the simulation set-up presented in Fig. 1. The 40 Gb/s information signal is combined with an orthogonally polarized 10 Gb/s impulse comb (control signal), positioned at a similar wavelength, and a continuous wave (CW). It is assumed that the pulses of the control signal have the similar temporal width as the impulses of the information signal. The resulting signal is injected in a highly non-linear fiber (HNLF) segment. For simplicity we have omitted amplification stages that would be required to adjust the signal power. After thc HNLF, the CW signal is extracted by an off-centered optical band-pass filter (OBPF). Assuming orthogonally polarized information and control signals allows assuming that the XPM induced on the CW signal is dependent on a linear sum of the intensities of the interfering signals. As such, in a first order approach the XPM induced on the CW takes the form:

$$\Delta\phi_{XPM}(t) \approx -2\gamma \left[\eta_s \cdot p_s(t) + \eta_c \cdot p_c(t) \right] * h_{xpm}(t) \tag{1}$$

where $p_s(t)$ and $p_c(t)$ are the instantaneous powers of the information and control signals and the symbol * denotes convolution. The terms γ, η_s and η_p are the HNLF non linear coefficient and the nonlinear coupling coefficients of the CW with respect to the information and control signals, respectively. $h_{xpm}(t)$ is the pulse response of the equivalent linear model for the cross-phase modulation. The XPM-induced instantaneous frequency shift may be obtained from (1) as $\Delta f(t) = d\phi_{xpm}/dt$. Fig.2 illustrates the XPM-induced instantaneous phase and frequency shifts at the HNLF output considering the control signal only, the information signal only and the combined signals.

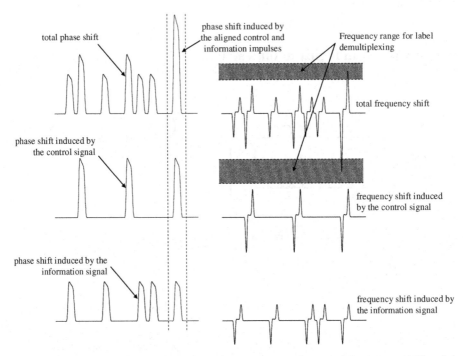

Fig. 2. Example of the XPM-induced instantaneous phase and frequency shifts of the CW signal, considering only the information signal (bottom), the control signal (middle) and both signals (top). The vertical axis of the XPM-induced phase shift has been reversed.

As shown in Fig.2, the alignment of an information impulse with a control impulse induces a particularly high phase shift. This generates a frequency shift significantly stronger that the phase shifts induced by the control or information signal alone. As such, one may define a range a frequencies within which the CW is positioned only when the control impulses superimpose the information pulses, as shown in Fig.2. Positioning the OBPF in this frequency range allows extracting the pulses from the original information signal. Note that nonzero dispersion and dispersion slope leads to asymmetric phase shifts. Therefore, in the presented example, the negative phase shifts are higher than the positive shifts. Fig.3 -a) and –b) present simulated traces and corresponding eye-diagrams of the signal at the OBPF output for two configurations of the latter. The simulated fiber parameters are presented in Table 1. We have assumed that the CW is co-polarized with the control signal. Furthermore, the OBPF output signal is detected using a simulated square-law receiver for 10 Gb/s, modeled by a 5-th order Bessel filter with a -3 dB cut-off frequency of 6.5 GHz.

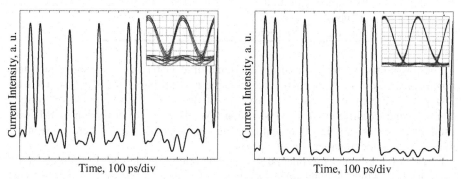

Fig. 3. Traces of the signal detected at the output of the OBPF. Left hand side: Filter off-center of -460 GHz and bandwidth of 250 GHz. Right hand side: Filter off-center of -460 GHz and bandwidth of 100 GHz. Insets present the corresponding eye diagrams.

Fig.3 a) and –b) show the extracted signal affected by interference due to the control and signal impulses, particularly on the space level. Reducing the filter bandwidth from 250 GHz to 100 GHz significantly improves the signal quality at the cost of reducing its power by a factor of ~10 dB. The low power efficiency of this process may be attributed to the residual dispersion of the fiber and dispersion slope, which limit the XPM-induced frequency shift.

3   Base Setup

Several effects and dynamics in the area of Cross Phase Modulation (XPM) between a CW and a modulated signal responsible for the Supercontinuum generation were explored in this work. The CW, see Figure 1, is placed outside the SC spectrum generated from the self phase modulation of one u2t tunable pico-second source (PMLLD) at 10GHz. This PMLLD, can be multiplexed by an Optical Multiplexer (OTDM) to rates of 20, 40, 80 and 160 GHz. The HNLF characteristics are presented in table 1.

Table 1. HNLF parameters.

Parameter	Value
Zero dispersion wavelength	1560 nm
Dispersion Slope	0.019 ps/nm^2/km
Nonlinear Coefficient	10 W^{-1} km^{-1}
Length	1500 m
Attenuation coefficient	0.76 dB/km

The possibility to obtain an all optical demultiplexer based on the above reported principles was also demonstrated. The principle behind this effect is the use of a 10 GHz pulsed stream at the same wavelength of the 40 GHz stream, and by correct alignment induce differentiated XPM in the CW. This, after conversion to IM, will result in a process similar to demultiplexing. In order to avoid coherence beating effects, two distinct polarizations were used.

Figure 4 reports the experimental set up. The output of the PMLLD source, at 10 GHz, has been amplified and subsequently split by means of an optical coupler. The same signal is, by means of this double arm structure, one part guided to an OTDM in order to obtain a 40GHz signal and the other send to a delay and amplitude adjustment arm. The signals in the two arms are polarization rotated to maximize the alignment with the Polarization beam combiner (PBC). Due to the fact that both signals originate from the same source they have the same working wavelength, 1556.8 nm. The orthogonally polarized signals arrive at the HNLF, after amplification, along with the auxiliary CW carrier (Figure 2A).

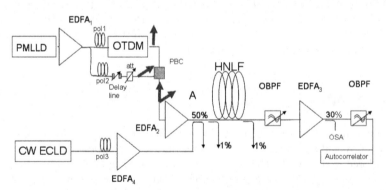

Fig. 4. All optical demultiplexing experimental Set-Up

In Figure 5, the HNLF input spectrum, CW and 40 GHz+10 GHz signals (A), and the Autocorrelator trace of the signal at 40 GHz (B) are been reported.

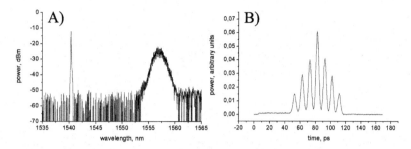

Fig. 5. A) HNLF input spectrum. B) Autocorrelator trace

In Figure 6 the output of the HNLF is reported. In particular, it can be noticed that when the 10 GHz signal is shut off, the 40 GHz modulation on the auxiliary CW carrier becomes more evident (Figure 3B).

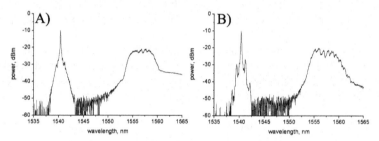

Fig. 6. Optical spectrum at HNLF output: A) with 40 GHz signal and 10GHz demultiplexing stream. B) the 40GHz signal alone.

Adjusting the power of the 10 GHz stream and moving the delay line is possible to observe the demultiplexing efficiency effect. From A) to D) (Figure 7) the delay line is moved always in the same direction. Figure 4C is reflecting what we can assume to be the deumultiplexing situation, since there is only one peak, which corresponds, due to the autocorrelator time window, to the perfect overlap between the 10 GHz demultiplexing stream and one of the 4 multiplexed pulses of the 40 GHz. All the other situations show the 40GHz residual stream (Figure 7A, B and D).

The concept behind the demultiplexing willing to be achieved with this setup, is the linear dependency of the nonlinear phase on the power of the XPM inducing signal. This higher power of the latter, the higher is the nonlinear phase shift suffered by the CW. Also, since coherence is a problem which could endanger the process, two polarizations were used. By formatting the filter shape and central position, one can control the XPM part which is extracted. In the experiments a simple off the shelf filter was used, therefore limiting the overall result. However, the same situation as described for 40 GHz, was achieved with a 80 GHz pulse stream. Another effect which will certainly be of interest are the time delays and amplitude difference between the 10 GHz demultiplexing stream and the stream to be demultiplexed. The advantage of having the two in the same wavelength is that the walkoff is minimized, increasing therefore the efficiency of the process.

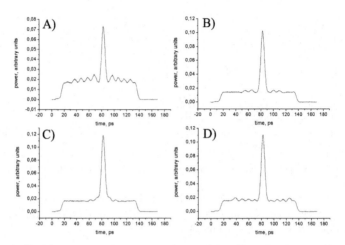

Fig. 7. A) B) C) D) Different positions of the delay line. In C) we have a good overlapping of the pulses and so a good demultiplexing from 40GHz to10GHz.

In Figure 8 are presented the results for several relative delays between the 80GHz stream to be demultiplexed and the 10GHz signal. A moderately good result was achieved, given the fact that the filter shape was not changed, therefore probably not optimum for demultiplexing at this frequency rate.

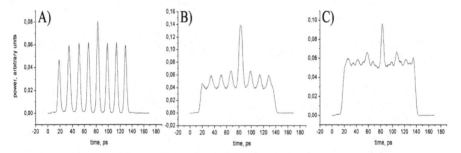

Fig. 8. A) 80 GHz input. B) and C) two different delays

In this process it is important to have a convenient choice of the powers of the two streams in order to maximize the effect.

4   Conclusions

Cross phase modulation occurring in high non linear optical fibers based on single wavelength orthogonal state of polarization has been used to achieve all optical demultiplexing. The results demonstrates that 40 GHz and 80GHz to 10 GHz demultiplexing can be easly achieved in a simple and scalable way. This technique furthermore suggests that good results could be obtained at even higher rated if tailored filters and suitable dispersion maps are used.

Acknowledgments. This work was supported by the project Cost 291. The authors would also like to thank the IREE for receiving us and the projects NoE e-Photon ONe + and CONPAC (POSC/EEA-CPS/61714/2004) for the complementary support.

References

1   Nolting, H.-P.; Sartorius, B. "Advanced devices and their features for future optical networks", _LEOS 2000. 13th Annual Meeting. IEEE_ Volume 2, 13-16 Nov. 2000 Page(s):627 - 628 vol.2

2   Russell, P. St.J.; "Photonic-Crystal Fibers" _Lightwave Technology, Journal of_ Volume 24, Issue 12, Dec. 2006 Page(s):4729 - 4749

3   Y. Ueno et al "Penalty-free error-free all-optical data pulse regeneration at 84 Gb/s by using a symmetric-Mach-Zehnder-type semiconductor regenerator," IEEE Photon. Technol. Lett., vol. 13, no. 5, pp. 469–471, May 2001.

4   S. Murata et alt "THz optical-frequency conversion of 1 Gb/s-signals using highly nondegenerate four-wave mixing in an InGaAsP semiconductor laser," IEEE Photon. Tech. Lett., vol. 3, pp. 1021–1023, Nov. 1991.

5   Bengt-Erik Olsson, Daniel J. Blumenthal et alt, " A Simple and Robust 40-Gb/s Wavelength Converter Using Fiber Cross-Phase Modulation and Optical Filtering", IEEE Phot. Technol. Lett., Vol.12, No.7, JULY 2000

6   P. V. Mamyshev, "All-optical data regeneration based on self-phase modulation effect," in Proc. Eur. Conf. Optical Communication (ECOC'98), Madrid, Spain, Sep. 20–24, 1998, pp. 475–476.

7   S. Taccheo and P. Vavassori, "Dispersion-flattened fiber for efficient supercontinuum generation", paper ThU5, OFC 2002, Anaheim, CA, USA (2002)

A Low Cost Migration Path Towards Next Generation Fiber-To-The-Home Networks

Reynaldo I. Martínez[1], Josep Prat[2], José A. Lázaro[2] and Victor Polo[2]

1: Universidad Simón Bolívar (USB), Dept. Electrónica y Circuitos, Sartenejas, Baruta, Edo. Miranda, 89000 (Venezuela)
2: Universitat Politécnica de Catalunya (UPC), Dept.of Signal Theory and Comm., c/Jordi Girona, 1-3, D4-S107, E-08034 Barcelona (Spain)
reynaldo.martinez.reyes@gmail.com, {jprat,jose.lazaro,polo}@tsc.upc.edu,

Abstract: A highly-scalable access architecture achieving high-user-density and enabling resiliency, centralized light-generation control, remote amplification and colorless ONU with Reflective Semiconductor Optical Amplifier (RSOA) is presented as the bridge between the already deployed Fiber-to-the-Home infrastructures and the advanced optical access networks. A techno-economical comparison of these optical access networks is done, depicting the proposed solution –SARDANA- as the most cost effective migration path towards the Next Generation Passive Optical Networks.

1 Introduction

Nowadays, there is no *technical* reason why any residential user could not have broadband internet access, gigabits/s of data to their home, using optical access systems already available in the market today – the obstacles are purely economics [1]. Therefore one of the fundamental scopes of today's optical communication research community has to be the cost reduction. Not only an effort in the reduction of the cost of the optical components or devices, but mostly in the creation of new architectures in order to make possible the delivering, in a cost effective way, of high bandwidth demanding integrated services through optical fiber up to the user premises.

Already standardized, the TDM-PON constitutes, without any doubts at all, the first step in the migration from cupper based access technologies to all-optical access networks (FTTH). Not only because of its low price but also because by means of its outside plant -splitters and fiber- it stands as the most simple and easy to maintain optical access architecture.

Once the first generations of FTTH networks are introduced, it is expected that capacity (in terms of speed and number of users) and scalability parameters will be the ones that are going to drive the design and the deployment of the Optical Access Networks (OANs). The user is going to start requiring more bandwidth demanding services with a strict QoS level, and here is where the WDM PONs will play a crucial role, because of the inability of the TDM PON in fulfilling those user demands. It is well know the efficiency of the WDM PONs in exploiting the optical fiber, achieving levels of capacity and scalability unreachable for their TDM counterparts due to their

I. Tomkos et al. (Eds.): ONDM 2007, LNCS 4534, pp 86-95, 2007.

lack of multiplexing in the wavelength domain. But as it was mentioned above, the cost of the components used in these next generation architectures remains high, making economically prohibitive for the operators the migration from their already deployed BPON/GPON/GE PON to the Next Generation Passive Optical Networks (NgPON).

This high initial capital expenditure required for the NgPON deployment compels network designers and operators to assure migration paths that guarantees fully future usage of the infrastructure investments, avoiding bottlenecks at any demand increase [2].

In this work we propose and demonstrate a low cost migration path towards future optical access networks topologies, based on a novel network topology that we have recently proposed [3] named SARDANA, which stands for Single-fiber-tree Advanced Ring-based Dense Access Network Architecture, aiming at offering: high user density, extended reach, flexibility, scalability and resiliency in a cost effective way.

In this document, a techno-economical comparison of the proposed network topology with other advanced optical access topologies is done, focusing in the Capital Expenditure (CapEx) per user on each network architecture. The total cost of each access solution is compared and also extrapolated in time.

This work is organized as follows: Section 2 is devoted to explain the proposed optical networks for analysis that can be found in the literature, the ones that are going to be compared with our solution.

Section 3 deals with the technical details about the architecture that we propose as the migration path in this work, the SARDANA solution. The analysis made in this section does not aim to fully explain the whole functionalities of the architecture. Here is just a general idea of what SARDANA is about, in order to situate the reader when making the comparison among the other FTTH solutions.

In Section 4 we present the results of the economical comparison between all the network topologies in terms of total cost per user and for different take rates. Finally in Section 5 we give the same economical comparison but using some future prices that the optical components within these networks are expected to have in some years from now.

2   Proposed Network Architectures for Analysis

In comparison with the SARDANA architecture we analyze five different already known FTTH solutions. They have been designed combining different multiplexing and optical techniques and are the following:

2.1 Point to Point – Two Fibers Network

As seen in Fig. 1, it has two fibers for every connection, therefore giving the complete network bandwidth for every direction [4].

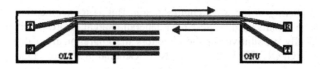

Fig. 1. P2P – 2 Fibers Network

2.2 Point to Point – Single Fiber Network

It is also a P2P Architecture but with a single fiber between the Optical Line Termina-
tor (OLT) and the Optical Network Unit (ONU), which reduce to the half the cost in
cabling [4]. It is shown in Fig. 2. It uses two different wavelength lasers for downlink
and uplink directions in order to avoid Rayleigh Backscattering impairments.

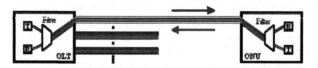

Fig.2. P2P – Single Fiber Network

2.3 TDM – PON

Shown in Fig. 3, it is the classical point to multipoint PON, where an optical splitter
divides the signal coming from the OLT in N identical signals, sending them to N
ONUs [4]. The users are multiplexed in time since they all receive the same signal.

Fig. 3. Single Fiber TDM PON

2.4 Coarse WDM – PON

It uses M Distributed Feedback Lasers (DFB) at the OLT to address through the Ar-
rayed Waveguide Grating (AWG) a specific power splitter, creating a virtual TDM
PON for every wavelength [4]. It uses a Reflective Semiconductor Optical Amplifier
(RSOA) as ONU, and it is represented in Fig. 4.

Fig. 4. Coarse WDM PON with Reflective ONU

2.5 Dense WDM –PON

Uses tunable lasers in the OLT to select the output port of the MxM AWG in the OLT as well as the 1xN AWG output port, which connects to a single ONU, where data is sent [5]. This architecture uses the Latin Routing characteristic of the AWG in order to have more flexibility when transmitting information to different users. It is shown in Fig. 5.

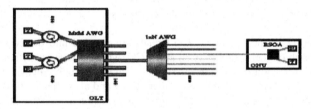

Fig. 5. Dense WDM PON with Reflective ONU

3 SARDANA: Single-fiber-tree Advanced Ring-based Dense Access Network Architecture

The proposed Architecture (SARDANA), shown in Fig. 6, is based on a WDM double-fiber-ring with single-fiber wavelength-dedicated trees connected to the main ring at the Remote Node (RN). Remote amplification is introduced at the RN by means of Erbium Doped Fibers (EDFs) for compensating Add/Drop and filtering losses. Pump for the remote amplification is provided by the pumping lasers located at the CO. The CO uses a stack of tunable lasers for serving the different tree network segments on a TDM basis.

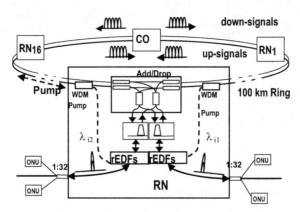

Fig. 6. Example of a 16RN-100Km Ring SARDANA Network Arquitecture – Double fiber completely transparent ring with wavelength dedicated trees

Downstream (λ^D_i) and upstream (λ^U_i) signals are coupled into the corresponding ring fiber through optical switches (OS) at the CO, which allows for dynamically adjustment of the direction of the transmission of each wavelength, always providing a path to reach all the RNs even in case of fiber failure and also offering traffic balancing capabilities. Pump can be coupled either to downstream or upstream fiber by pump/signal WDMs. The pump powers propagating through fiber ring to the RNs provides some extra Raman amplification. Because of this, upstream fiber is then preferred for pump propagation as upstream signal powers are usually weaker than the downstream ones. Two 1480nm pump lasers are required for bidirectional, balancing pumping and resilience against fiber failure.

At the RNs, a simplified OADM is accomplished by two fixed splitting ratio couplers (for the OADM function) and a 50/50 coupler (for protection function), for each of the two ring fibers. The fixed filters determine the dedicated wavelength of each Network Tree. Pump is previously demultiplexed and led to the EDFs for amplification of Up/Downstream of each tree. Two single fiber PON trees are connected to each RN by means of couplers; a 32 TDM PON is considered at each tree-PON network section. As a half-duplex system, the tunable laser of the CO is providing to the ONUs an optical carrier during half of the time for ASK upstream modulation by the RSOAs. Further details can be found in [3].

The SARDANA network can be scaled to reach a very large number of users; e.g. with 16 RNs and 32 wavelengths (as shown in Fig.6) it can serve up to 1024 ONUs. Finally, for a much more convenient network implementation with identical ONUs, we propose wavelength agnostic transmission devices. RSOAs are suitable devices due to their capabilities for re-modulation and amplification, as well as their wavelength independent nature.

4   Cost Analysis

The cost analysis in this work was done trying to keep similar conditions for all the networks topologies, in terms of number of users served, bandwidth delivered to each one and geographical conditions. This in order to be able to make fair comparisons between the total costs of the solutions. We obtained the equations of the CapEx of each one of the six different networks serving ≈5000 users, with ≈100 Mbps per user, in an urban area located in the city of Barcelona as depicted in Fig. 7, establishing those characteristics as our case of study.

In order to do so, we totalized the cost of all the components and lengths of fiber necessary in each part (OLT, Feeder, Distribution, and ONU) of each network architecture in order to deliver service in the proposed case of study (5000 users, 100Mbps per user, location of Fig.7).

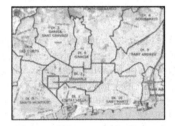 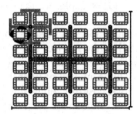

Fig.   7.   Left: Picture   of   Barcelona   and   the   considered   area   of   study
Right: Cabling Model used in the calculations on the cost of the outside plant-

We decided to choose the area of the Fig.7 because of its population density, which was similar to the potential client density. This situation, joined with the fact that the average OLT-Home distance in the selected area for coverage was similar to the distance between the OLT and the center of the area, allowed us to compute the average amount of feeder and distribution cable per home needed, and therefore compute the total cost in outside plant cabling of each architecture. The price of the components and devices employed in each architecture (AWG, splitters, RSOAs, lasers, receivers, circulators, etc) was averaged from the quotations of several vendors.

Figure 8 shows the total cost per user of each one of the studied networks with the obtained current prices for different take rates[1] (Installation and digging costs not included in the calculations).

---

[1]Take Rate: Relation between the number of homes connected to the network and the number of homes that can be connected to the network

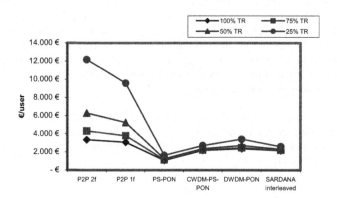

Fig. 8. CapEx per user with current prices for different take rates (TR).

The first thing that can be seen from Fig.8 is the incapacity of the point-to-point so-lutions to deliver the services in a cost effective way. In fact in some cases, the cost of the P2P approach is 8 times the total cost of the point-to-multipoint approach [6]. This is due to the huge amount of fiber that is needed in this architecture, as can be seen in Fig.9. The total cost of cabling scales directly with the number of users, because the P2P architectures require at least one fiber from each ONU up to the OLT. In a more efficient way, the point to multipoint architectures arrive to the served area with a sin-gle fiber from the OLT, and the segregation in one fiber per user is performed near the clients, reducing in that way the cabling cost.

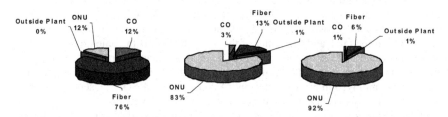

Fig. 9. Contribution to the total cost by each part of the network. Left: P2P Network, Center: TDM PON Network, Right: SARDANA

Another thing that can be seen from Fig. 8 is that with current prices, the TDM PON is the most adequate solution in terms of the CapEx for deploying a FTTH sys-tem. It is even a more cost effective solution than our proposed Arquitecture (SARDANA).

But while these already standardized TDM PONs are currently in deployment as the first step in all optical access networks, recent research is focused in the NgPONs as we claimed before, aiming at offering higher user density, extended reach, flexibil-ity, scalability, resiliency [7] among other characteristics, which are going to be mandatory for the future user demands, and which can not be achieved with TDM PON architectures.

By the time those next generation services start to appear in the market, we demonstrate in the next section that the SARDANA, an architecture able of delivering such services, will have a cost even lower than the –by then obsolete- pure TDM PONs, assuring a cost effective migration path to Advanced Optical Networks, like the Dense WDM PON, which by that time maybe would still have prohibitive costs for the operators.

5   Cost Analysis with Expected Prices

The cost of optical networks has decreased over the years as the underlying technology has advanced and manufacturing volumes have increased. The cost of the underlying electronic and optical technologies (lasers, optical fibres, ASICs, etc) is well-known to follow a cost reduction with volume known as a learning curve. A learning curve is defined as the percentage decline in the price of a product as the (cumulative) product volume doubles. Technologies typically follow an ~80% learning curve which means that the price of the product at volume 2V will be ~80% of the price at volume V [1].

Another important tendency curve is a law that has been in the literature for more than 40 years, called the Moore's Law[2]. It could be interpreted that the density of transistors doubles every year.

Combining these two ideas it would not be risky to expect that the prices of the components based in semiconductors will follow a similar learning curve, and therefore the access technologies based on semiconductors will follow a cost reduction as well.

It would be a huge mistake to expect that the total cost of the P2P solution for example, will be reduced following the mentioned learning curve because, as we concluded from the previous section and from the Fig. 9, most of its cost is due to the huge economical inversion in optical fiber cabling ($\approx$76%), and we can not assume that the price of the optical fiber is guided by the Moore's Law.

In the same Fig. 9 we see something very interesting: in SARDANA the 92% of the total cost of the network is due to the ONU, which is an RSOA as we explained in Section 3, a semiconductor which is expected to follow the cost reduction ideas explained above. So it should be no surprise that the prices of SARDANA, and the other networks topologies that employ components which follow the learning curves explained, might get down.

We can therefore use this historical learning curve and extrapolate the prices of the components which are affected by those curves. We did that and saw how the total cost of each one of the network topologies considered in section five changed through time, as it is shown in Fig. 10.

---

[2] " Moore's Law in Action at Intel" , Microprocessor Report 9(6), May 8, 1995.

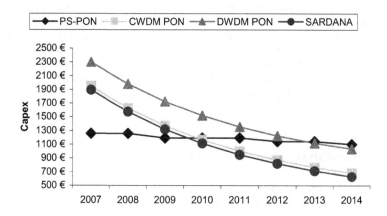

Fig. 10. Expected prices for network topologies considered in the study for the next decade (Take Rate=50%).

As it was expected, not all the Capital Expenditures experiment a reduction in the same way. This because not all the network architectures considered have the same amount of components that are affected by the learning curve (semiconductors). In fact, as it can be seen from figure 10, the PS-PON topology price barely varies through time. The SARDANA Architecture, having a total cost which is constituted in 92% by a semiconductor device, is the most affected architecture by the learning curve in the cost, and it is the first to beat the TDM PON topology in terms of the CapEx, having in addition the properties of capacity, speed, user density, scalability, resiliency and flexibility that allow this architecture to fulfill the future user demands.

6   Conclusions

Hybrid dual-fiber-ring with single fiber-trees access network topology with a wavelength agnostic ONU made of an RSOA demonstrates to provide a cost effective bridge between the already standardized and deployed Time multiplexed PONs and the Next Generation PONs, which have nowadays costs economically prohibitive for any telecommunication operator.

Further Economic studies have to be done in order to include the civil work costs (digging and installation) which in this work were not considered. Doing this the differences in the total investment between P2P and PON infrastructures would not be as high as the ones presented in this work. But the advantage of developing a PON infrastructure, due to their point-to-multipoint approach and resource sharing nature, is that it can take profit of all the dark fiber available in a city, minimizing the digging and installation costs in a way that a Point-to-Point approach can not.

The Techno-Economical comparison done in this work proves that SARDANA is, in terms of the cost and in terms of future usage of the existing infrastructure, the best migration path towards the truly broadband for all era.

Acknoledgments
This work was partially supported by the Spanish MEC, TEC2005-05160 (SARDANA) and the Ramon y Cajal Program.
Authors sincerely thank Raúl Sananes for his valuable contributions.

References

[1] J.Prat, P.Chanclou, R.Davey, J.M.Finochietto, G.Franzl, A.M.J.Koonen and S.Walker, "Long Term Evolution of Passive Optical Networks" in proceedings AccessNets 2006, Athens Greece
[2] J.Prat, C.Bock, J.A.Lázaro and V.Polo, "Next Generation Architectures for Optical Access", in Proceedings ECOC 2006, Cannes, Th2.1.3
[3] J. A. Lázaro, R. I. Martínez, V. Polo, C. Arellano and J. Prat "Hybrid dual-fiber-ring with single-fiber-trees dense access network architecture using RSOA-ONU," Optical Fiber Communications Conference, OFC/NFOEC'07, Proceedings, OTuG2, Anaheim, USA, March 2007.
[4] M. Nakamura, H. Ueda, S. Makino, T. Yokotani, and K. Oshima, "Proposal of networking by PON technologies for full and Ethernet services in FTTx," J. Lightwave Technol. 22, 2631–2640 (2004).
[5] C. Bock, J. Prat, J. Segarra, G. Junyent, A. Amrani: Scalable Two-stage Multi-FSR WDM-PON Access Network Offering Centralized Dynamic Bandwidth Allocation, in Proc. ECOC 2004 Stockholm, Tu4.6.6
[6] R.Sananes, C.Bock, S.Figuerola and J.Prat, "Advanced Optical Architectures Evaluation", in FTTH Conference 2004, Orlando, October 2004
[7] R.Davey et al, "The Future of Optical Transmission in Access and Metro Networks- an Operator's View", in Proceedings ECOC 2005, Glasgow, We2.1.3 (2005)

Securing Passive Optical Networks Against Signal Injection Attacks

Harald Rohde, Dominic A. Schupke

Siemens Networks
Otto-Hahn-Ring 6, D-81730 Munich, Germany
Corresponding author: Harald.Rohde@siemens.com

Abstract. Passive optical access networks are susceptible to intended attacks and unintended failures. This paper discusses intrusion by user-side signal-injection resulting in reduced network accessibility and it proposes possible countermeasures. The central function is that an intruding signal can be switched off when it is present.

1. Introduction

Any access network is subject to various intrusions, caused by intended attacks or by unintended failures (employed here analogously to "attacks"). Such attacks can target security items, e.g., information reaches a specific port to which it was not intended, or attacks may concern network accessibility, i.e., the possibility for a single user or a failing terminal to degrade (or even disable) the access network for other connected users. This paper discusses the latter case resulting in degradation or denial of service by signal injection in passive optical access networks. We also suggest countermeasures.

Shared-resources enable us to divide the expenditures of some of the network infrastructure (e.g., the medium) by the number of possible users. The throughput for a single user should not be significantly affected, if only a small fraction of the users at a given time access the resources (statistical multiplexing).

The following list presents some examples for shared-medium networks: Ethernet before the introduction of switches (the name "Ether" refers to the old perception of an ubiquitous medium), all wireless networks, cable modem networks, and within fiber optics Passive Optical Networks (PONs). Shared-resources, however, always raise security issues that have been sufficiently solved for the above network types (e.g., by "switched Ethernet"), except for PONs.

To the best of the authors' knowledge, the only paper addressing the issue is [1]. The system proposed in [1], however, uses optical fuses that have to be replaced after an attack. Our proposal allows for automatic switching-back to regular state once the attacking signal has been removed. We focus on the critical case of direct injection [2] of continuous signals, however, extension to sporadic signals appears realizable by sophisticated detection functions.

I. Tomkos et al. (Eds.): ONDM 2007, LNCS 4534, pp 96-100, 2007.

2. Modern passive optical networks

Figure 1 depicts a next-generation passive optical network. One single Optical Line Termination (OLT) handles a number of subscriber units (Optical Network Units, ONUs) on a split-fiber infrastructure. The optical splitter site is at best completely passive. A set of different architectures for this topic are currently under discussion [3-4]. However, apart from pure WDM-PONs, all future PONs have a high splitting-factor in common.

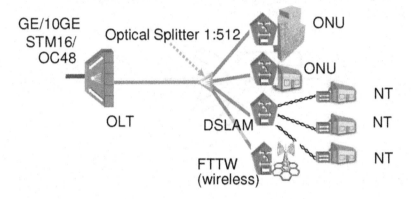

Fig. 1. Typical layout of a next generation passive optical network (PON).

As also depicted in Figure 1, a single OLT serves a number of different user-types. Possible users are private users, small/medium enterprises, wireless stations by Fiber To The Wireless (FTTW), and (outdoor) Digital Subscriber Line Access Multiplexers (DSLAMs) which feed a number of DSL users at the Network Terminations (NTs).

The larger the splitting factor becomes, the higher the danger of intended or unintended network disturbances will be. Because of the shared nature of the upstream data channel (upstream is here defined to be the direction from the ONU to the OLT), a single light source sending permanently or -even worse- casually light with the matching wavelength can stop the operation of the whole PON. While the downstream direction is optically unaffected, the lack of acknowledgment data packets from the single ONUs will immediately stop the operation of the PON.

Sending the light source can be done easily, including hacking into the ONU or using simple hardware. As enterprise users or wireless network operators attached to the ONU would refrain from relying on a network, which a single user could shut down, countermeasures have to be taken.

Today, a single PON has 16 to 32 users per fiber and because of this relatively small number a homogenous group of users (e.g., private-users groups, business-users groups) can be connected. Increasing split factors up to 512 or even higher makes homogenous groups increasingly difficult or even impossible, underlining importance of countermeasures against attacks.

3. Countermeasure against permanent signal injection

For the operator it is obviously interesting to identify the attacking port and to disconnect the attacker from the network. Applying a manual process for this can involve long PON outage durations, since the duration includes sending maintenance personnel to the OLT and checking ports one after another until the attacking port is found. Therefore we propose an automatic process for fast and administratively simpler reaction.

Figure 2 shows a generic architecture allowing to disconnect individual users by controlled optical switches at the splitter. These switches can also serve for other purposes such as for testing toward the ONU.

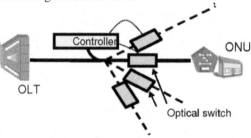

Fig. 2. : Architecture to disconnect attackers.

In the attack-case of sending a permanent signal from some ONU, the controller can firstly detect that such a continuous signal is sent (e.g., by detecting from a tap). Secondly, it identifies the port by briefly disconnecting the users, invoking the switches. This disconnection time should be very short, but still high enough to allow detection (e.g., ~5 ms). Once the attacking port is identified, i.e., when during a switch-off the permanent signal disappears, the corresponding port can be switched off and maintenance personnel can react on the malfunction at the ONU. The important point is that during this process the other users remain almost or even completely unaffected.

While active realizations are possible, for pure passive networks, we aim to realize the controller and the switches such that the splitter-site remains passive. For this, the controller can be placed at the active OLT, where it can detect the attacking signal. The attacking user is then associated with the corresponding switch that the controller addresses for disconnection. For switch activation, the controller can send a switching signal downstream from the OLT to the switches, allowing a passive realization of the switches. Hence, the switches are not invoked by a power supply, but by an optical signal filtered from the incoming OLT signal.

For cost-efficient components, we can use CWDM here. As the number of wavelength could then become exhausted, we can mitigate by addressing port groups (hence, a group of switches) over the invocation signal, that then disconnects a group of users. Such a solution compromises between cost and impact of attack.

In Figure 3 we present two potential technologies for the optical switches. In Figure 3a, only the data wavelength and a selected invocation wavelength can pass through the CWDM filter. Upon sending an invocation wavelength, the successive absorptive dye becomes opaque and thus can switch off the port, otherwise it remains transparent.

In Figure 3b, a part of the signal is tapped off and passed through a CWDM filter for the invocation wavelength. Upon sending the invocation wavelength, the successive photodiode (PD) generates a voltage that applies to the Mach-Zehnder Modulator (MZM) to switch the fiber optically off.

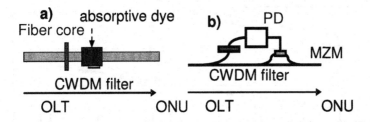

Fig. 3. Possible technologies for the optical switch.

4. Quantitative measure of the benefits

Figure 4 compares the Total Accumulated Outage Time (TAOD) which is defined as the Mean Time to Failure Recovery multiplied by the number of users for the three cases of a standard GPON, an unprotected enhanced PON, and a protected PON. The following assumptions hold: (i) half an hour to get a service technician ready and to get access to the splitter site, (ii) the distances to the splitter are 10 km for the GPON and 90 km for the enhanced PON, (iii) the average driving speed from the central office (where the OLT is based) to the splitter site is 50 km/h and (iv) 5 minutes for each ONU are needed to open the connection, check it and splice it again.

For an enhanced PON with 512 users the TAOD can be over 20 000 hours which can result in a tremendous financial penalty for the network operator. The TAOD for a protected PON with around one second delay until the protection sets in, is less than 10 minutes (512 seconds). The numbers of the example above may be varied, but in any case the TAOD of a protected PON will be 2 to 3 orders of magnitude below the TOAD of an unprotected PON for large user numbers.

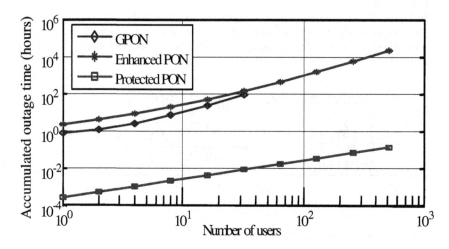

Fig. 4. Accumulated total worst case outage time for GPONs and Enhanced PONs

5. Conclusions

We have tackled a new network security issue which emerges together with the evolution of next-generation passive optical access networks. We discussed the important threat of user-side signal-injection and propose an efficient mechanism countermeasuring against this threat. We expect that in future passive optical networks such protection mechanisms will have to be installed, similarly to existing shared-resources networks.

References

[1] Shing-Wa Wong, Wei-Tao Shaw, Saurav Das, Leonid G. Kazovsky: "Enabling Security Countermeasure and Service Restoration in Passive Optical Networks", IEEE GLOBECOM, San Francisco, CA, USA, November 27 - December 1, 2006
[2] J-S. Yeom, O.K. Tonguz, "Security and Self-Organization in Transparent Optical Networks", Invited Paper, ACM AccessNets 2006, Athens, Greece, September 2006.
[3] Harald Rohde, "Modern PON Architectures", Asia-Pacific Optical Communications Conference (APOC), Shanghai, China, November 6-10, 2005..
[4] Harald Rohde, Sebastian Randel, "Project PIEMAN: A European approach to a symmetrical 10 Gbit/s, 100 km, 32 λ and 512 split PON", Asia-Pacific Optical Communications Conference (APOC), Gwangju, Korea, September 3-7, 2006.

Novel Passive Optical Switching Using Shared Electrical Buffer and Wavelength Converter

Ji-Hwan Kim[1], JungYul Choi[2], Jinsung Im[1], Minho Kang[1], and J.-K. Kevin Rhee[1]*

[1] *Optical Internet Research Center, ICU 119 Munjiro, Yuseong-Gu, Daejeon, 305-732, South Korea*
[2] *BcN Business Unit, KT Corp., 463-1 Jeonmin-Dong, Yuseong-Gu, Daejeon, 305-811, South Korea*

* rhee.jk@ieee.org

Abstract We propose a novel optical switching system using fast time-slotted passive switching with shared OEO buffers. Simulation results show that the proposed system significantly reduces the blocking probability below 10^{-6} at a 28% cost of a traditional system due to flexibility of contention resolution using electrical buffer and potentially requires very low power consumption. The proposed switching system is believed to be a technoeconomically feasible and implementable solution for optical packet and burst switching with current optical technologies.

Keywords: Active buffer, time-slotted switching, passive switch, cost analysis, optical packet/burst switching.

1   Introduction

In the next decade with fully featured multimedia network service requirement, the internet will revolve with newer technologies in order to provision the explosive growth of the capacity demand and delicate network QoS and traffic engineering functions. A single domain of a network may deal with petabit per second traffic demands. In order to support this trend, a disruptive technology employment such as optical packet switching (OPS) [1] and optical burst switching (OBS) [2] is considered. However, few technologies in this avenue yet look promising to achieve the industry goals to provide an ultimate bandwidth solution with practical application requirements of performance, cost, power consumption, and form factor. In order to re-investigate the fundamental understandings of the traditional cost-performance optimization problem of optical switch networks, we revisit the old but fundamental themes how to implement buffers in packet switched networks. In addition, because of the special WDM network constraints, the wavelength continuity requirement is also re-investigated. As a result, this paper proposes a novel optical switching system consisting of a passive switch with an active buffer module, which may open a new design concept of an optically transparent network system design.

I. Tomkos et al. (Eds.): ONDM 2007, LNCS 4534, pp 101-106, 2007.
© IFIP International Federation for Information Processing 2007

The OPS/OBS systems are mainly composed to two essential hardware functions, including the switch fabric and the buffer. The functions may utilize passive and active technologies as shown in Table 1. Most of active technologies utilize semiconductor optical amplifiers (SOAs) and wavelength converters (WCs) that offers capability of rather complicated optical signal processing, at the price of power for high data rates and complexity of control, and premium device technologies. Passive technologies, instead, can mitigate high power requirements and control complexity by far, but not capable of wavelength conversion that is important for providing wavelength continuity in WDM network systems and reducing high packet loss probability.

As one can notice from Table 1, the traditional switch architecture is 'active switches with passive buffers system,' where the wavelength continuity is fully provided [3] by a switch fabric. This design rule, however, may have been the critical road block against achieving low cost, low power, and low blocking probability (*BP*). This paper reports that '*passive switches with active buffers system*' is more effective in the respect of packet loss and overall system cost. Being capable of achieving extraordinary performance, the active buffers can be shared, which can reduce the system cost, power consumption, and form factor dramatically to achieve petabit switch solutions

Table 1.   Technology classification for optical switching. Bold-faced technologies suggest possible device solutions for the proposed node architecture design.

Technology	optical switch fabric	optical buffer
Passive (no WC)	**electro-optic, acouto-optic, ultra-fast MEMS, wavelength-selective switch**	fiber delay line, recirculating loop, slow-light photonic crystal
Active (WC capable)	electrical switch with OEO, AWGR/WC, SOA array	**electrical memory with OEO, bistable laser diode**

2   Passive switching with shared active buffering

Packet (or burst) switched DWDM networks require packet (or data burst) forwarding to the destined output fiber of a node with or without conversion of the wavelength. When only a fast passive switch fabric is used, the availability of the output fiber is limited to that of Aloha network because of no wavelength conversion. By the theory, several tens of percentiles of the input packets/bursts are lost due to contention. Only the contended packets are then sent to electrical buffers by O/E conversion (Fig.1).

As soon as any wavelength channel becomes available at the destined output fiber, the electrical data is converted to the available wavelength by a tunable E/O converter, i.e. a tunable transmitter. Because an electrical buffer provides buffering time flexibility and large buffer depth at low costs, *BP* can be reduced below 10^{-6} with a minimal investment of buffers (Fig.2). Here we define the sharing ratio, *SR*, as the ratio of the required number of buffers *B* to the total number of channels that is the product of the number of wavelength *W* and the number of fiber ports *F*.

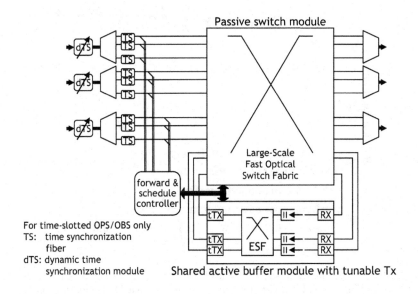

Fig. 1. Schematic node architecture of passive switching and shared active buffering

Time-slotted OPS/OBS [1] in a synchronized network can reduce BP performance of passive-only switching approximate by half in principle, reducing requirement of buffer SR by half or so, leading to an extreme savings of cost, power, and form-factor, as the OEO and buffer are the most cost critical subsystem, as compared in Fig.2. The combination of time-slotted packet switching and passive switching with shared active buffering (PSSAB) can achieve extremely high performance of $BP < 10^{-6}$ with tunable OEO buffer sharing ratio of only 15%. This offers major reduction of cost and power consumption because OEOs and WCs are most expensive and power-consuming part of the system.

The fundamental difference of the proposed idea is that the traditional method of wavelength switch with shared passive fiber-delay-line buffers (WSSPB) [3,4] attempt to resolve contention by wavelength conversion and the rest of contention is resolved by shallow buffering, while our scheme uses deep buffering first and then shared wavelength conversion. This difference bring several orders of magnitude enhancement in BP performance, especially for the case with a small number of system wavelengths, offering a practical deployment scenario of 'pay-as-you-grow'.

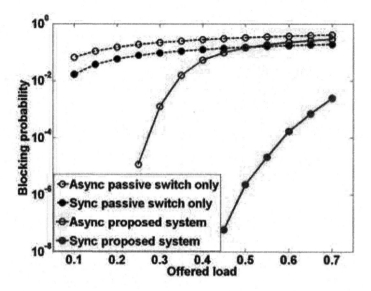

Fig. 2. Blocking probability versus offered load. The number of system wavelengths is 32 and the buffer sharing ratio is 15%. The proposed time-slotted ("sync") system shows orders of magnitude improvements.

3 Performance and cost analysis

The benefit of the proposed system is estimated by computer simulation. The blocking performance and system cost are found under the following system design parameters: the number of input/output fiber links (F) is 8, per-wavelength data rate 10Gbps, and the average optical packets size 100kbyte. These typical design parameters [5] require optical switches with sub microsecond response time. Optical packets are assumed to be uniformly distributed to all output fiber links. The electrical buffer is assumed to have a large enough depth of buffering, so there is no buffer overflow. Fig.3 presents the sharing ratio requirements for the shared OEO buffers as a function of an offered load in comparison between asynchronous and time-slotted synchronous cases. Two different time-slotted OPS/OBS systems are considered: a fixed-size synchronous packet ($N_{max}=1$), and a bi-step variable-size synchronous packet ($N_{max}=2$). Astonishingly, the time-slotted case requires a sharing ratio of only 12% in the case of 64 wavelength system, in order to provision for an offered load of 0.5. This is almost a factor of 10 improvement in the OEO requirement with respect to a current electrical cross connect (EXC) system.

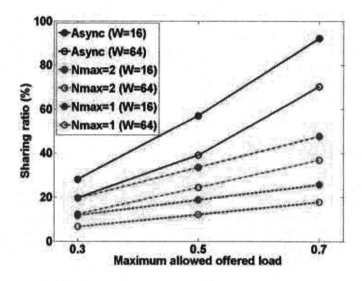

Fig. 3. OEO buffer sharing ratio requirement for blocking probability less than 10^{-6} for various conditions. N_{max} =1 and 2 correspond to fixed-size and bi-step-size time-slotted switching systems, respectively.

Fig. 4 shows overall system cost comparisons of the proposed system with WSSPB [6] and EXC systems for the same performance. The system cost of the proposed system is only 28% of the traditional EXC cost. Interestingly, the overall cost of the synchronous system is much less than that of asynchronous system.

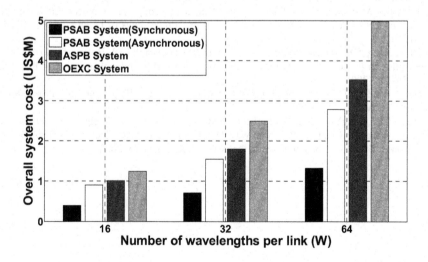

Fig. 4. Overall system cost comparison with constraint of target blocking probability at 10^{-6}

Packet (or burst) switched DWDM networks require packet (or data burst) forwarding to the destined output fiber of a node with or without conversion of the wavelength. When only a fast passive switch fabric is used, the availability of the output fiber is limited to that of Aloha network because of no wavelength conversion. By the theory, several tens of percentiles

4   Conclusion

We proposed a novel passive optical switching system with shared OEO packet buffers that may open a new paradigm for OPS/OBS practical applications. The use of shared electrical buffer can achieve several orders of magnitude improvement in blocking probability and potentially require very low power consumption. With practical system design parameters, only 12% of OEO buffers with respect to the total channels are required to achieve a 10^{-6} node blocking probability.

Acknowledgments. This work was supported in part by the KOSEF-OIRC project.

References

1. A. Bianco, E. Leonardi, M. Munafo, F. Neri, and W. Picco, "Design of optical packet switching networks," Proceedings of GLOBECOM '02, vol. 3, pp. 2752-2756 (2002).
2. S. Verma, H. Chaskar, and R.Ravikanth, "Optical burst switching: a viable solution for terabit IP backbone," IEEE Network, vol. 14, no. 6, pp. 48-53 (2000).
3. C. Develder, M. Pickavet, and P. Demeester, "Choosing an appropriate buffer strategy for an optical packet switch with a feed-back FDL buffer," Proceedings of ECOC '02, vol. 3, pp. 1-2 (2002).
4. C. M. Gauger, "Optimized Combination of Converter Pools and FDL Buffers for Contention Resolution in Optical Burst Switching," Photonic Network Communications, vol. 8, no. 2, pp. 139-148 (2004).
5. P. Bayvel and M. Dueser, "Optical burst switching: research and applications," Proceedings of OFC '04, vol. 2, pp. 4 (2004).
6. Ji-Hwan Kim, JungYul Choi, Minho Kang, and J.-K. Kevin Rhee, "Design of novel passive optical switching system using shared wavelength conversion with electrical buffer," IEICE Electronics Express, to be published in the issue of December 25, 2006.

160 Gbps Simulation of a Quantum Dot Semiconductor Optical Amplifier Based Optical Buffer

Maria Spyropoulou[1], Konstantinos Yiannopoulos[2], Stelios Sygletos[1], Kyriakos Vlachos[2], Ioannis Tomkos[1]

[1] Athens Information Technology Centre, 19.5Km Markopoulou Av.Peania 19002, Greece
{mspi,ssyg,itom}@ait.edu.gr
[2] Department of Computer Engineering and Informatics, and Research Academic Computer Technology Institute, University of Patras, Rio 26500, Greece
{giannopu, kvlachos}@ceid.upatras.gr

Abstract. We demonstrate the applicability of quantum-dot semiconductor-optical-amplifier based wavelength converters to the implementation of an ultra-high speed optical packet switching buffer. The buffer architecture consists of cascaded programmable delay stages that fully utilize the available wavelengths and thus minimize the number of wavelength converters that are required to implement the buffer. Physical layer simulations demonstrate error-free operation of the buffer with 3 cascaded Time-Slot-Interchager (TSI) stages at 160 Gbps in the 1.55um window.

Keywords: Optical buffers, Optical packet switching, Quantum Dot Semiconductor Optical Amplifiers, Wavelength Converters, Programmable Delays.

1 Introduction

Quantum dot semiconductor optical amplifiers (QDSOAs) are envisaged as the next generation optical signal processing devices in ultra high speed networks. This originates from their unique features such as high differential gain and ultra-fast gain recovery (~100fs) [1], which make them superior to conventional bulk or quantum well (QW) devices. Moreover, their enhanced properties make QDSOAs ideal candidates for signal processing applications at line rates that reach 160Gb/s.

In the current communication we demonstrate the applicability of QDSOAs in an optical buffer architecture that is suitable for ultra-high speed optical packet switching. Packet buffering is implemented in a multi-stage Time-Slot-Interchanger (TSI) that consists of QDSOA-based wavelength converters exploiting the cross-gain modulation (XGM) effect and feed-forward delay lines. Moreover, the TSI stages are designed to fully utilize the number of available wavelengths, which are defined in such a way to prevent spectral overlap in multi-hundred Gb/s line rates and also to fit within the spectral width of homogeneous broadening of a single-dot group. The use of many wavelengths is critical when considering the hardware cost of the buffer architecture and the signal quality degradation due to successive wavelength conversions. The buffer architecture is accompanied by physical layer simulations that derive the achievable buffer size in terms of cascaded TSI stages and illustrate its efficient performance in terms of output Q-factor and extinction ratio. Simulations reveal that up to 3 TSI stages may be cascaded for error-free operation at 160 Gbps.

I. Tomkos et al. (Eds.): ONDM 2007, LNCS 4534, pp 107-116, 2007.

2 Buffer Architecture and Control

2.1 Buffer Architecture

The proposed system architecture is presented in Fig. 1. It comprises cascaded programmable delay stages and each stage consists of two Tunable Wavelength Converters (TWCs) and two delay line banks. Each TWC provides w separate wavelengths at its output, and each wavelength is routed to the respective delay line of the delay line bank by means of a wavelength de-multiplexer. Adjacent TWCs and stages are connected by wavelength multiplexers.

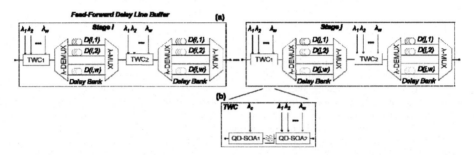

Fig. 1. (a) The structure of each delay stage. (b) The QDSOA tunable wavelength converter (TWC) setup. λ-MUX/DEMUX are the wavelength multiplexers and demultiplexers, respectively.

The delays that are introduced at each TSI stage are a design parameter of the proposed architecture and we will evaluate them from the timeslot transition graph (TTG) of the TSI architecture [2]. The TTG consists of nodes located at columns and rows; columns i and $i+1$ correspond to the input and output, respectively, of TSI stage i, and rows correspond to timeslots, each being occupied by a single optical packet. An optical packet that has arrived at the input of stage i and accesses one of the delay lines is placed in an output timeslot that appears later in time, and this action is represented as a straight line (time transition) connecting an input and an output node on the TTG.

Buffering a packet in the proposed design corresponds to a path on the TTG. The origin node represents the input-slot on which the packet has arrived and the destination node represents the output-slot where the packet leaves from. Taking into consideration that more that one packets arrive at the buffer inputs within a timeframe, it is evident that during each timeframe, an interconnection pattern which maps input to output nodes is formed on the TTG. The aim is to engineer the time transitions, or equivalently the delay times $D(i,j)$ at each stage, so that the interconnection pattern forms a $\log_n$-Benes graph as a subgraph. The $\log_n$-Benes graph is derived from the $\log_2$-Benes graph after replacing the 2x2 with nxn switches. Furthermore, it is considered a well-known re-arrangeably non-blocking interconnection topology.

The purpose of constructing the $\log_n$-Benes space-time graph is many-fold: the implementation requires a minimum number of serially connected stages that equals to

$$s = 2 \cdot m - 1 = 2 \cdot \lceil \log_n N \rceil - 1 \tag{1}$$

for buffering maximum N number of packets. Eq. (1) shows that by implementing the $\log_n$-Benes space-time graph, one can achieve a drastic reduction in the number of stages. This is of particular importance when considering the hardware cost of the implementation. Moreover, physical layer impairments aggravate the optical signal quality as the number of cascaded stages increases. An additional attribute of the Benes space-time graph is that, it is re-arrangably non-blocking resulting in the capability of the proposed design to store packets without suffering internal collisions. Finally, finding collision-free paths within the Benes graph is a well studied problem [3].

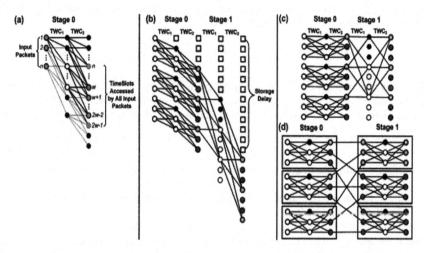

Fig. 2. (a) Formation of the $n{\times}n$ switch on the TTG. (b) Formation of the $\log_n$-Benes network on the TTG. (c) Representation without delays. (d) Equivalent representation. It is assumed that there are four available wavelengths in (b)-(d).

The building blocks of the $\log_n$-Benes graph are $n{\times}n$ switches, and thus the first step for constructing it, is to determine the switch size. In a previous communication, we had proposed the deployment of a single TWC per stage [4] for constructing the switches. This resulted in poor wavelength utilization, which amounted to approximately 50% of the available wavelengths, for the switch formation. Thus, the number of available wavelengths limited by the relation of the single-dot bandwidth and the detuning between adjacent wavelengths is of key importance. Therefore, it is proposed to double the number of TWCs that are required per stage, with objective to achieve almost 100% wavelength utilization when forming the $n{\times}n$ switches.

The switches are formed out of time transitions on the TTG, as shown in Fig. 2(a), which corresponds to the first stage (stage 0) of the buffer. A packet that has arrived at the input of stage 0 during the first timeslot may only access timeslots $\{1, ..., w\}$ at the output of the first TWC, since time transitions to previous timeslots are not allowed. In a similar fashion, the timeslots that are accessible by the aforementioned packet at the output of the second TWC are limited to $\{1, ..., 2w-1\}$. If n successive input packets are considered, then the timeslots at the output of the first and second TWC that are accessible by all input packets are $\{n, ..., w\}$ and $\{n, ..., 2w-1\}$, respectively.

The switch formation requires that the output timeslots that are accessible by all input packets exceed the number of packets themselves, so that the $n{\times}n$ switches may be always formed on the TTG. This is equivalent to

$$2 \cdot w - 1 - n + 1 \geq n \Rightarrow n_{\max} = w. \tag{2}$$

Moreover, the interconnection network that corresponds to the $n \times n$ switches must be non-blocking, so that packets do not arrive simultaneously at a TWC and are therefore not lost. This is satisfied by ensuring that there are at least two disjoint paths between all input and output timeslot nodes inside the switch. To ensure this, we take into account that the mid-nodes (nodes at the output of the first TWC) that are accessible to all input nodes are $\{n,...,w\}$. Provided that there are at least two fully accessible mid-nodes, shown in white colour in Fig. 2(a), there are always at least two disjoint interconnection paths towards the mid-nodes of the switch, and as a result

$$w - n + 1 \geq 2 \Rightarrow n_{\max} = w - 1. \tag{3}$$

Additionally, the existence of the two disjoint paths between the mid-nodes and the output nodes is assured when the output nodes are limited to $\{w, ..., 2(w-1)\}$. The switch is therefore formed (a) after selecting $n-2$ mid-nodes that are symmetrically located above and below the two fully accessible mid-nodes when n is even, or (1) after selecting $\dfrac{n-1}{2}$

and $\dfrac{n-3}{2}$ mid-nodes above and bellow the fully accessible mid-nodes, respectively, when

n is odd. Eq. (3) shows that almost full utilization of the available wavelengths has been achieved with the proposed buffer architecture.

The next step for constructing the $\log_n$-Benes graph is to determine the time transitions that form the graph's switches in the respective stages. The process is shown in Fig. 2(b)-(d) for the first and second stage of the buffer. The formation of the $\log_n$-Benes graph crossbars requires that at each stage i, time-transitions connect timeslots that are located n^i positions apart. This corresponds to setting the time delays, in timeslots, equal to

$$D(i, j) = j \cdot n^i, \quad i = 0,...,m-1, \quad j = 0,...,w-1. \tag{4}$$

The delays account for all time transitions on the TTG, even though not all transitions contribute to the formation of the virtual switches. The inactive transitions introduce a constant delay after which the output timeframe commences (white squares in Fig. 2(b)). At the output of each buffer stage, the delay equals

$$\Delta_i = n^{i+1}, \quad i = 0,...,m-1 \tag{5}$$

timeslots and as a result the delay that the packets experience when traversing the buffer is

$$\Delta = \sum_{i=0}^{m-1} n^{i+1} + \sum_{i=0}^{m-2} n^{i+1} = n \cdot \frac{n^m + n^{m-1} - 2}{n-1}. \tag{6}$$

Eq. (5) may be viewed as a constant buffer access time.

2.2 Buffer Control

Following the discussion of subsection 2.1, packets are buffered after being converted to the appropriate internal wavelengths and accessing the respective delay lines at each programmable delay stage. As a result, buffering requires that the state of wavelength converters has to be set prior to sending the packets to the buffer. From a TTG perspective setting the internal wavelengths is equivalent to calculating the state of the switches in all intermediate stages of the $\log_n$-Benes graph of in Fig. 2 (d) so that the input packet sequence is routed to the respective output sequence.

To perform routing in a $\log_n$-Benes graph, we have proposed a modified parallel routing algorithm [4] that extends the parallel routing algorithm on a binary Benes graph [3]. The algorithm involved setting the state of the outermost switches (at stages 0 and s-1) of the Benes graph given the respective packet sequences. The outermost switches are then omitted, and the remaining network is partitioned into multiple Benes graphs of reduced size. The algorithm is recursively applied on the resulting graphs until the state of all switches is set.

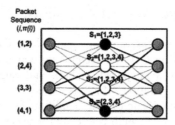

Fig. 3. Routing inside the *n*x*n* switches. It is assumed that there are five available wavelengths

Having determined the state of the switches, it remains to calculate the interconnection pattern inside the switches. We consider that each mid-node k of the switch is described by a set S_k that contains the input and output nodes it is connected to. Due to the symmetry of the switch, mid-nodes always connect to the same group of input and output nodes. Supposing that input node i must connect to output node $\pi(i)$, there are at least two mid-nodes that allow for this connection, since there are at least two disjoint paths between input and output nodes. The mid-nodes k that enable this connection satisfy

$$\left(i, \pi\left(i\right)\right) \in S_k. \tag{7}$$

An algorithm for selecting the mid-nodes so that all input-to-output connections are performed over disjoint paths (without collisions) is illustrated in Fig. 3. The algorithm involves the following steps:

    (a) For a given node pair (i, π(i)) find all available sets S_k that satisfy Eq. (7).
    (b) Select the set S_k with the smallest number of elements for the node pair.
    (c) Proceed to the next node pair (i+1, π(i+1)) and repeat.

The proposed algorithm calculates hops (i, k) and $(k, \pi(i))$ for all connections $(i, \pi(i))$ inside the switch. The calculated hops correspond to the wavelengths that are provided by the two TWCs of each respective stage. Following the discussion of subsection 2.1 on the switch formation, hop (i, k) corresponds to wavelength

$$\lambda_{\frac{n+1}{2}+k-i}, \quad n \ is \ odd$$

$$\lambda_{\frac{n+2}{2}+k-i}, \quad n \ is \ even. \tag{8}$$

In a similar fashion, hop $(i, \pi(i))$ corresponds to wavelength

$$\lambda_{\frac{n+3}{2}+\pi(i)-k}, \quad n \ is \ odd$$

$$\lambda_{\frac{n+2}{2}+\pi(i)-k}, \quad n \ is \ even. \tag{9}$$

3   Simulation of the Buffer Architecture

3.1   Description of the QDSOA based TWC subsystem

The proposed buffer architecture aims operation at 160Gb/s line-rate, therefore the TWC realization is based on QDSOA technology which can support high bit rates without patterning effects as opposed to bulk and quantum well SOAs. Fig.4a illustrates the TWC configuration which is based on the cross-gain modulation (XGM) effect of the saturated gain of each amplifier. Two cascaded QDSOA devices have been used to enable wavelength conversion between the modulating pump and continuous wave (cw) probe signals, respectively. Both pump and probe signals are assumed to co-propagate the device. The available wavelengths must fit within the single-dot bandwidth for efficient XGM to occur. Typical values of the homogeneous broadening of QDSOAs at room temperature are 10meV–20meV [5]. In the present study this value corresponds to 16meV, that is 31nm at the 1.55um window. At 160Gb/s the detuning frequency between adjacent wavelengths has been assumed 640GHz to prevent spectral overlap and it is equivalent to 5.1nm for operation at 1.55um. A set of four available wavelengths distributed evenly with respect to the center of the homogeneous broadening has been chosen. The latter coincides with the center of the inhomogeneously broadened gain profile of the QDSOAs located at $\lambda_c = $ 1.55um. It should be noted that, λ_c is used as a dummy wavelength for intermediate conversion between the two devices. The available wavelengths are related to λ_c as follows: $\{\lambda_1 = \lambda_c - 2\Delta\lambda, \lambda_2 = \lambda_c - \Delta\lambda, \lambda_3 = \lambda_c + \Delta\lambda, \lambda_4 = \lambda_c + 2\Delta\lambda\}$, where $\Delta\lambda = 5.1$nm corresponds to the detuning between adjacent wavelengths including λ_c.

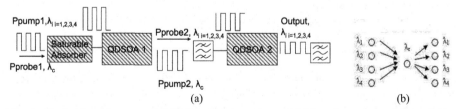

Fig. 4. (a) Configuration setup of TWC consisting of two QDSOAs in cascade. (b) Available wavelengths within homogeneous broadening of single-dot gain.

The input signal represented by P_{pump1} carries the binary data at one of the available wavelengths and is used as a pump signal to modulate the carrier density and subsequently the gain of the first QDSOA (QDSOA1). A (cw) probe signal represented by P_{probe1} and located at λ_c, enters the same device experiencing the gain modulation. Effectively, the binary information carried by P_{pump1} is copied to the (cw) signal with inversed polarity. The output of QDSOA1 constitutes the pump signal that modulates the gain of QDSOA2 and it represented by P_{pump2}. A second (cw) signal enters the second QDSOA to achieve wavelength conversion from λ_c to one of the available wavelengths. The output of the second QDSOA constitutes the output of the TWC. It is noteworthy that, tunable filters are placed at the output of each QDSOA in order to cut off the unnecessary input pump signals. For the specific simulations the device has been assumed as a simple passive element of 2dB loss. Additional losses of 6 dB have been included for the MUX/DEMUX elements giving rise to 8 dB losses in total.

A saturable absorber is used at the input of each TWC to scale down the extinction ratio degradation along the cascaded subsystems in the buffer architecture. The specific element is also based on QD technology and therefore exhibits very fast response time (~1.5psec) [6] which enables support of signal processing applications at line-rates of 160Gb/s. A simple static transfer function is needed to represent the reshaping characteristic of the device [7]. The latter is introduced through the loss parameter $\alpha(t,P)$ that tracks the signal envelope according to $\alpha(t,P)=\alpha_0/(1+P(t)/P_{sat})$, where P_{sat} is the saturation power α_0 and is the steady state loss being -0.1dB in this case. Extensive simulations have indicated an optimum value of +20dBm for the P_{sat} parameter.

The rate equation model presented in [8] has been employed, to evaluate the performance of the QDSOA devices in this configuration setup. This is a generalization of the approach used to model bulk/QW devices based on four main assumptions. Firstly, the quantum dots are spatially isolated exchanging carriers only through the wetting layer. Secondly, the dots were grouped together by their resonant frequency in order to treat the spectral hole in the inhomogeneously broadened gain spectrum. Thirdly, the wavelength separation between the pump and probe signals depends on the width of homogeneous broadening of the single-dot gain. Finally, carrier dynamics within the dots is described by four rate equations corresponding to the energy levels of the wetting layer, the continuum state, the excited state and the ground state.

3.2 Results and discussion

This section presents the results of the physical layer dimensioning of the proposed buffer architecture. In depth optimization analysis has been carried out to determine the maximum number of stages that can be cascaded. To evaluate the signal quality along the cascade two different metrics as figure of merit functions have been used, corresponding to the Q-factor and the extinction ratio. The Q-factor ratio is directly related to the actual Bit-Error-Rate of the system, only when the signal degradation follows Gaussian statistics. Although this does not happen in the present case, this metric can still be used to reflect the efficiency of that process, due to the regenerative capabilities of the TWCs.

The operating conditions for the TWC subsystem have been decided after extensive simulations where various parameters have been optimized such as the active length and the current density of each QDSOA [9]. The main objective is to achieve ultrafast XGM operation and to prevent bit patterning phenomena. For the purpose of the present work, the simulations have been performed assuming input current density 36kA/cm^2 to ensure high carrier population at the upper energy states of the quantum-dots, which act as a reservoir of carriers for the lower energy states. Also, the QDSOAs have been assumed 10mm long and the average power of the pump channel has been +27dBm. The corresponding pulse stream consists of 2-psec 1st order Gaussian pulses, modulated by a 2^7-1 PRBS bit pattern at 160Gb/s. Also, a limited amount of amplitude distortion corresponding to Q_{in} =10, has been assumed at the input, whilst the extinction ratio is 13dB. A detailed set of the QDSOA simulation parameters used in this study, can be found in [8].

Important design issue for the subsystem of the TWC is the identification of the optimum power levels for the probe wavelengths at the input of the first and the second QDSOA represented by P_{probe1} and P_{probe2}, respectively. The goal of such an optimization process has three aspects. The first is to maintain the extinction ratio at the output at the same level as that of the input. The use of the saturable absorber has a significant impact on this. Secondly, the operation of the TWC should be regenerative, which practically indicates improvement of the Q-factor ratio. Finally, the output signal should maintain a high power level to ensure an efficient XGM performance at the following TWC.

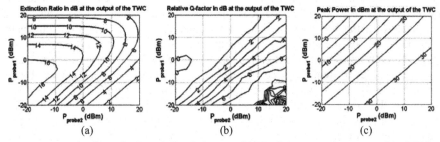

Fig. 5. (a) Extinction ratio (b) Relative Q-factor and (c) Peak power, at the output of the TWC

Fig. 5 (a-c) illustrates the extinction ratio, the relative Q-factor improvement and the peak power of the output pulse-stream as a function of P_{probe1} and P_{probe2}, respectively. It is clear that, extinction ratio is severely degraded as P_{probe2} increases, which has also been demonstrated for SOAs in XGM operation [10]. On the other hand, high power level of P_{probe1} is required to achieve sufficient pump power at the input of the second QDSOA and satisfactory extinction ratio at the output of the TWC. The regenerative efficiency of the subsystem is related to the Q-factor which shows opposite behaviour to the extinction ratio. In particular, increased level of P_{probe2} will accelerate the recovery of the saturated gain of QDSOA2, which in turn results in reduction of patterning effects of the output pulse stream. It has already been mentioned that the output power level of the TWC is very important because it feeds the next TWC. Fig. 5c illustrates that, the peak power of +30dBm is maintained around a region where the extinction ratio has decreased to around 8 dB whereas, the Q-factor has increased about 6 dB with respect to the input value ($Q_{in} \approx 10$).

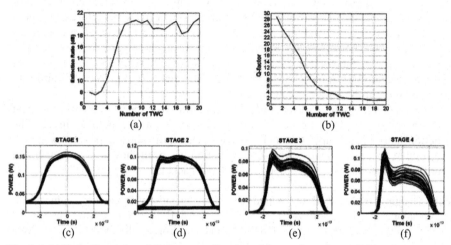

Fig. 6. (a-b) Extinction ratio and Q-factor as a function of the number of cascaded TWCs (c-f) eye diagrams at the output of the 1st, 2nd, 3rd and 4^{rth} stage.

It is noteworthy that, the operating conditions change when the TWCs operate in cascade. The worst case scenario has been considered here, in which conversion from λ_1 to λ_4 and λ_4 to λ_1 occurs successively at each stage. Fig. 6 illustrates the extinction ratio and the Q-factor as a function of the number of cascaded TWCs as well as the eye diagrams after the first four stages of the buffer architecture. The operating conditions have been identified {-15dBm, 0dBm} for P_{probe1} and P_{probe2}, respectively. The performance of the subsystem in terms of extinction ratio has improved along the cascade owing to the use of the saturable absorber which suppresses the amplitude variation at the spaces. Furthermore, the extinction ratio increases towards a steady level of 20dB, 7dB above the extinction ratio at the input of the first converter (13dB).

However, the performance of the cascade is limited by the degradation of the Q-factor at the output of each converter. The eye diagrams of the output pulse streams illustrate that the pulse width increases along the cascade introducing a duty cycle distortion. Moreover, overshooting is experienced by the leading edge of the pulse which is related to the fact that the leading edge saturates the amplifier so that the trailing edge experiences less gain than the leading edge. This effect becomes more severe along the cascade and limits the performance of the buffer to 3 stages. In addition, the output peak power along the cascade is reduced and the Q-factor decreases. In particular, at the output of the third stage, that is, at the output of the sixth TWC, the Q-factor is equal to its input value indicating error-free operation. It is clear that, the information at the output of the fourth stage is distorted. The number of packets that the proposed buffer architecture can support by use of four available wavelengths and s = 3 stages is 9, as derived from equation (1).

4   Conclusion

In this communication we have demonstrated the applicability of tunable wavelength converters (TWC) based on QDSOA technology to the implementation of an ultra-high speed optical packet switching buffer. The buffer architecture consists of cascaded stages of two TWC in series and makes use of all available wavelengths located within the homogeneous broadening of the single-dot gain of the QDSOA device. Effectively, the number of cascaded TWC needed to serve a certain number of input packets is significantly decreased, reducing the cost in terms of hardware implementation of the buffer architecture.

The cascadability results of the TWC configuration have been studied in terms of extinction ratio and Q-factor to ensure good quality of the output signal. The extinction ratio at the output of the TWC has been significantly improved by use of the saturable absorber. Finally, error-free operation has been shown up to 3 stages of the buffer architecture and thus, support of 9 input packets.

Acknowledgments. The work was supported by the Operational Program for Educational and Vocational Training (EPEAEK), PYTHAGORAS II Program and by EU via the IST/NoE e-Photon/ONe+ project, COST 291 and the project TRIUMPH (IST-027638 STP).

References

1. Borri, P., Langbein, W., Hvam, J.M., Heinrichdorff, F., Mao, M.-H., Bimberg, D.: Ultrafast gain dynamics in InAs-InGaAs quantum-dot amplifiers. IEEE Photon. Technol. Lett. 12 (2000) 594-596
2. Hunter, D.K., Smith, D.G.: New architectures for optical TDM switching. IEEE/OSA J. Lightwave Technol. 11 (1993) 495-511
3. Lee, T.T., Liew, S.Y.: Parallel Routing Algorithms in Benes-Clos Networks. IEEE Trans. Commun. 50 (2002) 1841–47
4. Yiannopoulos, K., Varvarigos, E., Vlachos, K.: Multiple-Input Buffer and Shared Buffer Architectures for Optical Packet and Burst Switching Networks. IEEE/OSA J. Lightwave Technol. submitted
5. Sugawara, M., et al.: Effect of homogeneous broadening of optical gain on lasing spectra in self-assembled InxGa1-xAs/GaAs quantum dor lasers. Physical Review B 61 (2000) 7595-7603
6. Massoubre, D.: High Speed Switching Contrast Quantum Well Saturable Absorber for 160Gb/s Operation. Conference on Lasers & Electro-Optics (CLEO), 2005, CThD3
7. Audouin, O., Pallise, E., Desurvire, E., Maunand E.: Use of Fast In-Line Saturable Absorbers in Wavelength-Division-Multiplexed Solitons Systems. IEEE Photon. Technol. Lett. 10 (1998) 828-830
8. Sugawara, M. Akiyama, T., et al.: Quantum-dot semiconductor optical amplifiers for high-bit-rate signal processing up to 160 Gb/s and a new scheme of 3R regenerators. Meas. Sci. Technol. 13 (2002) 1683-1691
9. Spyropoulou, M., et al.: Study of Multi-wavelength Regeneration based on Quantum Dot Semiconductor Optical Amplifiers. IEEE Photon. Technol. Lett. submitted
10. Lee, H., et al.: Theoretical Study of Frequency Chirping and Extinction Ratio of Wavelength-Converted Optical Signals by XGM and XPM using SOA's. IEEE J. of Quantum Electronics 35 (1999) 1213-1219

SIP Based OBS networks for Grid Computing[*]

A. Campi, W. Cerroni, F. Callegati

Department of Electronics, Informatics and
Systems, University of Bologna
Via Venezia 52 – 47023 Cesena ITALY
{fcallegati,acampi,wcerroni}@deis.unibo.it

G. Zervas, R. Nejabati, D. Simeonidou

Department of Electronic Systems
Engineering, University of Essex
Colchester - CO4 3SQ, UK
{rnejab, dsimeo, gzerva }@essex.ac.uk

Abstract. In this paper we discuss the use of the Session Initiation Protocol (SIP) as part of the control plane of an application aware Optical Burst Switching (OBS) network, able to support Grid computing applications. The paper presents the possible alternatives for the architecture of such control plane and reports of an experiment in an existing OBS test-bed where this approach was successfully tested in practice.

1   Introduction

In a Grid computing environment the network becomes part of the computation resources, therefore it plays a key role in determining the performance. Optical networks promise very high bandwidth interconnection and are a good candidate to support this sort of services [1].

Mostly optical circuit switching based on wavelength routing has been considered to this end but, if we imagine a future scenario with a broad range of user communities with diverse traffic profiles and connectivity requirements, providing network services with wavelength granularity is a solution that has drawbacks in terms of efficiency and scalability. In this case Optical Burst Switching (OBS) [2] is a candidate to implement a more scalable optical network infrastructure to address the needs of emerging Grid networking services and distributed applications [3]. The feasibility of a programmable or application aware OBS network, to support very high speed and "short lived" grid sessions, has recently been proposed and investigated [4].

In this paper we address the problem of implementing the control plane for an application aware optical network. In spite of its importance the research on this topic is rather limited and again mainly related to the wavelength routing scenario [5]. In this paper we propose an architecture that is based on three layers of standard protocols: the Job Subscription Description (JSDL) [6] language used by the application to communicate their service needs, the Session Initiation Protocol (SIP) [7] used to map the jobs requests into communication sessions and the Generalized Multiprotocol Label Switching (GMPLS) [8] used to control the networking

[*] This work is partially supported by the Commission of the European Union through the IST e-Photon+ project.

I. Tomkos et al. (Eds.): ONDM 2007, LNCS 4534, pp 117-126, 2007.
© IFIP International Federation for Information Processing 2007

functions. We will show that, by properly interfacing these layers this architecture proves to be effective, flexible and scalable.

The paper is organized as follows. In section 2 we discuss the architecture of the control plane. In section 3 we describe the experimental set-up used to prove the feasibility of this concepts. In section 4 we describe the experiment results and in section 5 we draw some conclusions.

2   Application Aware Control Plane based on SIP

The Job Submission Description Language (JDSL) specifies a general syntax for documents used by the applications to notify to the infrastructure the requirements of the jobs they intend to submit. The result is the definition of an XML schema for documents to be exchanged among grid entities. In broad term JDSL can be considered the communication language used by the control plane components of the Grid at the highest application level.

At the same time network elements have their own control plane to guarantee proper forwarding. IP based routing is not easy to implement at very high speed. Forwarding based on a simpler function than full IP header processing is therefore desirable. Moreover traffic engineering and dedicated protection/restoration are not easy to implement with a pure connectionless best-effort network protocol like IP. This is what motivated the introduction of MPLS [8] and GMPLS [9] that are efficient telecom-oriented solution for the fast and automated provisioning of connections across multi-technology (IP/MPLS, Ethernet, SDH/SONET, DWDM, OBS etc.) and multi-domain networks, enabling advanced network functionalities for traffic engineering, traffic resilience, automatic resource discovery and management.

Indeed JDSL is a pure application protocol that is not meant to deal with networking issues while GMPLS is not natively designed to support service oriented functionalities. *As a consequence the technical challenge rises from the inter-working of the application platforms and of their resource management systems with the underlying next generation optical networks powered with GMPLS network control plane.*

The abovementioned problem is not peculiar to grid computing and is rather typical of emerging networking scenarios. Next generation networks will very likely deal with applications that require "communication services" within "communication sessions" rather independently with respect to the networking environment they are using. Such a problem is addressed by the IMS specifications [10]. In broad terms the goal of the IMS is to define a service control architecture that enables all sorts of multimedia services to be carried over an IP based networking infrastructure.

The IMS has the SIP protocol as the basic language to establish and control communication sessions. SIP is a session control protocol well known but indeed its adoption as a cornerstone of the IMS architecture has indeed largely increased the interest into it. SIP is a rather general protocol that can support almost any "user defined" functionality by means of suitable payloads. It provides all the messages that are necessary to control a long lasting communication session, including authentication, presence discovery, re-negotiation and re-routing etc.

The proposal presented in this paper conceptually follows the IMS approach applied to the Grid over optics scenario. A service boundary is kept between the user and the network and all the exchanges of information related to the management of the specific application service needs are done within SIP communication sessions. In such a view the SIP sessions become the key element for the application aware optical network. Sessions can be established after proper user authentication, suspended, continued, re-routed and their service profile can be modified adding or taking away resources or communication facilities according to the needs. In broad terms the introduction of the session provides the network with the status information that can be used to manage the data flows (retrieve, modify, suspend, etc.) both according to the network need and with reference to the application requirements.

Basically SIP will enable end-to-end dynamic service provisioning across a global heterogeneous optical network infrastructure. SIP will give to the middleware the capacity to exploit the network-oriented features of GMPLS on one side and the rich semantic of application oriented languages such as JDSL on the other.

SIP interworking with OBS

The OBS network is a transport cloud with edge routers at the boundaries. Edge routers provide access from legacy domains into the OBS network by performing burst assembly and by creating the burst control packets, while the OBS core routers are devoted to pure optical switching of the data bursts.

SIP is used to challenge, negotiate and maintain application sessions and brings the notion of session into the application aware OBS network. This notion provides a significant new spectrum of opportunities in terms of quality of the communication. Generally speaking the network layer deals with data flows. Since a communication session at the application layer may involve several data flows (either parallel or sequential or both) the introduction of SIP on top of the OBS transport layer enables the possibility to manage the communication requirements for the entire application session and not only for the individual burst/packet or stream.

In this paper we propose and demonstrate the use of the SIP protocol to support Grid networking over an OBS network. We will refer in the following to a Grid-aware SIP Proxy (GSP) as a network component that is able to satisfy the communication requests of Grid applications by exploiting the SIP protocol over the OBS network.

The GSPs are an integral part of the control plane of the Grid-aware OBS network. They works coupled with OBS edge and/or core routers by means of a middleware that process the SIP message payloads and talks with the OBS layers by means of suitable APIs. GSPs have SIP signalling functionalities to manage the application sessions and exploit the OBS network to provide the required communication facilities. This means that GSPs send the requests to open communication paths (for instance GMPLS Label Switched Paths or LSPs) to the optical nodes according to the applications requirements and leave to the network layer all the network related problems such as flow control, QoS guarantee etc.

The middleware between the GSPs and the OBS control plane is made of 3 main parts:

1. Interface with SIP servers to interpret the applications requests related to the sessions supporting Grid computing requests (for instance parsing and processing JSDL documents);

2. interface to the control plane (for instance GMPLS) of the OBS network;

3. internal engine which translate the application requests into network related communication instances (for instance JSDL request into GMPLS LSPs requests).

Sessions established by means of SIP signaling can be mapped on the network layer, by means of the middleware, according to two well known approaches, outlined in the following [5].

Overlay approach

The overlay approach keeps both physical and logical separation between the SIP (session) layer and the optical network. The non-network resource (i.e. Grid resource: computing and storage) and the optical layer resources are managed separately in an overlay manner. An IP legacy network carries the SIP signals and the OBS network carries the Grid data: the slow network (based on legacy technology) is used for the signalling while the fast network (optical OBS) is used only for data transmission.

The GSPs are placed into the edge routers only and use a legacy electronic connection to forward the SIP signals to the other GSPs (at other edges of the OBS cloud). GSPs negotiate the session and are composed by registrar, location and proxy servers in order to provide all the functions of a SIP network. The users (i.e. the application) use the SIP protocol to negotiate the Grid communication session. When the session is set the middleware is responsible to request a data path between the edge routers involved in the session to the optical network control plane. Then the session data cut through the OBS network and the SIP layer is not involved any more.

The main advantage of this approach is to use the various technologies for what they do best. The well-established legacy technology based on the current Internet is used to carry the signalling, i.e. low speed and rather low bandwidth data transmissions while the high speed OBS network is used to carry bulk data.

Integrated approach

The integrated approach enriches the optical control plane with SIP functionalities, to realize a pure OBS network that works controlled by the SIP protocol to negotiate the application (grid) sessions and by GMPLS to control the connection management functions. No legacy networks are into play any more and signalling and data share the same networking infrastructure. Since the signalling and the data plane share the same network infrastructure, all OBS node must at least have the capacity to read and forward a SIP message.

The GSPs in the edge nodes are full functional and logically identical to those mentioned before. On the other hand the GSPs in the core nodes can be equipped with a subset of functionalities, to satisfy the best performance/complexity trade-off. On one side the SIP functionalities could be limited to a light proxy with forwarding functions only. In such a way most of the intelligence of the SIP layer can still be segregated at the boundaries of the OBS network. On the other side the OBS core nodes could be equipped with full functional GRID proxies and therefore could take part in the operations of managing the application requests, for instance actively participating in the resource discovery process.

Resource discovery and reservation

As an example of application of the concepts presented in the previous sections, here we discuss how SIP can be used to implement resource discovery and reservation, a meaningful case to discuss.

We assume that Grid resources and users are divided into domains, an organization that is particularly indicated in a Grid network oriented to consumer applications where a large number of resources and users can be present. A domain can be the set of users connected to a single GSP as well as a subset or superset. SIP can be used to implement resource discovery and reservation in two different ways.

The former follows a single phase approach where both resource discovery and reservation are performed at the same time. The user with data to be processed remotely (e.g. Grid user, e-Science) sends a request to the GSP at its ingress edge router to the OBS cloud in the form of an INVITE SIP message. Encapsulated into the INVITE there is the JSDL document which describes the job requirement. The SIP server passes the job requests (JR) to the its underlying middleware that performs a resource discovery algorithm to find out whether there are enough computing resources available within its known resource. If the answer was "yes" the proxy would start establishing the session, if it was "no" it would forward the message to the other proxies in the network to look for the requested resources until the message arrive to an available resource or it is dropped. If a resource is found the INVITE is acknowledged and a session between user and resource is created, thus reserving the resource usage. At the same time a data path is created in the OBS network to carry the job related data.

In the latter approach two phases are used, resource discovery at first with a notifications mechanism and then direct reservation by the client. The basic concept behind this idea is the understanding that the resource management in a Grid network is similar to the presence notifications of a SIP network. A resource is associated with SIP address (i.e. a user of the SIP network) and has a set of proprieties with a state. The presence notifications messages of the SIP network can then be used to update and notify, support presence, messaging, state change detection, etc. Using messages like PUBLISH, SUBSCRIBE and NOTIFY the various management functions of a resource can be provided. Fig. 1. shows two simple call flows for resource discovery management.

- The scenario represented in Fig. 1.a describes a localised publishing approach. In this approach the Grid resources are published only to the attached SIP proxy with a PUBLISH message. A user requesting resources sends a SUBSCRIBE message to the nearest GSP. The GSP 1 checks the status of its own resources and if they can satisfy the request then notifies the client. Otherwise, the

SUBSCRIBE message is propagated to the other known GSPs either by utilizing sequential or parallel forking in order to discover the requested resources. The GSP with available resources sends a NOTIFY message back to user. The NOTIFY is used to communicate to the user the availability and location of the resources (i.e. address of the end point or GSP or domain).

- Fig. 1.b portrays the scenario in which the availability of Grid resources is distributed to all GSPs on the same domain by utilising PUBLISH message. Then the SUBSCRIBE message sent by the user to describe its request is sent to the nearest GSP. This GSP is aware of all resources of the domain and discovers the requested ones can send a NOTIFY message back to the user. Otherwise propagates the SUBSCRIBE message to other domain(s).

After the resource discovery the user knows the location (SIP name or network address) of the resource and can attempt a direct reservation by an INVITE message. The detailed description of the reservation process is presented in experimental description section.

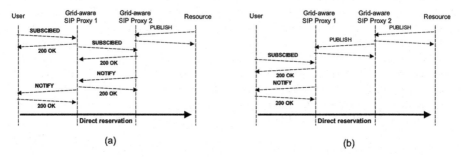

(a)   (b)

Fig. 1. Example of resource discovery and state notification. (a) Localized approach. (b) Distributed approach.

3   Experimental set-up

In this section we describe an experimental set-up that demonstrates the feasibility of using SIP as the application signaling component in a GRID over OBS network. It refers to the overlay approach with one way reservation that has been tested at first, since it does not require to embed the SIP functionalities into the OBS nodes and to encapsulate SIP messages into optical bursts. Nonetheless the experiment here reported is, in our view, significant to prove the feasibility and effectiveness of the proposed concept. Moreover this choice is also motivated by the fact that the overlay approach will likely be the first choice in the actual deployment of such solution. The experiment exploits an existing OBS test-bed whose main characteristics have been already reported in the literature [4].

In the experiment the end user intending to submit a job request to the Grid environment prepares a JDSL document. It registers to its service access point, i.e. a

GSP, and sends its request by encapsulating the JSDL document into a SIP message. The message is processed at the GSP in the ingress edge node. By analyzing the JDSL the GSP execute the resource discovery and then forwards the SIP request to the SIP proxy connected to the egress edge node leading to the computational resources. Upon acceptance of the job request, the OBS network is used to carry the job submission (application data) over optical bursts.

The overlay network testbed architecture utilizes SIP protocol on a higher physical and logical layer to the OBS network testbed and it is illustrated in Fig. 1. The testbed operates at 2.5 Gbps both for the Optical Burst Ethernet Switched (OBES) control channels [11] as well as the out-of-band data channels which both operate in asynchronous mode. It also provides full-duplex communication between two Edge Routers and one Core Router. Both edge and core OBS routers are equipped with an embedded network processor, implemented with a high-speed Field Programmable Gate Arrays (FPGA). The data plane generates variable sized bursts with variable time intervals and operates in bursty mode.

The GSP are based on a SIP stack called pjsip [14] that runs on a PC. In the experiment we implemented a GSP for any edge router. In the specific case of the testbed this means two GSP are coupled with the two edge routers.

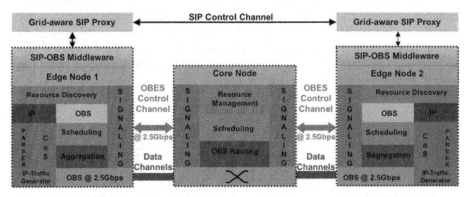

Fig. 2. Overlay network architecture utilizing SIP over OBS. It incorporates two OBS Edge Routers equipped with Grid-aware SIP Proxies on top and one Core Router. The Testbed operates in full-duplex mode.

Both edge routers operate as ingress and egress nodes by utilizing Xilinx high-speed and high-density VirtexII-Pro FPGA prototype boards. At the ingress side, the edge routers utilize one fast and widely tunable SG-DBR laser each for wavelength allocation agility. Both OBS edge nodes are connected to the GSP running in the two PCs. The PCs are connected by a normal Ethernet network, thus realizing the separate signaling interconnection for SIP messages between remote proxies.

The core router integrates an FPGA and a 4×4 all optical switching matrix (OXS). The FPGA controls the OXS. It extracts the incoming BCHs, processes it and in turn drives the OXS with appropriate control signals to realize the switching path required by the BCH. Finally it reinserts a new BCH towards the egress edge router. The 4×4

OXS matrix has extremely low crosstalk levels of < -60 dB and fast chip switching time of ~1.5 ns have been achieved [11].

4   Experiment Description

The network concept demonstrated in the test-bed provides application layer resource discovery and routing of application data or user's jobs to the appropriate resources across the optical network. The OBS control plane architecture comprises a resource discovery stage and a traditional OBS signalling stage using Just-In-Time (JIT) bandwidth reservation scheme [2].

The user with data to be processed remotely (e.g. Grid user, e-Science) sends a request to the GSP of the ingress edge router to the OBS cloud. The request carries the job specification and resource requirements (i.e. computational and network) in the payload, in the form of a JSDL document. The SIP server classifies the incoming job requests (JR) and then either process it or forward it to the next proxy which has knowledge of the resources available.

In this experiment, due to the very simple network topology, the distribution of resources is static. Client applications are connected to the edge router 1 while resources are connected to the edge router 2. Therefore all messages with requests are sent by the clients to GSP 1 coupled with edge router 1. Since GSP 1 does not have any available resource, the requests are forwarded to the GSP 2, coupled with edge router 2, from where the resources are reachable.

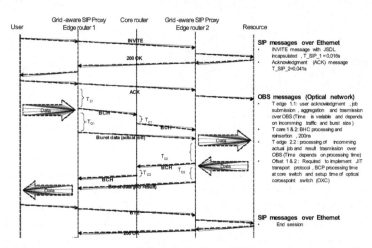

Fig. 3. Session control messages and data flows in a grid job transfer over the OBS network. The figure shows the messages exchanged between edge router 1 and edge router 2 via the OBS core router, by means of either Ethernet (SIP) or the OBS testbed (grid job data).

The message flow is presented in Fig. 3. The user sends the job specification with an INVITE message to the proxy of the edge router. Encapsulated into the INVITE there is the JSDL document which describes the job requirement. The proxy reads the INVITE and performs a resource discovery algorithm and forward the message to the GSP2. After the INVITE is processed at GSP 2 the user is informed about the result of the resource discovery, with a positive reply message (200 OK), that is received by GSP 1 T_{S1} = 0.016 seconds (experimentally measured) after sending the INVITE message. In this case GSP 1 forwards the OK message to the client that will then set up the application to sent the job before acknowledging the session establishment. In this case a computational resource reservation signalling (ACK) message is sent to GSP 2 after T_{S2} = 0.041 s. The time elapsed between the arrival of the OK message and the departure of the ACK is due to the signalling between GSP 1 and the client (not reported in the figure) and of the set up of the application to transfer the data referring to the communication session under negotiation. Then the user sends the actual job to edge router 1 for transmission to the reserved resources attached to edge router 2.

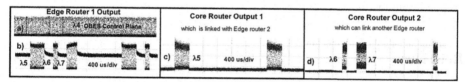

Fig. 4. Experimental results, a) OBES Control Plane, b) Asynchronously transmitted variable sized bursts, c) burst routed through core router towards edge router 2 and d) bursts routed towards emulated router.

Edge router 1 aggregates the job into optical bursts that are sent to the reserved resources by utilizing JIT bandwidth reservation scheme. The edge router is able to send data from different users to different reserved resources across the network as shown in Figure 4.b. Before that, a BCH has to be transmitted in order to setup the path (Figure 4.a) The first burst is routed through core router output port 1 (Figure 4.c) to be sent to edge router 2. The following two bursts are routed through core router output port 2 (and Figure 4.d) which is used as an additional link to emulate an additional edge router (e.g. edge router 3). The offset time required between BCH and Burst Data is 10.2 μs in order to incorporate core router processing time plus switch setup time (T_{C1} = T_{C1} = 200 ns) and optical crosspoint switch switching time (10 μs, Figure 4.d). Upon reception of the job, edge router 2 requires T_{E2} to read the incoming burst and emulate the forwarding to the resource and the processing time. Then it sends an emulated job result back to the user through the core router and edge router 1 (Figure 4.e).In this experiment, the data plane transmitted variable length optical bursts (from 60 μs up to 400 μs) with their associated BCHs over OBES control plane (Figure 4.a), on three different wavelengths (5=1538.94 - 6=1542.17nm- 7=1552.54 nm).

5   Conclusions

This paper proposes to introduce the concept of session in the control plane of a Grid enabled OBS network. To this end the Session Initiation Protocol (SIP) is used on top of the OBS network control layer (for instance based on GMPLS). SIP is used to establish application sessions according to the application requirements and then the network control plane is triggered to create the high bandwidth data paths, by properly controlling the OBS nodes. This approach in principle enriches the network with a number of application oriented features thanks to the rich session oriented semantic of SIP.

The paper analyzed the various architectural approaches of such solution and its possible development. Then an experimental set-up of an overlay architecture was implemented on an OBS test-bed to validate this approach. The results of the experiment reported in this paper prove the feasibility and effectiveness of the proposed solution for the control plane.

References

1. D. Simeonidou, et al., "Optical Network Infrastructure for Grid" Open Grig Forum document GFD.36, August 2004.
2. C. Qiao, M. Yoo, "Optical Burst Switching - A new Paradigm for an Optical Internet", Journal of High Speed Networks, vol. 8, no. 1, pp. 36-44, Jan. 2000.
3. D. Simeonidou, et al., "Grid Optical Burst Switched Networks (GOBS)," Global Grid Forum Draft, May 2005.
4. D. Simeonidou et al. "Dynamic Optical-Network Architectures and Technologies for Existing and Emerging Grid Services", IEEE Journal on Lightwave Technology, Vol. 23, No. 10, pp. 3347-3357, 2005.
5. R. Nejabati, D. Simeonidou, "Control and management plane considerations for service oriented optical research networks", IEEE 8th International Conference on Transparent Optical Networks, June 18-22, 2006 - Nottingham, UK.
6. A. Anjomshoaa et al., "Job submission description language (JSDL) specification v. 1.0", Open Grig Forum document GFD.56, November 2005.
7. J. Rosenberg et al., "SIP: Session Initiation Protocol", IETF RFC 3261, June 2002.
8. E. Rosen, A. Viswanathan, R. Callon, "Multiprotocol Label Switching Architecture", IETF RFC 3031, January 2001.
9. E. Mannie (Ed.), "Generalized Multi-Protocol Label Switching (GMPLS) Architecture", IETF RFC 3945, October 2004.
10. M. Poikselka, et al., "The IMS: IP Multimedia Concepts and Services", 2nd ed., Wiley, 2006.
11. Georgios Zervas, et.al., "QoS-aware Ingress Optical Grid User Network Interface: High-Speed Ingress OBS Node Design and Implementation", OFC 2006, paper OWQ4, Anaheim, California, USA.
12. G. Zervas, et. al., "A Fully Functional Application-Aware OpticalBurst Switched Network Test-Bed", to be published in OFC 2007, paper OWC2, Anaheim, California, USA.
13. Zhuoran Wang; Nan Chi; Siyuan Yu, Lightwave Technology, Journal of,Volume 24, Issue 8, Aug. 2006 Page(s):2978 – 2985.
14. http://www.pjsip.org

Job Demand Models for Optical Grid Research

Konstantinos Christodoulopoulos[1], Emmanouel Varvarigos[1],
Chris Develder[2], Marc De Leenheer[2], Bart Dhoedt[2]

1: Dept. of Computer Engineering and Informatics, and
Research Academic Computer Technology Institute
University of Patras, 26500, Patras, Greece
2: Dept. of Information Technology (INTEC), Ghent University – IBBT,
G. Crommenlaan 8 bus 201, 9050 Gent, Belgium
{kchristodou, manos}@ceid.upatras.gr
{chris.develder, marc.deleenheer, bart.dhoedt}@intec.ugent.be

This paper presents results from the IST Phosphorus project that studies and implements an optical Grid test-bed. A significant part of this project addresses scheduling and routing algorithms and dimensioning problems of optical grids. Given the high costs involved in setting up actual hardware implementations, simulations are a viable alternative. In this paper we present an initial study which proposes models that reflect real-world grid application traffic characteristics, appropriate for simulation purposes. We detail several such models and the corresponding process to extract the model parameters from real grid log traces, and verify that synthetically generated jobs provide a realistic approximation of the real-world grid job submission process.

Keywords: Optical Grids, Job demand models, Profiling, Expextation Maximization algorithm, Probabilistic modeling

1. Introduction

Today, the need of network systems for storage and computing services for scientific and business communities are often answered by relatively isolated islands, known as clusters. Migration to truly distributed and integrated applications requires optimization and (re)design of the underlying network technology. This is exactly what Grid networks promise to offer: a platform for cost and resource efficient delivery of network services to execute tasks with high data rates, processing power and data storage requirements, between geographically distributed users. Realization of that promise requires integration of Grid logic into the network layers. Given the high data rates involved, optical networks offer an undeniable potential for the Grid. An answer to the demand for fast and dynamic network connections could lie in the (relatively) new switching concepts such as Optical Packet Switching (OPS) and Optical Burst Switching (OBS) [1].

Delivering the Grid promise implies answering a series of fundamental questions [2]: (re)design the architecture of a flexible optical layer, development of the

I. Tomkos et al. (Eds.): ONDM 2007, LNCS 4534, pp 127-136, 2007.
© IFIP International Federation for Information Processing 2007

necessary design techniques for e.g. dimensioning, algorithms for routing and control offering both QoS and resilience [3] guarantees. It is this—to a large extent unexplored [4]—area of fundamental research that is the subject of the supporting studies within the Phosphorus project.

Many of the answers to these research questions are addressed through simulations. A necessary prerequisite to obtain useful results is an adequate model of the traffic (i.e. jobs) that will be submitted to the Grid. Although, a great deal of work has appeared in literature on job characterization and modeling for single parallel supercomputers [5], similar work in the area of (optical) Grids is quite limited. Medernach [6] analyzed the workload of an LCG/EGEE cluster, proposing a 2-dimensional Markov chain for modeling single user behavior in a Grid. Li et al. [7] used the LCG Real Time Monitor to collect data from the global EGEE Grid, and proposed models at three different levels: Grids, Virtual Organizations and regions. They conclude that Markov Modulated Poisson Processes (MMPP) with sufficient number of states can reflect the real world job arrival processes. In the work presented here, we introduce a new model, referred to as the Pareto-Exponential model and compare it with the previously proposed model. We show that, despite its more compact parameter set, the Pareto-Exponential model is a valid alternative. Finally, we also propose a model for the job execution times that is based on the hyper-exponential distribution.

We start in Section 2 by briefly outlining the Phosphorus project and how the job demand modeling work fits in the whole concept. In Section 3 we introduce the candidate traffic models considered, and discuss how we fitted them to real world traffic traces at different aggregation levels in Section 4. Finally, Section 5 summarizes our conclusions.

2. The Phosphorus project

As indicated in the introduction, a new generation of applications is emerging, coupling data and high-end computing resources distributed on a global scale. These impose requirements such as determinism (e.g. guaranteed QoS), shared data spaces, large data transfers, that are often achievable only through dedicated optical bandwidth. High capacity optical networking can satisfy bandwidth and latency requirements, but software tools and frameworks for end-to-end, on-demand provisioning of network services need to be developed in coordination with other resources (CPU and storage) and need to span multiple administrative and network technology domains.

In response to the above requirements, the European IST project Phosphorus will address some of the key technical challenges to enable on-demand end-to-end network services across multiple domains. The Phosphorus network concept and testbed will make applications aware of the Grid environment, i.e. the state and capabilities of both computational and network resources. Based on this information, it is possible to make dynamic, adaptive and optimized use of heterogeneous network infrastructures connecting various high-end resources. The testbed will involve European NRNs and national testbeds, as well as international resources (GÉANT2,

Internet2, Canarie, Cross Border Dark Fibre infrastructures and GLIF virtual facility). A set of highly demanding applications will be adapted to prove the concept.

In the Work Package 5, "Supporting Studies", architectural and algorithmic questions will be addressed. These include research in the area of job routing and scheduling algorithms (decide where to execute a given job and how to reach that destination, referred to as the anycast routing problem [8]), examine techniques that jointly reserve computation and communication resources, and compare packet versus circuit switching technologies.

3. Job Demand Models

To objectively evaluate the performance of e.g. job scheduling and routing algorithms, it is desirable that the job submission model accurately reflects the characteristics of real world grid jobs. When simulation is used, an analytical model is preferred over actual job traces, since this approach allows the different job parameters (e.g. average load) to easily be adjusted. In this section we present such analytical models for the job arrival/processing times.

The classical Poisson process model, in which the inter-arrival times are exponentially distributed, forms the basis for most of these more advanced models.

3.1. Non-Homogeneous Poisson Process (NHPP)

As can be easily intuitively accepted, the job arrival rate can exhibit time dependent behavior, especially on e.g. national scales. On a daily scale, day/night differences can be observed, and also the difference between week days and weekends may be visible. Hence, the arrival rate of a Poisson process can be considered to be a function of time $\lambda(t)$, leading to a non-homogeneous Poisson process. More specifically, the number of arrivals $N(t)$ in the interval $[0, t]$ follows the distribution shown in (1).

$$\Pr[N(t) = n] = e^{-m(t)} \frac{(m(t))^n}{n!}, n \geq 0 \text{ and } m(t) = \int_0^t \lambda(s)ds \qquad (1)$$

In fitting this model to real life job traces (described in Section 4), we will consider the job arrival rate $\lambda(t)$ to be a stepwise function. In this case, the job generation model can be considered as a state process, where the system evolves from one state to the next while maintaining a fixed arrival rate λ_i that depends only on the state i.

3.2. Phase-type process

An m-phase type distribution represents a random variable (in our case e.g. job inter-arrival times) whose values are the transition times until absorption of a continuous-time Markov chain with m transient states and one absorbing state. In general, any inter-arrival process can be approximated by a phase-type distribution provided enough states are introduced. A special case of the general phase-type distribution is the *hyperexponential* distribution (HE), which has two or more non-identical phases that occur in parallel (i.e. each of the phases only has a non-zero probability to transit to the absorbing state). The probability density function (pdf) of a hyperexponentially distributed variable X is given in (2). This corresponds to the weighted sum of m exponentially distributed random variables Y_i (with average $1/\lambda_i$).

$$f_X(x) = \sum_{i=1}^{m} p_i f_{Y_i}(y) \quad \text{where} \quad \sum_{i=1}^{m} p_i = 1 \tag{2}$$

3.3. Markov Modulated Poisson Process (MMPP)

The Markov modulated Poisson process (MMPP) is a doubly stochastic Poisson process [9], characterized as a (finite state) continuous time Markov chain with m states. Each state i is a Poisson process (arrival rate λ_i) in itself, and state transitions are defined by a state transition matrix Q. Thus, the system is fully defined by a matrix Q, as defined in (3), and a vector $\Lambda = [\lambda_1 \lambda_2 \ldots \lambda_m]$.

$$Q = \begin{bmatrix} -\sigma_1 & -\sigma_{1,2} & \cdots & -\sigma_{1,m} \\ -\sigma_{2,1} & -\sigma_2 & \cdots & -\sigma_{2,m} \\ \vdots & \vdots & \ddots & \vdots \\ -\sigma_{m,1} & -\sigma_{m,2} & \cdots & -\sigma_m \end{bmatrix} \quad \text{where} \quad \sigma_i = \sum_{j=1, i \neq j}^{m} \sigma_{i,j} \tag{3}$$

3.4. Pareto-exponential model (PE)

In the Pareto-exponential model busy periods, in which jobs arrive, succeed each other. Each busy period has an exponentially distributed duration (with mean of $1/\mu$ seconds), and within a busy period jobs arrive according to a Poisson process (at a rate of λ jobs/second). The times between the start times of a busy period are distributed following a truncated Pareto distribution with shape parameter α, minimum value X_{min} and maximum value X_{max}. An intuitive interpretation of the busy periods can be that these correspond with job submissions from a particular virtual organization (VO) participating in the Grid.

4. Real life measurements

To validate the suitability of the various models for the job arrival process and their execution times, we have collected traces from operational Grid environments. We then fitted the aforementioned models to the traces, and used the parameter values found to drive a simulator generating the IATs according to the respective models with the parameter values found by the fitting algorithm.

4.1 Measured infrastructure

Since the Phosphorus test-bed is still under construction, we gathered traces on the LCG/EGEE infrastructure [10]. The Enabling Grids for E-sciencE (EGEE) project offers an always-on Grid computing infrastructure, geographically distributed across the globe. The worldwide LHC Computing Grid project (LCG) was created to offer computing infrastructure processing and analyzing the data of the Large Hadron Collider (LHC) experiments at CERN. The LCG and EGEE projects share a large part of their well established infrastructure; hence we refer to it as the LCG/EGEE infrastructure. Currently, it comprises 207 cluster sites from 48 countries. In the observation period, we recorded the presence of 39,697 CPUs (of which on average 31,228 were active) and 5 Petabytes of storage.

The job lifetime comprises various phases, as illustrated in Fig. 1. Users submitting jobs are part of a VO (Virtual Organisation), which is a dynamic collection of individuals and institutions sharing the same permissions etc. In order to submit jobs to the Grid, a user has to log in to a user interface (UI) and provide the job specification in a JDL (job description language) format. This job submission is then forwarded to the corresponding Resource Broker (RB), which will schedule the job for execution taking into account information provided by the JDL as well as information service (e.g., the VO, global traffic load information). The job, wrapped in an input sandbox, is eventually sent to a Computing Element (CE) at a particular site, where a local resource management system will assign it to a Worker Node (WN).

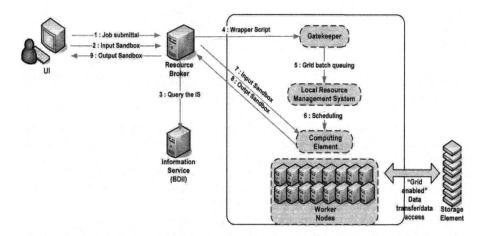

Fig. 1. Job flow in the Grid environment.

As indicated, a job evolves through various states. Of particular interest to optical grid job modelers are (i) the job inter-arrival times (IATs), and (ii) the time spent in the "Running" state, which amounts to actual execution time of a job, including the I/O time. In the following we will establish suitable models by fitting them to the measured data. The subsequent subsection details the fitting methodology used.

4.2. Trace fitting methodology

In order to fit the distribution parameters to measured data samples, we use a maximum-likelihood estimation (MLE) technique, and specifically the Expectation-Maximisation (EM) algorithm. Further details can be found in [11].

4.3. Jobs at the Grid level

Using the LCG Real Time Monitor [11], we collect data for jobs submitted to all Resource Brokers (RBs) participating in the EGEE project. The RTM records the times at which user jobs are submitted, the way they are distributed to the sites, the times at which the jobs complete the different states of their processing, and finally depending on the successful or not execution it also presents the times of delivery of the execution outcome to the corresponding user. Of main interest here were the job IATs and running time. For this, we collected job arrival data during 1-31 Oct. 2006 (totaling 2,228,838 jobs).

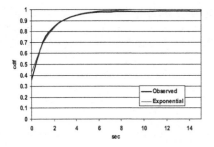

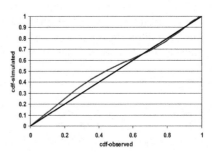

Fig. 2. Empirical cdf and Poisson process fit for IAT at the Grid level.

Fig. 3. Empirical vs Poisson P-P plot for IAT at the Grid level.

The empirical cumulative distribution function (cdf) is shown in Fig. 2. Given the observed standard deviation being close to the mean IAT, and the absence of a heavy tail, it is clear that a Poisson process can be a suitable model. Using MLE fitting, we found a Poisson process with mean IAT $(1/\lambda)$ of 1.6077. Note that since our measurement data has a resolution of 1 second, we actually converted the Poisson process IATs by rounding them to the closest integer.

In Fig. 3, the probability-probability (P-P) plot (composed by pairing percentiles corresponding to the same value) is shown for the rounded Poisson model versus the empirically observed data. This plot being close to the line between (0,0) and (1,1) we may conclude the Poisson process model is adequate.

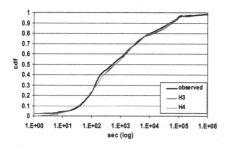
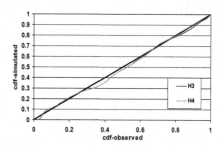

Fig. 4. Empirical cdf and HE fit for job execution time at the Grid level.

Fig. 5. Empirical vs HE P-P plot for job execution time at the Grid level.

- The WN execution times exhibit peaks at certain time values, reflected in a sharp rise in the cdf as depicted in Fig. 4. Thus, we resorted to fits with the hyper-exponential model, one with 3 phases (H3) and one with 4 phases (H4), based on the observation that the cdf curve exhibits 3-4 "steps". We used the EMpht utility [12] implementing the EM algorithm described in [14].

To assess the quality of these fits, we again generated time values by simulation and obtained the cdfs (Fig. 4) and P-P plots (Fig. 5). Since the accuracy of H3 and H4 fits is similar, we can conclude that the 3-phase exponential process is sufficient to model the WN execution times at the Grid level.

4.4. Jobs at the Grid Site level

In addition to the Grid level traces, we also collected information for jobs submitted to individual Grid sites (computing elements in the scheme of in Fig. 1).

4.4.1. Kallisto site

The Kallisto node located in Patras is part of Hellasgrid and has been a production site since 1 Feb. 2006. It comprises 64 Intel Xeon CPUs at 3.4GHz and 2Gb RAM, of which 60 are actual Worker Nodes (the 4 other service nodes comprise the EGEE core servers), using g-Lite middleware and running scientific Linux v3.

For the measured job IATs, we considered the four different models discussed in Section 3:

- *Non-Homogeneous Poisson Process (NHPP):* We defined a stepwise function for $\lambda(t)$, being constant over 1 hour intervals, with hourly values of λ obtained by averaging over all days in the observation period.
- *Hyper-Exponential Model:* We considered (i) a 2-phase (H2), and (ii) a 3-phase (3H) hyper-exponential model, based on the observation of 2-3 steps in the empirical cdf. Fitting was done using the EMpht software.
- *Markov Modulated Poisson Process (MMPP):* We considered (i) a 3-state (3MMPP), and (ii) a 4-state (4MMPP) MMPP.
- *Pareto-Exponential Model:* In this case, we chose a truncated Pareto distribution with X_{max}=10800 sec, since this was the (deterministic) periodicity of site

functional test. For the other parameters, we fitted: λ=18 arrivals/sec for busy periods, mean duration $1/\mu$=22.5 sec of the busy periods, a=0.48 and X_{min}=32 sec.

To evaluate the applicability of the models, we generated synthetic job traces using a simulator implementing the models, resulting in the cdf graphs of Fig. 6 and the corresponding P-P plots shown in Fig. 7.

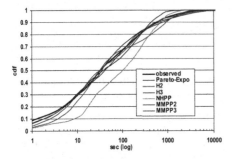 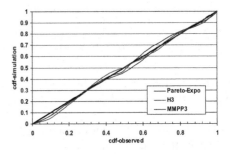

Fig. 6. Empirical cdf and various models fitting for IAT at Kallisto.

Fig. 7. Empirical vs various models P-P plot for IAT at Kallisto.

From the above graphs, we can conclude that conclude that the proposed Pareto-Exponential model generates traces that are very close according to the P-P plot to those observed in our cluster. H3 and 3MMPP models also simulate satisfactorily the job arrival process. However, the Pareto-Exponential model is simpler, more concise and more intuitive than the other proposed models, since it is based on a smaller number of parameters, and seems to correspond to actual VO behavior. With respect to the HE and MMPP models we observe, as expected, that the fit improves by increasing the number of phases (states for MMPP).

We have also computed the Hurst parameter for the four models (as a measure of long range dependency, aka self-similarity). Only the Pareto-exponential and the MMPP models experience long-range dependence (H=0.58 for PE, H=0.62 for 2MMPP and H=0.64 for 3MMPP with confidence levels higher than 99%). The value in the measured data amounted to H=0.68.

With respect to the job execution times (page limitations prevent us from showing the details), we observed similar behavior as on the Grid level, and fits using the 3H and 4H models proved to produce adequate results. On a quantitative level, we observed some differences which are due to the smaller number of VOs served by the Kallisto node compared to the complete EGEE (11 vs 75 VOs, with the most active VO in Kallisto –ATLAS VO- being the 3rd on a global scale).

4.4.2. BEGrid site

The BEGrid site located in Ghent is part of the Belnet grid initiative, originating in 2003. It comprises 41 dual Opteron (1.6 GHz) worker nodes, and another 15 dual dualcore Opteron (2 GHz) nodes, all having 4Gb RAM, and 5 service nodes, using g-Lite middleware and running scientific Linux v3.

In contrast to the Kallisto and global EGEE results, we found a far less smooth, more step-wise empirical cdf of the job IAT for the BEGrid measurements. We found a particularly steep increase, corresponding to a peak in the probability density function (pdf) around 8-15 seconds. The reason for this is that a large group of BEgrid users commits their jobs using scripting, with script submission overhead resulting in a job IAT of the order of 10s. Obviously, neither the exponential distribution nor the more complex functions succeed in reproducing this abrupt cdf. Hence the poor P-P plots.

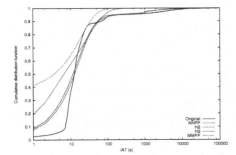
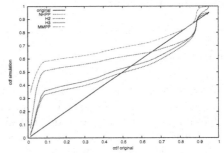

Fig. 8. Empirical cdf and various models fit for IAT at BEGrid site.

Fig. 9. Empirical vs various models P-P plot for IAT at BEGrid site.

With respect to the job execution times, the observed cdf (omitted because of space limitations) was of a similar slightly step-wise shape as the EGEE and Kallisto measurement data. Again, we found 3- and 4-phase hyper-exponential models to generate the most satisfactory results.

5. Conclusions

Due to the high equipment cost involved in the research of optical grids if actual hardware were to be used, simulation techniques are often put forward as a viable alternative. To warrant accurate and useful results, it is important that a realistic grid job load is used as input for the simulation. To this end, we presented analytical job models, and the methodology to extract model parameters from actual grid log traces. This approach guarantees a very flexible, analytical job submission model, yet providing a very realistic approximation of the real life grid job submission pattern.

Using real life measurement data, gathered at different aggregation levels in a Grid environment (local site vs global Grid), we judged the usefulness of various models fitted to that data. This was achieved by implementing the models in simulation software. From this study, we concluded that:

- Job inter-arrival times on the observed Grid level can be successfully modeled by a Poisson process, but on the Grid site level (eg. Kallisto traces) the long range dependency needs to be taken into account and HP, MMPP or Pareto-Exponential models need to be used.
- For the job execution times, we achieved the most satisfactory results with a (3 phase) hyper-exponential process.

Acknowledgements

The work presented in this paper was supported by the EU through the IST Project Phosphorus (www.ist-phosphorus.eu) and e-phton/ONe+ (www.e-photon-one.org). C. Develder thanks the Research Foundation – Flanders (FWO – Vlaanderen) for his post-doctoral fellowship. M. De Leenheer thanks the IWT for financial support through his Ph. D. grant.

References

1. C. Develder, et al., "Delivering the Grid Promise with Optical Burst Switching" (Invited), Proc. Int. Workshop on Optical Burst/Packet Switching (WOBPS East 2006), at the Joint Int. Conf. on Optical INternet and Next Generation Networks (COIN-NGN 2006), Jeju, South Korea, Jul 2006.
2. D. Simeonidou, et al., "Dynamic Optical Network Architectures and Technologies for Existing and Emerging Grid Services", IEEE Journal of Lightwave Technology, 23(10):3347-3357, Oct 2005.
3. J. Zhang, and B. Mukherjee, "A review of fault management in WDM mesh networks: basic concepts and research challenges", IEEE Network, 18(2):41-48, Feb 2004.
4. K.H. (Kane) Kim, "Wide-Area Real-Time Distributed Computing in a Tightly Managed Optical Grid - An Optiputer Vision" (Keynote), Proc. 18th IEEE Int. Conf. on Advanced Information Networking and Application (AINA'04), Vol. 1, pp. 2–11, Fukuoka, Japan, Mar 2004,.
5. D. Feitelson, "Workload modelling for computer systems performance evaluation", http://www.cs.huji.ac.il/~feit/wlmod
6. E. Medernach, "Workload analysis of a cluster in a Grid environment", Proc. 11th Int. Workshop Job Scheduling Strategies for Parallel Processing (JSSPP), Cambridge, MA, USA, Jun 2005.
7. H. Li, M. Muskulus, and L. Wolters, "Modeling Job Arrivals in a Data-Intensive Grid", Proc. 12th Workshop on Job Scheduling Strategies for Parallel Processing, Saint-Malo, France, Jun 2006.
8. T. Stevens, et al., "Anycast Routing Algorithms for Effective Job Scheduling in Optical Grids", Proc. European Conference on Optical Communication (ECOC), Cannes, France, Sep 2006.
9. W. Fisher, and K. Meier-Hellstern, "The Markov-modulated Poisson process (MMPP) cookbook", Performance Evaluation, 18(2):149-171, Sep 1993.
10. http://public.eu-egee.org/
11. G. McLachlan, and T. Krishnan, "The EM Algorithm and Extensions", Wiley Series in Probability and Statistics, 1997.
12. Real Time Monitor, http://gridportal.hep.ph.ic.ac.uk/rtm
13. S. Asmussen, O. Nerman, and M. Olsson, "Fitting phase-type distributions via the EM algorithm", Scandinavian Journal of Statistics, 23(4):419-441, 1996.
14. W. Roberts, et al., "On Ryden's EM Algorithm for Estimating MMPPs", IEEE Signal Processing Letters, 13(6):373-376, Jun 2006.

Experimental implementation of Grid enabled ASON/GMPLS Networks

Eduard Escalona[1], Jordi Perelló[1], Salvatore Spadaro[1],
Jaume Comellas[1], and Gabriel Junyent[1]

[1] Optical Communications Group, TSC Dept.
Universitat Politècnica de Catalunya (UPC), Jordi Girona, 1-3
08034 Barcelona, Spain
{escalona, jperello, spadaro, comellas, junyent}@tsc.upc.edu

Abstract. Current GMPLS protocols can be used in a Grid enabled environment by properly defining a new interface between Grid and Transport layers. However, the asymmetric nature of Grid traffic makes GMPLS inefficient when establishing bidirectional connections. The User to Grid traffic is unlikely to be the same as the Grid to User traffic; whilst the location of the processing sites may be of no importance to the user, the destination address of the Grid to User downstream is the same for all the different flows akin to the same job. This may result in asymmetric resource allocation throughout the network. We propose and experimentally evaluate an implementation for a middleware interface as well as a novel scheme that reduces blocked job requests due to lack of network resources in such environment.

Keywords: GMPLS, Control Plane, Grid, Asymmetry, Testbed.

1 Introduction

As applications demand for higher processing and storage capacity, the concept of Grids become more and more useful. Single machines cannot cope with the amount of capacity required for next generation applications and most small to medium enterprises (SME) cannot afford specialized hardware to provide such services. Moreover, supercomputers are a very expensive solution and are not scalable. Grid architectures allow the division of huge tasks into smaller jobs to process them separately in remote locations and then reassemble them to obtain the resulting information.

The main problem arises when the amount of data to be transmitted is large. Optical networks have made large bandwidths available. Nevertheless, if Grids networks are to better utilize this capacity, mechanisms to efficiently manage the optical resources are required.

The emerging Grid scenario requires a closer cooperation of the service and transport layers as bandwidth requirements of applications increase. Lately, several studies have been done to improve the performance of optical networks and considerable effort made to define optimal ways to manage and control network

I. Tomkos et al. (Eds.): ONDM 2007, LNCS 4534, pp 137-145, 2007.

resources. To this effect, the Internet Engineering Task Force (IETF) has developed a set of protocols defined as the Generalized Multi Protocol Label Switching (GMPLS [1]) paradigm. GMPLS, being an extension of MPLS, supports different types of switching; it has mechanisms that allow the establishment of optical channels in a fast, effective and dynamic way, acting as the control plane of the automatically switched optical networks (ASON [2]). Grid networks can take advantage of GMPLS capabilities as it fits the requirements detailed by the standardization organisms [3].

To implement the required procedures that enable the establishment and deletion of connections across heterogeneous networks, the Optical Internetworking Forum (OIF) defines a User to Network Interface (UNI). The UNI describes the service control interface between user and transport network addressing the procedures required for its interoperation based on the adaptations of two GMPLS signaling protocols, Label Distribution Protocol (LDP) and Resource ReserVation Protocol with Traffic Engineering extensions (RSVP-TE).

In order to extend the services offered by the UNI to facilitate on-demand access to Grid services, the Open Grid Forum (OGF) settled the concept of the Grid UNI (G.UNI) abstractly defining procedures and functionality considerations [3]. As the implementation of the G.UNI is open, we have designed an overlay architecture based on Web Services that will be further described in Section 2.

However, the deployment of Grid Services in a GMPLS-based transport network discovers that the asymmetric nature of the Grid traffic requires a more efficient management of the resources. When a job request is split into different connection requests the data to be processed is forwarded to diverse locations. In turn, the processed information should be sent back to the originating user or to another specific address. In such environment, unidirectional, symmetric and asymmetric bidirectional connections may be required when handling distributed processing. Within this scenario, the resource allocation becomes asymmetric, which means that, in a network link, available bandwidth may not be found for both directions.

Current GMPLS protocols support the establishment of unidirectional and bidirectional connections. However, when creating bidirectional connections, GMPLS is constrained to use the same path for downstream and upstream. Furthermore, the upstream label is chosen using local information. In Lambda Switched Capable networks (LSC), this results in a high blocking probability when wavelength conversion is not available as the same wavelength has to be maintained along the whole path.

In this paper, a novel connection scheme for GMPLS networks is proposed in order to reduce the rate of blocked job requests due to the lack of available network resources. It is based on calculating routes for both downstream and upstream connections separately, allowing the diversification of routes for bidirectional connections. It is beneficial when no upstream resources are available along the same path of the downstream connection. The proposed scheme has been experimentally tested over a real ASON/GMPLS network testbed, CARISMA [4].

This paper is organized as follows. Section 2 defines the interface between Grid and the optical control plane as well as our specific implementation in the CARISMA testbed. Section 3 addresses the asymmetry issue in GMPLS networks and Section 3.1 describes a novel connection scheme to improve its efficiency. Some experimental results are shown in Section 3.2 for performance evaluation. Finally, Section 4 concludes the paper.

2   Interfacing Grid and GMPLS

The Open Grid Forum (OGF) defines the requirements for an optical infrastructure for grid and describes the characteristics of a Grid User network Interface (G.UNI) [3]. The G.UNI is the interface between the Grid users and the network to dynamically allocate and provision bandwidth and Grid resources on request.

The OGF groups G.UNI functionalities in signaling and transport. Signaling deals with bandwidth allocation, scheduling, AAA (Authentication, Authorization and Accounting) information, resilience and propagation of service events. Transport deals with traffic classification, traffic shaping, grooming and security.

The first step in the introduction of the GUNI is the integration of the optical control plane following an overlay model in which Grid user applications will not have knowledge of the network topology. The next step is a peer model in which users and grid clusters are also elements of the network. This paper deals with the implementation, deployment and evaluation of the overlay model at our research premises in the CARISMA testbed.

2.1   The CARISMA Testbed

The CARISMA project was initiated in 2002 as an initiative to build a high performance Wavelength Division Multiplexing (WDM) based network to be used as a field-trial for the integration and evaluation of the current emerging innovative technologies. It is intended to provision bandwidth on demand while ensuring Quality of Service (QoS) between IP networks. The CARISMA testbed implements the ASON architecture based on a GMPLS control plane.

The control plane has been built associated to the transport plane, thus having a Signaling Communication Network (SCN) formed by Ethernet point-to-point links, following the same architecture as the transport plane. In such point-to-point architecture, only one control channel between Optical Connection Controllers (OCCs) is supported over each interface. Particularly, the protocols implemented are RSVP-TE [5] for signaling, OSPF-TE [6] for IP routing and information dissemination and Link Management Protocol (LMP) for discovery and control channel maintenance. These protocols have been implemented using Linux-based routers (Pentium 4 – 2GHz) running the GMPLS control plane and forming the OCCs.

CARISMA supports both soft-permanent and switched connections as defined by ASON. The management plane, responsible of the establishment of soft-permanent connections and the monitoring of the network topology and equipment status, uses the Simple Network Management Protocol (SNMP) for the communication with the transport and control planes. For the establishment of switched connections, OIF UNI 1.0 has been implemented.

2.2 G.UNI Implementation

The design of the G.UNI in the CARISMA network allows the Grid application to decide itself the type of connection required for a better performance. For the implementation we have used Java and Globus Toolkit 4 (GT4) to build the G.UNI as a WSRF (Web-Service Resource Framework) Grid Service. The result is a Web Service capable of accepting requests for optical channel establishments. Initially just three basic services have been defined to be supported by the G.UNI: Connection Setup, Connection Tear-Down and Connection Status Query. The functional block of the G.UNI that handles the requests is called Lightpath Provisioning Server (LP Server). The overlay G.UNI implementation should support different service invocation reference configurations likewise to the OIF UNI, namely Direct and Indirect Invocation Models. Under the direct model, the Grid User (or Grid Broker) invokes the transport services directly over the G.UNI (Fig. 1).

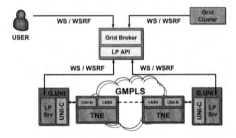

Fig. 1: G.UNI Direct Invocation Model

In the Indirect Invocation model a G.UNI proxy acts on behalf of one or more clients (Fig. 2). The interface between the G.UNI and the GMPLS control plane is open to any implementation. Specifically, our design accepts SNMP (soft-permanent connections) and OIF UNI 1.0 (switched connections) requests and uses the Indirect Invocation model.

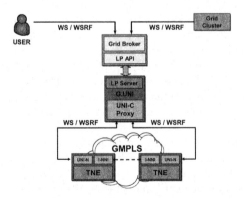

Fig. 2: G.UNI Indirect Invocation Model

2.3 Preliminary Results

In order to evaluate the performance of the designed G.UNI a Grid client has been implemented to emulate a Grid application requesting for optical connections. Results regarding setup latency prove the effectiveness of our implementation (Table 1).

Table 1: Connection Setup Time Flow using G.UNI

Action	Signaling Protocol	Delay
Client → G.UNI (LP Server)	SOAP (WS)	32 ms
G.UNI (LP Server) → UNI-C Proxy	UNIX Socket	13 ms
UNI-C Proxy → UNI-N (Source)	RSVP UNI 1.0r2	8 ms
UNI-N (Source) ←→ UNI-N (Dest)	RSVP-TE	27 ms
UNI-N (Dest) → UNI-C Proxy	RSVP UNI 1.0r2	8 ms
TOTAL		88 ms

3 Grid asymmetry impact over GMPLS

If we think about typical Grid applications like video streaming, Grid computing or storage we can easily see that the traffic generated is mainly asymmetric. In a general network with Grid applications over an optical control plane, in which unidirectional, symmetric and asymmetric bidirectional connections have to coexist, the distribution of network resources becomes also asymmetric (i.e. available bandwidth is not the same for the upstream and downstream on the same link).

When using current GMPLS standards, bidirectional connections are established adding the RSVP Upstream Label (UL) object to the Path messages so upstream resources are reserved hop by hop using local wavelength information [5]. With sparse (or lacking) wavelength conversion capabilities it results in very high blocking rates. Previous studies propose methods to select, if available, any free upstream resources (e.g. wavelengths) for the chosen path using an Upstream Label Set [7], or Global Wavelength information [8] (GW). Despite providing better performance than the UL scheme, they are constrained to use the same path for upstream and downstream connections. In case of asymmetric allocation of the network resources, connection requests are then blocked if not enough bandwidth is available on the upstream direction of the connection. As a further contribution, in this paper we propose a novel connection scheme for GMPLS networks to face the impact of the Grid traffic asymmetry.

3.1 UR Connection Establishment Scheme

The scheme proposed (UR) consists in calculating a new upstream route taking into account the actual available resources also for the upstream connection. Our scheme allows choosing a disjoint route when a bidirectional path can not be established because of the lack of resources in one of the directions. The signaling schemes are the same used in [9]. The upstream tunnel signaling is automatically triggered after the egress node has calculated an available upstream route. Although this kind of signaling doubles the setup time, this is not the predominant issue in a circuit-switched environment as flooding resource information updates take substantially longer.

In addition, we have also implemented a routing algorithm that takes into account the actual link usage, routing the traffic through the less congested links. Flooding capabilities of the OSPF-TE protocol are used to propagate the link usage information by means of Link State Advertisements (LSAs) assigning a metric field to the data links (wavelengths). This field has value zero when all data links on a TE link are free. When a connection is established and some wavelengths are reserved, the metric value is increased for all the data links belonging to the same TE link than the reserved one. This information, after disseminated, is used to calculate less congested routes using (1) and (2) for every wavelength following the wavelength continuity constraint and applying the Dijkstra algorithm. Note that, in (1) Cj is the cost for the route j, Hj is the number of hops for the route j, DLi is the number of data links in the link i (total or free) and n is the weight given to the new optical metric value added. The higher value of n, the more importance is given to the optical metric. The resulting route is the one with minimum cost of all the computed routes. In our studies we have used a value of n equal to 3 in order to assign more weight to the optical metric than to the distance without making the number of hops negligible.

$$C_j = H_j + \sum_{i=link_1}^{i=link_{H_j}} \left(DL_i^{TOT} - DL_i^{FREE} \right) \cdot n \qquad (1)$$

$$C_F = \min\{C_j\} \qquad (2)$$

Studies with different weight values should be done to observe the effects over the network performance. It is important to note that no extra packet or byte overhead is added as the metric is included in the metric field of the OSPF opaque LSAs and routes are updated upon route changes. We call this algorithm the Optical Metric algorithm (OM).

3.2 Performance Evaluation

We have carried out some experiments to evaluate the performance of the proposed scheme for Grid networks. The meshed-based network scenario shown on Fig. 3 composed by 5 nodes and 4 bidirectional wavelengths per link (8 data links) has been considered. To emulate the asymmetric allocation of the network resources (typical in a Grid network context), we assume that three permanent unidirectional connections using one wavelength each (OCC2-OCC1, OCC3-OCC2 and OCC4-OCC5) are initially established.

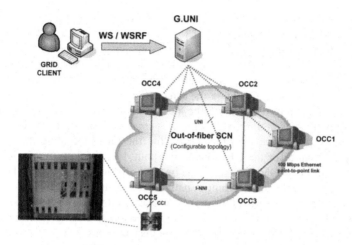

Fig. 3. CARISMA testbed configuration

Three different schemes to establish bidirectional connections have been compared experimentally as a function of the traffic load, considering blocking probability as the figure of merit: the standard GMPLS scheme with a First Fit wavelength assignment (UL), the GW scheme and the UR scheme. We have also evaluated the improvement of the performance of the UR scheme with the OM algorithm (UR+OM). Grid connections requests arrive with an exponentially distributed inter arrival time (IAT) of mean 20s. The connection holding time (HT), also exponentially distributed, is varied to generate different traffic load. The requests are set up between pairs of nodes chosen randomly so the traffic load is distributed uniformly among the nodes.

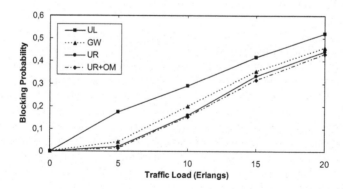

Fig. 4. Blocking probability

As a sample of the obtained results, Fig. 4 shows the resulting curves for the four compared schemes (UL, GW, UR, UR+OM). These results have been obtained using a statistical relevant number of connections (10000). GW improves the results of the standard GMPLS implementation (UL) as expected, as it uses global wavelength information to maintain the wavelength constraint.

Nonetheless, in the considered network scenario, UR shows better performance than the GW scheme, since resource allocation is not symmetric. In fact, when no upstream resources are available on the same route, the UR scheme is able to select a diverse one. The improvement of the performance on the average in the tested scenario is about 4% which should increase in conditions of more traffic asymmetry.

It can be observed, on the other hand, that the OM algorithm allows to further improve performance of the UR scheme in about 4% with high traffic loads. Please note that UR and UR+OM have the same behavior for low traffic loads. This is due to the fact that UR+OM may use longer paths than UR to avoid links that are already carrying traffic for a better load balance. However, in low loaded networks these links are not as prone to be saturated as with high traffic loads. This effect compensates the gain obtained with the wavelength occupancy distribution.

Results regarding average lambda occupancy per link have also been taken. Lower differences between maximum and minimum occupancy values for the UR+OM scheme can be seen, which means that this scheme offers a more compensated traffic distribution among links than GW. Whereas the GW scheme makes some links more loaded than others, OM tries to compensate these values, shortening the difference and as a consequence, providing a better load balance.

4   Conclusions

The amount of bandwidth required for the emerging intensive grid applications makes necessary to build a service-aware optical transport network. As GMPLS seems to fulfill the requirements needed for a Grid applications correct performance, a Grid middleware has to be defined and implemented to merge service and optical transport environments.

Experimental results show that the obtained setup delays are suitable for the requirements of Grid applications.

Moreover, a novel connection scheme for bidirectional connections in multi-service networks has been presented to address the asymmetric nature of Grid networks. It is based on calculating a separate route for the upstream connection allowing the selection of a diverse path when no upstream resources are available along the same path used by the downstream connection. The results obtained show that the UR scheme outperforms the already proposed GW and UL schemes and it is further improved if a routing algorithm that takes into account the global wavelength availability is also considered for optimizing the load balance. Ongoing further studies will consider larger networks and varied topologies.

Acknowledgements

The work reported in this paper has been partially supported by the Spanish Science Ministry through Project "Red inteligente GMPLS/ASON con integración de nodos reconfigurables (RINGING)", (TEC2005-08051-C03-02), and by the i2CAT Foundation (www.i2cat.cat).

References

1. E. Mannie: Generalized Multi-Protocol Label Switching (GMPLS) Architecture, RFC 3945, Oct. 2004.
2. ITU-T Rec. G.8080/Y.1304, Architecture for the Automatically Switched Optical Network (ASON), November 2001
3. D. Simeonidou et al.: Optical Network Infrastructure for Grid, GFD-I.036, March 2004
4. J. Perelló et al.: Control Plane protection using Link Management Protocol (LMP) in the ASON/GMPLS CARISMA network, IFIP Networking 2006, May 2006.
5. L. Berger: GMPLS Signaling RSVP-TE Extensions, RFC3473, January 2003.
6. D. Katz et al.: Traffic Engineering (TE) Extensions to OSPF Version 2, RFC3630, September 2003
7. E. Oki et al.: Bidirectional Path Setup Scheme Using an Upstream Label Set in Optical GMPLS Networks, IEICE Trans. Communic , vol E87-B.
8. R. Martinez et al.: Experimental GMPLS-based Dynamic Routing in All-Optical Wavelength-Routed Networks, Proc. ICTON 2005, Barcelona.
9. E. Escalona et al.: Establishing Source-Routed Bidirectional Connections over the Unidirectional ASON/GMPLS CARISMA Testbed, Proc. ICTON 2005, Barcelona.

Reservation Techniques in an OpMiGua Node

Andreas Kimsas, Steinar Bjornstad, Harald Overby and Norvald Stol

Norwegian University of Science and Technology (NTNU)
O. S. Bragstads plass 2B, 7491 Trondheim, Norway
e-mail: akimsas@item.ntnu.no

Abstract. An OpMiGua node integrates a packet switch for low priority traffic and a circuit switch for high priority traffic. Both traffic classes share the same input and output ports using time division multiplexing, but absolute priority is given to circuit switched packets. The circuit switched packets do not experience contention at output ports and are not subject to delay jitter; hence a guaranteed service class is created. Previous studies of OpMiGua nodes have used one of two reservation techniques to assure priority; either a time-window approach or a preemptive approach. This article introduces two new reservation techniques and investigates advantages and drawbacks associated with the four techniques. It is shown that each reservation technique is associated with specific loss mechanisms and methods to reduce their influence are proposed. Simulation results demonstrate that the choice of reservation technique is highly influenced by the relative share and length of high priority packets.

Keywords: OpMiGua, hybrid, time-window, preemption, packet loss

1 Introduction

Different hybrid Optical Burst/Circuit Switched (OBS/OCS) and Optical Packet/Circuit Switched (OPS/OCS) networks have been introduced over the last years. A survey of these networks is given in [1], where OpMiGua[1] [2] belongs to the "integrated" type of hybrid networks. Common for the integrated hybrid network approaches is the ability to share the bandwidth of a wavelength between packet switched and circuit switched traffic. The time-scale for resource sharing is on a packet-by-packet basis and is different from hybrid schemes where resources are reserved for timescales equalling circuit switched connection times. This capability is used to increase the resource utilisation as compared to pure OCS networks.

The OpMiGua concept was developed to create a network architecture that is capable of achieving high resource utilisation *and* capable of providing a service class with guaranteed delivery. Guaranteed service class traffic (GST) is created by employing optical circuit switched paths to avoid contention in transit nodes and to eliminate delay jitter. To increase the bandwidth utilisation, statistically multiplexed (SM) traffic is injected in the voids between consecutive GST packets. Reservation techniques are used to assure that the insertion mechanism does not interfere with the GST packets. It will be shown that the reservation techniques used to assure priority

---

[1] Optical Migration Capable Network with Service Guarantees

I. Tomkos et al. (Eds.): ONDM 2007, LNCS 4534, pp. 146–155, 2007.

for GST packets are associated with loss mechanisms that influence the performance of the SM traffic. With the final goal of reducing the blocking probability of SM traffic, this article investigates the loss mechanisms and performance of different reservation techniques.

Earlier articles on OpMiGua suggested using reservation techniques based on a time-window approach [2] or based on a preemptive approach [3]. In addition to the two previous techniques, two new reservation techniques are introduced; an improved time-window approach and a combined time-window and preemptive approach. Comparison of the four reservation techniques will provide insight into the loss mechanisms involved and the performance of the two new reservation techniques can then be compared to the existing ones. With a focus on reservation techniques buffering for contention resolution is not considered.

Each technique is evaluated via simulations of an OpMiGua node, whose operation is briefly explained in section 2. Section 3 discusses the logical behaviour of the four reservation techniques and identifies the loss mechanisms involved in each method. The simulation scenario and results are treated in section 4 and the main findings are then summarised in the conclusion.

2   The OpMiGua Node

An OpMiGua core node can be decomposed into four main parts. A packet separator (PS) detects if the incoming packets belong to the GST class or to the SM class by analysing its optical label. Possible labelling methods include use of orthogonal polarisations [4] or subcarrier modulation [5]. Regardless of the separation technique in use, the packet separator will forward GST packets to an optical cross connect (OXC) and SM packets to an optical packet switch (OPS). Performance of a blocking OPS module was studied in [3], but in this article both OXC and OPS modules are assumed strictly non-blocking.

The wavelength routed GST packets are directed to its output port according to the current configuration of the OXC, while the SM packets can use any idle output wavelength which is part of the correct output fibre. The insertion of SM packets is performed by an optical packet switch inside the OpMiGua node. Two questions then arise; how can the arrival of GST packets be detected and signalled, and how can absolute priority be ensured for GST traffic? The answer to the first question is given below, while the second question is treated in section 3.

The packet combiner (PC) is responsible for signalling the presence of incoming GST packets to the control unit of the packet switch and re-inserts the optical label if necessary. The physical detection of GST packets is enabled by an optical splitter, which is linked to the control unit. After passing the optical splitter, the GST packets enter a fibre delay line (FDL) whose purpose is to create a time-window for the control unit. The size of the time-window is determined by the length of the FDL and its actual length depends on the reservation techniques in use. Having knowledge of future GST arrivals, the control unit can make the necessary decisions to avoid simultaneous transmission of GST and SM packets on the same shared output port.

3 Logical behaviour of reservation techniques

The time-window technique [6] [7] is used in OpMiGua to predict and avoid simultaneous transmission of GST and SM packets on the same wavelength and is a *proactive* scheme. The preemptive technique [8] [9] avoids simultaneous transmission by interrupting SM transmission after a conflict is detected and is a *reactive* approach. As will be explained in this section, each method has their advantages and disadvantages with respect to network layer efficiency and implementation complexity.

3.1 Time Window Techniques

Upon arrival of an SM packet at the OPS switch, the control logic is aware of all GST arrivals within the time $t_{arrival}+\Delta$, where Δ is the size of the time-window and equals the transmission time for the longest SM packet. Provided that SM packets are scheduled to an output port whose time-window is empty there is no risk of contention with GST packets. Figures 1 and 2 illustrate the logical behaviour of the two time-window techniques considered in this article; the simple time window (STW) scheme and the length aware time window (LATW) scheme. For simplicity, only one output wavelength is shown. Also, contention among SM packets is supposed to be handled by the OPS module which explains the absence of SM packets that overlap in time.

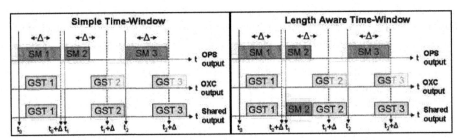

Fig. 1 left (*STW*) and Fig. 2 right (*LATW*): The blocking probability is higher for the STW technique, but its implementation is simpler as it does not take SM packet length into account.

The STW technique does not consider the length of SM or GST packets when scheduling SM packets. The control-logic will schedule SM packets on the output port as long as GST packets are not detected within the time-window. Referring to the figure, at time t_0 a scheduling decision has to be made for SM-1. The control logic has knowledge of the arriving GST-1 packet which should be given priority. SM-1 is then inserted into a buffer, scheduled on an alternative available wavelength or discarded. The same occurs for SM-2 at time t_2 and SM-3 at time t_3. However, as exemplified by SM-2 in the figure, there are inefficiencies related to this reservation technique. Although SM-2 is sufficiently short to be scheduled without disturbing GST-2 it was not feasible since the control-logic does not take packet length into account. This inefficiency has been named reservation induced blocking (RIB) [2] and adds to blocking caused by contention of simultaneously arriving SM packets. Thus, at the expense of RIB absolute priority to GST packets is achieved.

The LATW technique is suggested to reduce the RIB. For each output wavelength, the time to arrival of the GST packets is continuously updated so that the control-logic can compute whether there is sufficient time to schedule the SM packet on one of the outputs. Either it can schedule the packet on the first wavelength that has sufficient space, or it goes through all wavelengths and schedules the packet on the wavelength which creates the smallest gap between the SM and GST packet. On average, the first option requires less computation, but the second option makes better use of each wavelength resource. Our simulations use the minimum gap method, but whichever case being chosen the RIB is eliminated because SM packets are only rejected when there is contention at the output port. Also, the blocking probability for any packet length using the LATW scheme will be inferior or equal to the STW scheme. This scheme can therefore be seen as the optimal choice with respect to performance, but requires access to the SM packet length and involves more computation. Additionally, in contrast to the STW technique blocking probability increases for longer SM packet lengths which introduce unfairness.

3.2 Preemptive and Combined Techniques

The preemptive (PRE) technique depicted in figure 3 does not rely on a time-window, but will always attempt to transmit the SM packet if the shared output wavelength is currently available. At time t_0, the control logic has no knowledge about a possible conflict between SM-1 and GST-1, so SM-1 is scheduled on the common output wavelength. Once the GST packet is sensed on the input, transmission of SM-1 is immediately terminated and SM-1 is said to be preempted. Fragments of preempted packets will cause an increased load at the subsequent nodes, unless precautions are taken to detect and remove the fragments.

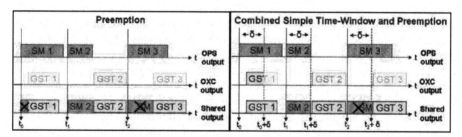

Fig. 3 left (PRE) and Fig. 4 (COMB): The preemptive technique allows SM packets to be scheduled on any free output port at the risk of being preempted. A cross indicates preempted packets. With the combined scheme SM packets which are longer than the time window δ risk preemption, while shorter packets may be blocked by incoming GST packets.

In contrast to the STW technique, SM-2 is scheduled at time t_1 and is successfully inserted between GST-1 and GST-2. Although not suffering from RIB, the performance of the preemptive scheme is affected by a different loss mechanism. If several outputs wavelengths are free, the control logic must schedule the packet on one of these outputs without knowledge about future GST arrivals. Hence, it is possible that the selected output port results in preemption, even though it could have been successfully transmitted on a different output port. This inherent inefficiency for the preemptive scheme can bee seen as a loss of multiplexing gain and is termed preemption induced blocking (PIB).

When considering the length of SM-2 it should be clear that the probability of a successful transmission decreases when the packet-length is increased. Again this is exemplified with SM-3, which would not have been preempted if it was of the same length as SM-2. Hence, the preemptive scheme introduces unfairness with respect to packet length, where shorter packets are favoured.

The combined (COMB) scheme use elements from the STW and PRE techniques. With this method we try to find a compromise between the amount of RIB and PIB, as well as reducing the number of preempted packets. The scheme use a time window of size δ, with $0 < \delta < \Delta$. It should be noted that the combined scheme equals the STW scheme for $\delta = \Delta$ and the preemptive scheme for $\delta = 0$. As shown in figure 4, the time window is sufficiently large to predict collision between SM-1 and GST-1. When compared to the STW scheme, the successful scheduling of SM-2 illustrates a reduced RIB. Since Δ is the minimum length required to assure collision avoidance, the shorter time window implies that all scheduled packets with length between δ and Δ may interfere with incoming GST packets. If interference is detected at the output, then absolute priority to GST packets is obtained by preemption, illustrated by SM-3 in the figure.

4 Simulation Scenarios and Results

4.1 Simulation Scenarios

The performance of each scheme is assessed via simulation of an all-optical OpMiGua node, with its main parameters listed in table I. Strictly non-blocking OPS and OXC modules is assumed to get the full benefit of statistical multiplexing for SM packets and eliminate blocking probability for GST packets. Both SM and GST packets are generated according to a Poisson arrival process. All simulation scenarios in this article assume that the input traffic is buffered in ingress nodes, i.e. packets that overlap in time will be transmitted back-to-back. The simulation points were obtained running ten independent replications to establish 95% confidence intervals and the simulation length was adjusted to produce confidence intervals which are smaller than 10% of the average value. To improve the readability of the figures the confidence intervals are not represented, but their values are included in the discussion whenever necessary.

According to the connection-oriented nature of GST packets, all packets generated at an input port are directed to an output port which is not used by other GST sources. The destination fibre for SM packets is uniformly distributed among the four fibre outputs. While SM packets are transmitted one-by-one, earlier studies [2] have shown that aggregating GST packets into bursts at the ingress nodes have positive effects for the SM packet loss rate (PLR). Consequently, most of the simulations set the burst length to $100 \cdot SM_{MEAN}$, which is thought to be a good compromise between performance and aggregation delay. The aggregation process itself was not simulated, as a thorough study of different GST burst aggregation schemes will be investigated in a separate paper.

Table I: Parameters for Simulation Study. Brackets indicate a list of possible values.

Component / Quantity	Parameter	Parameter value	Unit
Bitrate per wavelength	Rate	10	[Gbps]
Fibre inputs / outputs	F	4	
Wavelengths per fibre	W	32	
Normalised load	A	$0.6 \leq A \leq 0.9$	
GST share of total load	S	$0 \leq S \leq 100$	[%]
SM packet length	SM_{LENGTH}	Uniform 40-1500	[bytes]
SM average packet length	SM_{MEAN}	770	[bytes]
SM maximum packet length	$SM_{MAX\ LENGTH}$	1500	[bytes]
GST burst length	GST_{LENGTH}	$[1,10,100,1000] \cdot SM_{MEAN}$	[bytes]
Time-window	Δ	$SM_{MAX\ LENGTH} \cdot 8$ / Rate	[s]

4.2 Influence of GST share on the different reservation techniques

This section aims at quantifying the effect of the different loss mechanisms related to the SM class, particularly in presence of varying GST shares. Being statistically multiplexed, the SM packets will experience contention at the output ports whenever all possible output wavelengths are occupied by SM or GST packets. Throughout the paper this loss mechanism is termed contention induced blocking (CIB) and is present for all reservation techniques studied here. The figures 5-8 show the SM PLR for different loads and for different shares of GST traffic.

Time Window Techniques
When considering the STW technique, the total loss is the result of the CIB and the RIB. The RIB equals zero in the absence of GST traffic, which implies that the system is purely packet switched, the curve for GST share S=0% in figure 5 is then used as reference for the curves with non-zero GST share.

Using experience for purely packet switched systems, the introduction of two service classes normally introduces inefficiencies, hence an increase in PLR could have been expected [8]. In figures 5 and 6 SM PLR clearly benefits from high GST shares for loads inferior to 0.8, thus an OpMiGua node with two service classes performs better than a single service class OPS node. So, for low and moderate loads, the negative effect of RIB is more than compensated for by the reduced CIB. When the load exceeds 0.8 this is no longer true. Take the points at A=0.9 as examples; the upper value for the S=0% curve equals 0.062 while the lower value for S=90% equals 0.081. Here the output ports are so congested that blocking due to RIB overshadows the reduced CIB. A confirmation of this analysis is found in figure 7, where the RIB is reduced by employing the LATW technique. The curves around A=0.9 almost overlap within the confidence intervals.

The LATW technique was proposed to eliminate the effect of RIB. As expected this leads to better performance regardless of the load. When comparing the curves of equal GST share in figures 5 and 6 the difference is small, but still present. For

S=30% at load 0.65 the PLR is improved with 4.7 %, but for S=70% the improvement is 24.4 %. Similarly, at load 0.9 the gain increase from 5.1% to 18.2%. The trend is clear: higher GST share gives higher improvement. This is in accordance with the fact that with the STW technique the RIB will be more and more prevalent as the amount of GST traffic increases, so reducing the RIB will favour the simulation scenarios with high GST share.

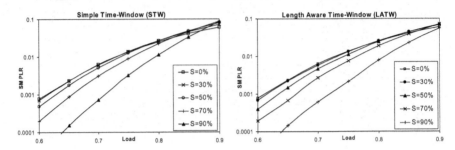

Fig. 5 left *(STW)* and Fig. 6. right *(LATW)*. $GST_{LENGTH} = 100 \cdot SM_{MEAN}$. For loads A<0.85 the SM PLR benefits from high GST shares. The SM PLR for the LATW technique is improved by up to 20% for the curves represented here.

The effect of the time window approaches can then be summarised. First, the SM traffic in an OpMiGua node with moderately aggregated GST traffic performs better than a purely packet switched node with a single service class. Second, since the GST traffic bypass the OPS switch, the OPS switch will experience a reduced *internal* load as compared with a purely packet switched node. Hence, the SM class in OpMiGua is less influenced by internal blocking when compared to low priority traffic in a purely packet switched node. This property can be exploited by using a cost-effective OPS module with internal blocking, rather than the more expensive and technologically immature non-blocking alternatives [3]. Finally, a relatively low PLR is obtained for the SM class, even without the use of buffering for contention resolution. This is especially true for high GST shares. Hence, it could be beneficial to let part of the SM traffic be sent as GST packets if the traffic demand on an optical lightpath is low. However, the performance and implementation of this last issue is outside the scope of the article, and is left for future studies.

Preemptive and Combined Techniques
The SM PLR represented in figure 7 shows a behaviour that is opposite to the behaviour using time-window techniques; the performance deteriorates with increased GST share. Again, the total PLR is the result of two effects, CIB and PIB. The analysis for the time-window techniques showed that the CIB decreases as more GST traffic is inserted into the system. Since the simulation scenarios only differ with respect to the reservation technique, the same should apply here. Even more, with a higher packet loss the average occupancy probability of the output port will decrease, so the CIB should be even lower than for the time-window techniques. Consequently, the increased PLR is caused by the PIB.

At low loads, the SM PLR for curves with high GST shares is very different from the case without GST traffic. One may conclude that for high GST shares the loss mechanism is primarily influenced by the PIB, while for low GST shares the primary loss mechanism is the CIB. As the total load increases towards A=0.8 the curves almost overlap regardless of GST share. Therefore, in this regime the CIB is the dominant loss mechanism. At even higher loads the curves split, with the highest loss for the highest GST shares. Both CIB and PIB increase, but the PIB increases faster than the CIB.

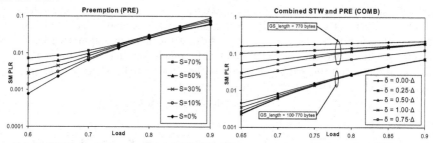

Fig. 7 left (PRE) and Fig. 8. right (COMB): The preemptive technique for different GST traffic shares (S) and the combined scheme with different time-windows δ and S=30%.

Some general observations can be made concerning the performance of the preemptive technique. For all loads it is outperformed by the LATW technique, and for loads inferior to 0.9 it also is less performing than the STW technique. However, the difference is not considerable for GST shares inferior to 30%. At load 0.7 and S=30%, the PLR for the preemptive technique is $8,2 \cdot 10^{-3}$ compared to $6,2 \cdot 10^{-3}$ for the STW technique.

The combined technique is evaluated by varying δ in the interval [0, Δ]. For δ=0 the reservation technique is entirely preemptive and for δ=Δ the scheme corresponds to the STW technique. When the GST packets are aggregated into burst of 100, then all curves for the combined scheme in figure 8 are confined within the limits defined by the curves for δ=0 and δ=Δ. If packet fragments are not considered, the choice of reservation technique is reduced to finding a compromise between the SM packet loss rate and added GST delay at the intermediary nodes. Experience from the preemptive technique and the STW technique indicates that the difference between the curves will be reduced for lower GST shares and increased for higher GST shares. At higher loads the difference between the reservation techniques will be small, regardless of the GST share and regardless of the scheme being used.

When considering the case without GST packet aggregation, cf. the five upper curves in figure 8, the performance of the combined scheme with δ = 0.75·Δ outperforms the STW technique. It is not the case for other values of δ. So, by allowing a well chosen ratio of packets to risk preemption, the performance is improved.

The analysis illustrates the complexity of the packet loss mechanisms in an OpMiGua node. Hence, a large parameter space is required to get a good understanding of the system. Furthermore, the two set of simulations in figure 8 also illustrate the sensitivity on GST burst length, an issue which is studied in next section.

4.3 Effect of Burst Length

If the STW technique is used, an arriving GST burst will contribute the RIB during the time interval defined by the time of entry into the FDL and the arrival time at the packet combiner. The length of the burst is of no importance when considering RIB, better performance can then be expected if the average number of burst arrivals is reduced. For a given GST share, this may be achieved by increasing the burst length. Indeed, significant improvement in the SM PLR is observed for the STW technique in figure 9, especially when moving from no aggregation to an aggregation length of 10 GST packets. The effect drastically decreases after an aggregation length of 100 GST packets, almost nothing is gained when shifting to an aggregation length of 1000 packets. In other words, the RIB is already much smaller than the CIB for a GST_{LENGTH} equal to 100 packets. Unless buffers for contention resolution are used, extending the burst length will only imply added delay.

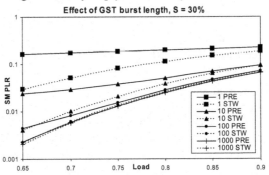

Figure 9: Effect of burst length for the preemptive technique (*PRE*) and the simple time-window technique (*STW*). S=30%.

The same trend applies for the preemptive technique, but the difference between the simulation scenarios is larger than for the STW technique. As explained in the section on the preemptive approach, PIB is the dominating loss mechanism at low and moderate loads. Reducing the number of GST burst arrivals reduce the influence of PIB. For the 100 STW scenario the RIB is not dominant compared to the CIB, but for the preemptive technique this is not the case. A substantial improvement can still be observed when increasing the aggregation length to 1000 GST packets, illustrating that the preemptive technique needs more GST aggregation than the STW technique to achieve the same SM PLR.

5 Conclusion

In this paper we have identified and proposed reservation mechanisms for the OpMiGua hybrid optical node. The novel length aware time window and the combined preemption/time-window schemes have been thoroughly investigated. Circuit switched guaranteed traffic (GST) has no packet loss. Blocking probability of low priority traffic is used as performance parameter.

We found that regardless of reservation technique, the blocking probability is heavily influenced by the share of GST traffic, and improves as the GST burst length increases. Hence, a compromise between GST aggregation delay and packet loss must be found when designing an OpMiGua network.

For traffic loads less than 0.8, the STW and LATW techniques outperform the preemptive and combined techniques. However, for GST shares inferior to 30% the difference between the techniques is small, e.g. less than 10% for a GST share of 10% and a load of 0.7.

For traffic loads beyond 0.8 the difference in performance decreases, for all traffic loads beyond 0.8 the difference in performance decreases, for all schemes and for all GST traffic shares.

Furthermore, a higher performance than for an OPS with a single service class is found using both the STW and the LATW techniques, given a load below 0.8 and a GST shares above 30%. This illustrates the potential of the OpMiGua concept and motivates a future study of scheduling approaches where SM packets are transmitted as GST traffic on circuits with low GST share.

Acknowledgment: Telenor, the NRC and COST action 291 are recognised for funding research activities in the field of optical networking.

References

1. Gauger C.M., Kühn P.J., Breusegem E., Pickavet M., Demeester P.: Hybrid Optical Network Architectures: Bringing Packets and Circuits Together. IEEE Communications Magazine (August 2006), vol. 44, pp. 36-42.
2. Bjornstad S., Hjelme D.R., Stol N.: A Packet-Switched Hybrid Optical Network with Service Guarantees, IEEE JSAC, Supplement on Optical Communications & Networking, (August 2006), vol. 24, pp. 97-107.
3. Kimsas A., Bjornstad S., Overby H., Stol N.,: Protection Using Redundancy in a Hybrid Circuit/Packet Node Design. Proc. ECOC (2006), We3.P.143.
4. Tuft V.L., Bjornstad S., Hjelme D.R.: Time Interleaved Polarization Multiplexing for Polarization Labelling, Proceedings of the 7th International Conference on Transparent Optical Networks (July 2005), vol. 1, pp. 47-51.
5. Breusegem E. et al.: A B Broad view on Overspill Routing in Optical Networks: a Real Synthesis of Packet and Circuit Switching?", Journal of Optical Systems and Networking (2005), vol. 1, no. 1, pp. 51-64.
6. Hu G., Gauger C.M., Junghans S.: Performance of MAC Layer and Fairness Protocol for the Dual Bus Optical Ring Network (DBORN), Conference on Optical Network Design and Modeling (February 2005), pp. 467-476.
7. Qiao C., Yoo M., Dixit S.: OBS for Service Differentiation in the Next-Generation Optical Network. IEEE Communications Magazine (February 2001), vol. 39, no. 2, pp. 98–104.
8. Overby H., Nord M., Stol N.: Evaluation of QoS Differentiation Mechanisms in Asynchrounous Bufferless Optical Packet Switched Networks. IEEE Communications Magazine (August 2006), vol. 44, no. 8, pp. 52-57.
9. Phuritatkul, J., Yusheng J.: Analysis of Bandwidth Allocation with Wavelength Preemption Scheme for Supporting Relative Service Guarantees in OBS networks. Broadband Networks (October 2005), vol. 2, pp. 1300 – 1309.

R & Ds for 21st Century Photonic Network in Japan

Ken-ichi Kitayama

Osaka University
2-1 Yamadaoka, Suita, Osaka 565-0871, JAPAN
kitayama@comm.eng.osaka-u.ac.jp

Abstract. Government-funded R&D initiatives have been playing crucial roles for developing photonic networks. The R&D project is typically carried out for a relatively long period of five-years, and the goal is set so high that no single private sector can afford the risk. In this paper, challenging government-funded R&D initiatives of photonic networks in Japan is introduced. They have been launched in 2006, which aims to establish the *photonic platform* to provide abundant bandwidth on demand, at any time and from anywhere, to promote bandwidth-rich networked applications. In addition, other government-funded R&D activities relevant to the photonic networks in Japan is also presented, including photonic devices and subsystems and large-scale optical interconnect for ultra-high performance computing.

Keywords: R&D initiative, photonic network, optical communication, network testbed, government fund, Japan

1   Introduction

Government R&D initiatives have been playing crucial roles to evolve voice-oriented legacy telecommunication networks to IP-based NG networks [1]. The R&D programs are characterized by participants, the period, and the budget size as well as the goals. A consortium is first formed consisting of both academic and industrial partners, the R&D is typically carried out during a relatively long period of four- or five-years, which is financially well supported, and goals are set so high that no single private sector can afford the risk.

In this paper, challenging government-funded R&D initiatives of photonic networks in Japan is introduced. They have been launched in 2006, which aims to establish the *photonic platform* to provide abundant bandwidth on demand, at any time and from anywhere, to promote bandwidth-rich networked applications. In addition, other government-funded R&D activities relevant to the photonic networks in Japan is also presented, including photonic devices and subsystems and large-scale optical interconnect for ultra-high performance computing.

I. Tomkos et al. (Eds.): ONDM 2007, LNCS 4534, pp. 156–165, 2007.

2   Photonic Network R&D Initiatives for Builiding Photonic Platform

2.1   Overview

Four R&D projects of photonic networks have been launched in 2006 under the financial support of an agency of the Japanese government, the National Institute of Information and Communications Technology (NICT). The outline is summarized in Table 1.

Table 1.   R&D projects supported by NICT

Project	Period	Participants
Photonic node with multiple granularity switching capability	2005-2009	NTT Comm., Fujitsu, NTT, Nagoya Univ.
λ Access	2006-2010	NTT, NEC, NTT Com., Mitsubishi Electric, Hitachi, KDDI Labs., Univ.Tokyo, Keio Univ.
λ Utility	2006-2010	NEC, Fujitsu, Mitsubishi Electric, Oki Electric, Osaka Univ.
Optical RAM for all-optical packet switching	2006-2010	NTT, NEC, Osaka Univ., Kyushu Univ.

2.2   Photonic Node with Multi-granularity Switching Capability

The objective of the project is to develop fundamental technologies for 100 Tera-bps class photonic nodes with multiple granularity switching capability and for the design and control of backbone networks with Peta-bps class throughput utilizing the multi-granularity nodes. As shown in Fig.1 the reesearch themes are the followings;

- The architecture of large-scale optical switches with several nano-second switching time for handling optical burst data. Materials, devices, module fabrication and control techniques for large-scale high-speed optical switches.
- The architecture of a multi-granularity photonic node that can handle optical bursts to/from waveband paths. Key technologies for realizing the node hardware such as multi-wavelength light sources, optical virtual concatenation, waveband conversion, and stable optical modulation/demodulation with highly spectral efficiency.

- The architecture and design of a photonic network that employs hierarchical optical paths. The control technology of combined optical switches with different switching granularities.

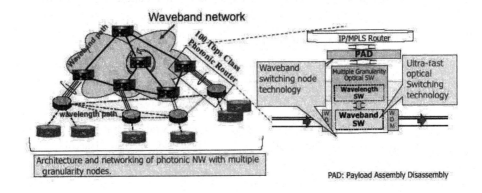

Fig. 1 High performance photonic node with multi-granularity switching capability

2.3 λ Access

Multiple-lambda optical networks envision virtual Terabit LAN in geographical scale, and two of its key technical challenges are to provide user network interface (UNI) with scalability and controllability. The "Lambda Access" is a new five-year project funded for these challenges in Japan by NICT (Nation Institute of Information and Communications Technology), that focuses on network access at 100Gbps and beyond with single or multiple lambda interfaces in cooperation with the related project "Lambda Utility". As shown in Fig. 2, the Lambda Access Project includes the following two themes;

- WDM Seamless Access Technology

This theme aims at the establishment of Terabit LAN network interfacing technology that allows an end-user to access network with multiple lambdas at 100Gbps and more. It includes user-controlled link bonding for scalable access, new-generation MAC with Mega-Byte-size frames, and network access protocols for seamless user-side controllability.

- Frame-multiplexed Ultra High Speed Access Technology

This theme aims at the establishment of Ethernet frame aggregator and its uplink single-lambda 100Gbps interface.   It includes statistic Ethernet frame aggregation at 100G bps and beyond, 100Gbps single-lambda modulation/demodulation techniques as well as its physical layer implementation, and frame-base protocols for end-to-end OAM.

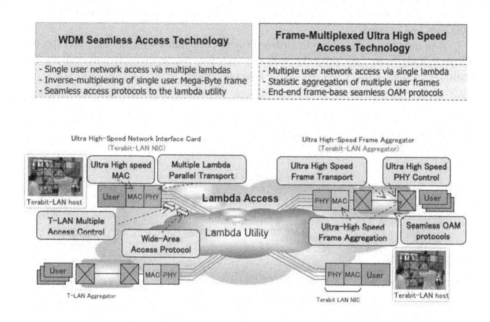

Fig. 2. Outline of λ Access project

2.4   **λ Utitity**

The Lambda Utility Project cooperating with the Lambda Access Project performs technical development which realizes the terabit-class LAN environment over wide area networks.   The goal of these projects is to provide each end-user stress-free 100Gbps-class bidirectional communication with various services, such as a super high definition video conference, GRID computing, and so on, as shown in Fig. 3. The Lambda Utility Project, focusing on technologies in a wide area network, includes the following three themes;

- Borderless optical path control and management technology

This theme aims at the establishment of photonic service gateway technology which utilizes the optical path over multiple wide area networks with scalability of more than 1000 nodes.

- Highly-spectral-efficient transmission link technology

This theme aims at the establishment of optical link technology at 100Gbps and beyond. It includes the multi-level modulation and demodulation technology for pursuing high spectral efficiency. The high coding gain forward error correction code is also included in this theme to cope with the higher signal to noise ratio requirement for the multi-level modulation format.

• Modulation-format-independent all-optical 3R technology

This theme aims at the establishment of all-optical 3R technology which is capable for both phase modulation format signal and intensity modulation format signal at 100Gbps and beyond.

It is noted that the Lambda Utility Project cooperates with the Lambda Access Project, and those two projects will conduct the interoperability experiment for realizing a terabit-class LAN over wide area networks.

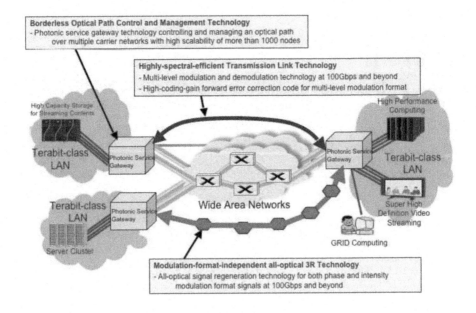

Fig. 3. Outline of λ Utility project

2.5   Optical RAM for All-optical Packet Switchings

This project aims at developing basic technology for all-optical packet switch. A main focus is on optical random access memory (RAM). Three research topics of this project includes are the following;

• Nano-structured optical devices for a bit memory based upon ultrafast bistability.

- Basic technology for the fabrication process of nano-structured optical devices and their materials.
- Optical interface and subsystem for optical RAM focusing on the optical addressing to the bit memory and the serial-to-parallel conversion of the bit-serial optical packet.
- Architecture of all-optical packet switch and buffer scheduling as well as its transmission control protocol, preferably adopted for small buffering capacity.

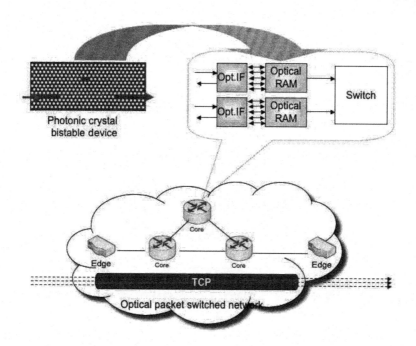

Fig. 4. Outline of optical RAM project.

2.6   Network Testbeds

The configuration of the Japan Gigabit Netowrk (JGN) II [2] and the experimental network is shown in Fig.5. JGNII's L2 service used for the GMPLS control layer (C-Plane) made it possible to achieve interoperation among Keihanna Center, Otemachi, and Koganei. The C-Plane is used to enable GMPLS nodes to exchange control protocols. For the transport layer (D-Plane), on the other hand, JGNII's STM-64 (standard for 10 Gbps synchronous transmission) service was used. STM-64 was employed under constant connection, with the GMPLS nodes at both ends controlled as a lambda path (LSC-LSP). A multi-layer control experiment was conducted by controlling the STM16 (standard for 2.5-Gbps synchronous transmission) path, with GMPLS as a TDM path (TDM-LSP) and connecting to the MPLS network (MPLS-LSP). In the meantime, settings for a virtually wider control network (total extension distance for signaling: 1,320 km) ere attempted by dividing the control layer in two and causing signals to perform a round trip between Koganei and Keihanna. As for the multi-vendor environment, verification was carried out by combining different items of equipment from nine different vendors and generating several types of paths.

JGNII also is provides with optical testbeds both in Tokyo and Keihana. Field experiments of data-granularity-flexible reconfigurable OADM with wavelength-packet-selective switch [3] and optical quantum cryptosystem [4] have been conducted.

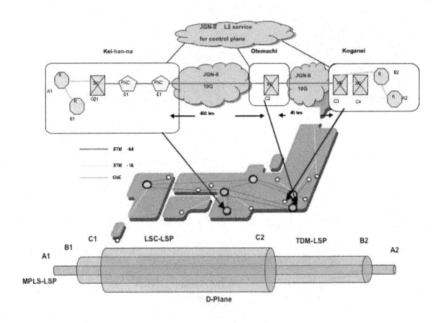

Fig. 5. GMPLS nation-wide interoperability test on Japan Gigabit Netowrk (JGN) II

3   Photonic Devices and Subsystems R&D Initiative

There has been a five-year long R&D project (2002-2006) for devices and subsystems, "Photonic Network Technology Development Project," funded by New Energy and Industrial Technology Development Organization (NEDO). The project goals are to develop advanced photonic components and demonstrate the feasibility of optical label-switched networking. The consortium was organized by Optoelectronics Industry and Technology Development Association (IOTDA). In the consortium University of Tokyo, KDDI Laboratory, Fujitsu, Mitsubishi Electric, Furukawa Electric, Hitachi, and Hitachi Cable in the consortium has demonstrated 40/10Gbps bit-rate transparent optical burst switching (OBS) [5]. In the experiment the bit-rate transparent switching and wavelength contention resolution as well as the optical label processing were performed. Using a PLZT-based 5x5 optical matrix switch, a 2 wavelength, 2x2 node was developed. A monolithically integrated MZI-SOA wavelength converter was optimized for single-arm 40Gbps operation, and was used for contention resolution in a shared loop-back configuration. A fast tunable DBR laser diode was used to switch wavelengths within a μs. The wavelength converter supported both 10 and 40Gbps PRBS 2^7-1 payloads under same bias conditions to simulate a network supporting multiple bit-rate payloads. Error-free (BER $<10^{-9}$) wavelength conversion and switching was achieved for less than 1dB and 4.8dB power penalty for 10 and 40Gbps payloads, respectively. The results underline the feasibility of bit-rate transparent fast optical networking.

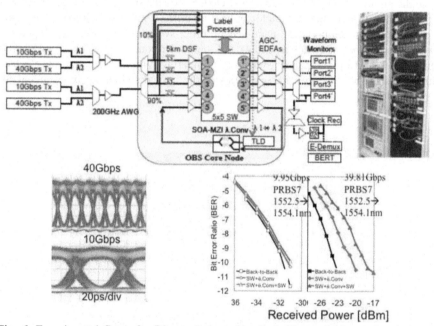

Fig. 6. Experimental Setup for Bit-rate Independent Switching and photo of optical router prototype, in addition to the BER measurement results for contention resolution.

4   Large-scale optical interconnect for UHPC

Peta-scale system interconnect (PSI) project [6] is one of the national projects on elemental technologies for ultra-high-performance computing systems, funded by the Ministry of Eduction, Culture, Sport, Science and Technology (MEXT). It is a six-year long project in the period of 2006-2011. The project works intensively on WDM optical packet interconnect to enables peta-scale supercomputers with over 10,000 compute-nodes. For example, 1 peta-flops system consists of 100 computing node groups, where each group has 200 CPUs and 10 Tflops capability, and the groups are interconnected by 1 T Byte/s class interconnection link. Thus, 100-Tbps class (100 port x Tbps class link) interconnection is required for the peta-flops system. The data granularity ranges from 64-bytes-length short data to several mega-byte bulk data, and the latency with in microsecond-order latency are necessary to obtain the high simulation speed. One of the big issues facing realization of such interconnection systems is the huge amount of interconnecting elements, especially switches, cables, and OE/EO modules. In the proposed network archtecture [7], an electric leaf switch is used for the switching of the intra-rack signals and the aggregation of the inter-rack signals by using packet queues for high bandwidth data interconnection. The aggregated signals are converted to optical packet signals and interconnected by the optical spine switch among computing node groups under the control of the arbiter. WDM packet switching is used to increase the bandwidth per switching port and reduce the number of OE/EO modules and cables. Broadcast and select type of optical switch with multi-gate semiconductor optical amplifier (SOA) switch to realize both large port counts (~256port) and nano-second order switching has been developed by Fujitsu. The number of switches, cables and the power consumption are significantly reduced by WDM optical packet switching comparing with conventional all-electric swithing system because the characteristics of the optical switch are independent of the signal bit rate.

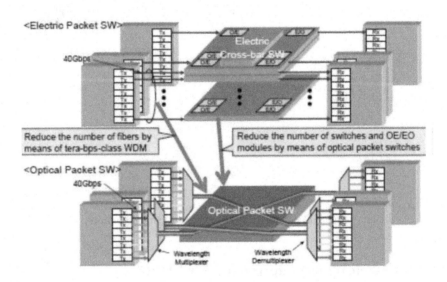

Fig. 7. WDM packet switched interconnect for UHPC.

5   Conclusion

Newly launched government-funded R&D initiatives of photonic networks in Japan has been presented. They aim at establishing the *photonic platform* to provide abundant bandwidth on demand, at any time and from anywhere, to promote bandwidth-rich networked applications. In addition, other government-funded R&D projects relevant to the photonic networks, including photonic devices and subsystems and large-scale optical interconnect for UHPC have also been described.

Acknowledgments. The author would like to thank S.Araki of NEC, O.Ishida of NTT Labs., T.Morioka of NICT, Y.Nakano of Univ. Tokyo, H.Onaka of Fujitsu, and A.Takada of NTT Labs. for providing materials for the manuscript.

References

1. K. Kitayama, T. Miki, T. Morioka, H. Tsushima, M. Koga, K. Mori, S. Araki, K. Sato, H. Onaka, S. Namiki, and T. Aoyama, "Photonic network R&D activities in Japan – Current activities and future perspectives," *IEEE J. Lightwave Technol.*, Vol.23, No.10, pp.3404-3418, 2005.
2. http://www.jgn.nict.go.jp/
3. N.Kataoka, N.Wada, F.Kubota, K.Sone, Y. Aoki, H.Miyata, H.Onaka, and K. Kitayama, "Field trial of data-granularity-flexible reconfigurable OADM with wavelength-packet-selective switch," *IEEE J. Lightwave Technol.*, vol.24, No.1, pp.88-94, 2006.
4. T.Hasegawa, T.Nishioka, H.Ishizuka, J.Abe, K.Shimizu, and M.Matsui, "Field experiments of quantum cryptosystem in 96km installed fibers," CLEO/Europe-EQEC, EH3-4-WED, Munich, 2005.
5. A. Al Amin, K. Shimizu, M. Takenaka, T. Tanemura, R. Inohara, K. Nishimura, Y. Horiuchi, M. Usami, Y. Takita, Y. Kai, Y. Aoki, H. Onaka, Y. Miyazaki, T. Miyahara, T. Hatta, K. Motoshima, T. Kagimoto, T. Kurobe, A. Kasukawa, H. Arimoto, S. Tsuji, H. Uetsuka and Y. Nakano, "40/10Gbps bit-rate transparent burst switching and contention resolving wavelength conversion in an optical router prototype," ECOC2006, Th.4.1.6 (Cannes, Sep.2006).
6. http://www.psi-project.jp/
7. H.Onaka, Y.Aoki, K.Sone, G.Nakagawa, Y.Kai, S.Yoshida, Y.Takita, K.Morito, S.Tanaka, and S.Kinoshita, "WDM Optical Packet Interconnection using Multi-Gate SOA Switch Architecture for Peta-Flops Ultra-High-Performance Computing Systems," ECOC2006 Tu4.6.6(Cannes, Sep.2006).

Optical Burst Switching Network Testbed

Wu Jian, Zhang Wei, Wang Minxue

Key Laboratory of Optical Communication & Lightwave Technologies, Ministry of
Education

Beijing University of Posts and Telecommunications, Beijing 100876, P.R.China

jianwu@bupt.edu.cn

Abstract: In this paper, several key technologies of optical burst
switching (OBS) network are implemented and experimentally verified.
The TCP performance over OBS is experimentally investigated on this
testbed and multi-QoS traffic transmission on the testbed are
demonstrated as well. Currently, an interconnection experiment
between OBS and GMPLS network are also reported.

Keywords: optical burst switching, TCP, QoS

1. Introduction

Optical burst switching (OBS) [1] has been receiving increased attention worldwide
as a promising technology for building the next-generation network. In OBS networks,
a control packet is sent ahead of time on control channel to reserve resource at
intermediate nodes for data burst, and then the associated data burst consisting of
multiple assembled IP packets are transmitted and switched along the configured
route all-optically in OBS networks.

To implement the OBS network, a lot of challenges must be solved. In physical
layer of OBS network, high-speed assembling and scheduling technologies are
required as well as burst-mode transmitter and receiver [2-4]. In higher layer, OBS
technology is designed for supporting future IP over wavelength division
multiplexing (WDM) network, and due to the domination of TCP traffic in current IP
traffics, it is very important to investigate the performance of TCP over OBS network.
Some simulation results show that the TCP performance is very sensible to burst loss
[5-6]. However, there are other different results showing that the TCP performance
over OBS network is much more robust even in the case of burst loss probabilities as
high as 1% [7].

Quality of service (QoS) provision issue becomes more and more important in
OBS networks with the emergence of applications with different demands. In general,
QoS can be provided by introducing differentiation at some point in the network, such
as differentiated offset times, differentiated burst assembly, and differentiated
scheduling [8-11].

In this paper, an OBS testbed is developed to provide a platform for investigating
above mentioned key technologies in OBS network [12]. Burst assembling and
scheduling are implemented with high-speed FPGA technology and burst-mode
transmitter and receiver are demonstrated as well with fast clock and data recovery

I. Tomkos et al. (Eds.): ONDM 2007, LNCS 4534, pp 166–175, 2007.

(CDR) time and high receiving sensitivity. Based on this OBS testbed, the TCP performance over OBS is experimentally investigated [13]. Experimental results show that the performance of TCP traffic is greatly degraded by burst loss probabilities. Moreover, attaching higher QoS level to TCP traffic is also contributive to the improvement of TCP performance. A multi-QoS traffic transmission is also demonstrated on the OBS network testbed [14], and puts an extra effort on TCP transmission over OBS to investigate its end-to-end performance.

2. OBS network testbed

Fig.1 shows the OBS network testbed. It consists of three edge nodes and four core nodes. Each edge node is connected with core node through a pair of fibre links in which three WDM data channels and one dedicated control channel are included. Bit-rate for all channels is 1.25Gb/s. Edge node provides user interface including 8 Fast Ethernet ports and 2 Gigabit Ethernet ports to access IP-based traffic. Input packets are classified according to their egress nodes and QoS demands, and then assembled into data bursts based on assembly algorithm. Assembled bursts are launched into OBS network following a control packet by an offset time. Core node functions as electrical processing of control packets and configuration of optical switching matrix according to the information provided by control packet. Finally, IP packets are disassembled out of bursts at egress edge node and dispatched to their corresponding destinations.

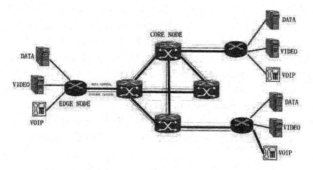

Fig.1. Optical burst switching network testbed

With the consideration of flexibility and scalability, burst assembler and scheduler in the testbed are implemented with embedded processor and high speed FPGA respectively. When latest available unused channel with void filling (LAUC-VF) [15] algorithm is employed, maximum processing delays of scheduler at core node and edge node are 2.5us and 10us respectively. Moreover, an optical switching fabric with switching speed less than 100ns is also developed [12].

3. Experimental Verification

When FF-VF algorithm is employed, Fig.2 shows scheduling result for bursts with three kinds of QoS demands (their offset times are 1750us, 900us and 50us respectively). The upper of this figure shows the BHPs on control channel, and the corresponding bursts are showed in the two different data channels at the bottom of this figure. Due to the fact that the load is different for traffic with different QoS demands, bursts with different lengths are generated from assembler. Only bursts with lower QoS demand, namely with shorter burst length, are assigned to low priority channel $\lambda 1$.

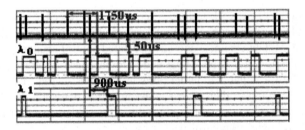

Fig.2. Scheduling results at edge node

Fig.3(a) shows the measurement results of the burst-mode receiver. In this OBS testbed, there are preambles of 200ns attached to the header of each data burst for fast recovery of clock and data at receiving end. It can be seen from this figure that it actually takes less than 80ns to recover the clock and data correctly form incoming data burst, which means that the synchronization preambles can be reduced further to improve the link utilization efficiency. Even in the case of the 200ns time interval of two neighbouring bursts, the two data bursts also can be received correctly, shown in Fig.3(a), that indicates the 10us guard band for data bursts is enough and also can be greatly reduced to improve transmission efficiency. In addition, the clearly opened eye-diagram of bursts at receiving power of -22dBm, which is shown in Fig.3(b), means the high receive sensitivity of the designed burst-mode receiver.

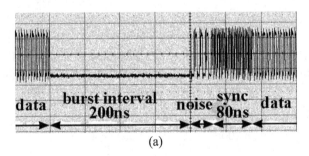

(a)

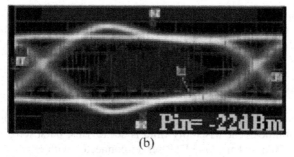

(b)

Fig.3. Receiving Bursts (a) and Eye-diagram(b)

4. TCP Transmission Experiments

The experimental setup for TCP transmission is shown in Fig.4. Only two edge nodes and one core node are used for simplicity. Standard interface such as Fast Ethernet and Gigabit Ethernet are provided to access IP-based traffic. In this experiment, burst assembly algorithm with fixed time and length threshold is applied, and for burst scheduling algorithm LAUC-VF is used.

As shown in Fig.4, a TCP server is connected to one edge node and the client is connected to the other one, by which a TCP connection can be set up through the OBS network. Four Gigabit Ethernet ports of the IP data quality analyzer are connected to the two edge nodes respectively, by which controllable traffic load is provided to simulate TCP traffics with different QoS requirement in OBS network. In the TCP transmission, a maximum congestion control window of 64KBytes and maximum segment size of 512Bytes are employed in the TCP server.

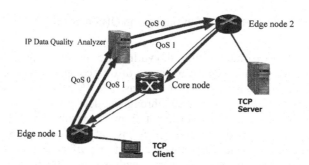

Fig.4. TCP transmission experiment in OBS testbed

5. Experimental results and discussions

In this paper, the TCP performance over OBS is experimentally investigated on OBS testbed [5][6]. Experimental results show that the performance of TCP traffic is influenced in the case of burst loss, both for ACK burst and data burst. But ACK burst loss affect less than data burst loss on TCP throughput performance, caused by TCP slide Window control mechanism. Moreover, considering that QoS provision is provided in OBS network, contention resolution mechanism of dropping ACK burst when network is busy is proposed in further investigations.

5.1 TCP throughput

In TCP transmission experiment, the network is configured to dropping bursts at certain loss probabilities. The assembly time is set to 1ms, burst length is 90kbytes and offset time is fixed to 1ms. One FTP server connected with edge node 2 is set to send data to TCP client on edge node 1 through OBS testbed. The burst loss probability varies from 1% to 0.1% to investigate the TCP transmission performance in OBS network.

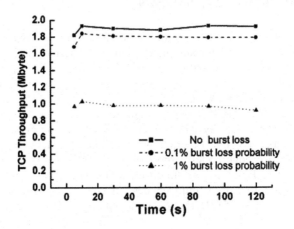

Fig.5. TCP throughput in OBS network

According to the experiment results shown in Fig.5, burst loss probability of 0.1% only causes an average throughput decreasing of 6%, compared with the case of no burst loss. As the burst loss probability increased to 1%, the performance of TCP throughput degrades greatly. The average throughput decrease to 50% compared with the case of no burst loss. This is different from the result in [7], in which TCP throughput only decreased 5% in average in case of burst loss probability of 1%. Further experiments are made about how the burst loss affect TCP throughput as the burst length changed. In this case, the burst length is set to 200kbytes and 400kbytes respectively.

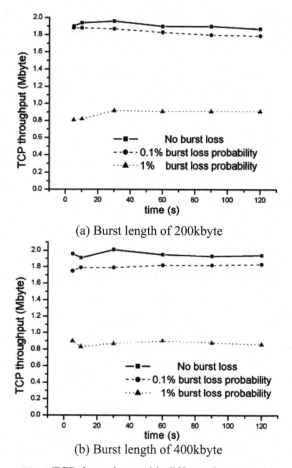

(a) Burst length of 200kbyte

(b) Burst length of 400kbyte

Fig.6. TCP throughput with different burst length

Fig.6.(a) and (b) show that as the burst length increased to 200kbyte and keeping constant burst loss probability of 0.1%, only 3.7% TCP throughput decreasing is achieved which means better TCP performance is obtained. As increasing the length to 400kbyte, the average decreased throughput is only 8%. It can be explained that there is a tradeoff between burst lengths when transmitting TCP flows. Bandwidth efficiency can be improved as the burst becomes longer, but TCP sources synchronization will decrease when lose large burst.

From this experiment it can be concluded that OBS network is supposed to be robust in TCP traffic transmission under burst loss probability below 0.1%.

For TCP transmission over OBS network, TCP setup, ACK packets and data segments are all buffered and then assembled into bursts at the edge node, and this causes to the increasing of round trip time (RTT) in TCP layer and results in the decrease of available TCP bandwidth, that is so called delay penalty. This conclusion is proved by both of experimental results and simulation results. As shown in Fig.7, with the increasing of assembling time, the RTT in TCP layer increases linearly and causes the decreasing of TCP bandwidth from 86Mb/s to about 25Mb/s for one TCP thread. Even double the TCP threads, the TCP bandwidth still decreases to less than 50Mb/s. In this experiment, the RTT is roughly measured with Ping command.

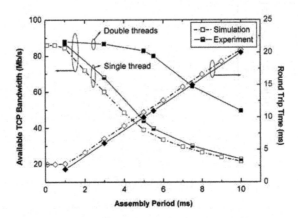

Fig.7. TCP bandwidth decreasing due to delay penalty

5.2 QoS provision

In QoS provision experiment, two classes of QoS traffics are set in IP data quality analyzer, namely QoS0 and QoS1, and QoS0 is a higher QoS class. Transmission performance is recorded for each of traffics, and comparison is given in Fig.8(a) and (b).

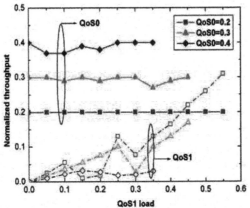

(a) Impact of low QoS traffic to high QoS traffic

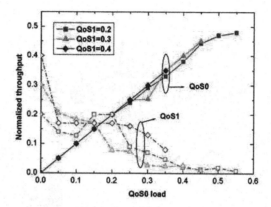

(b) Impact of high QoS traffic to low QoS traffic

Fig.8. QoS provision transmission in OBS network

In Fig.8(a) the normalized throughput of QoS0 is set to 0.2, 0.3 and 0.4, and then the throughput of QoS1 is increased slowly. Results show that low QoS traffic doesn't affect the throughput of high QoS traffic. On the contrary, it is shown in Fig.8(b) that high QoS traffic will take bandwidth away from low QoS traffics. Results from this experiment show that by configuring to high class QoS, TCP throughput can be guaranteed.

6. Interconnection Experiment between OBS and GMPLS network

Generalized Multi-Protocol Label Switching (GMPLS) is one of enablers to provide flexible control and management for carrier's optical transport networks in near future. In the meantime, Optical Burst Switching (OBS) has become a promising technology for the next generation optical network with advances in ultra-fast switching technologies and optical burst control techniques. Considering such current situations, it is very important to investigate the network architecture interconnected between OBS domains via GMPLS-controlled core transport network [16].

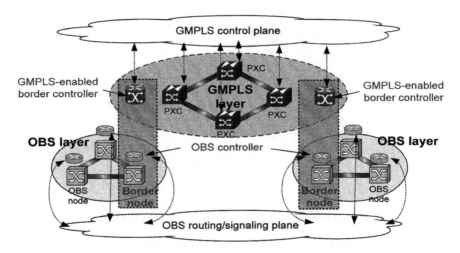

Fig.9. Scenario of interconnection between OBS and GMPLS network [16]

During the submission of this paper, the experiment is still under developing. The result will be reported on the meeting.

7. Conclusions

In this paper, several key technologies of OBS network are addressed. Burst assembling and scheduling are implemented with high-speed FPGA technology with high processing speed. Burst-mode receiver with high receiving sensitivity and fast clock and data recovery time (less than 80ns) is implemented. All other modules are implemented and experimentally verified in this testbed, which means a flexible and scalable OBS testbed is demonstrated and can provide a flexible platform for further research works.

Moreover, the TCP performance over OBS is experimentally investigated on this testbed and multi-QoS traffic transmission on the testbed are demonstrated as well. Experimental results show that the performance of TCP traffic is greatly influenced in the case of 1% burst loss probabilities. With the decreasing of burst loss probabilities to 0.1%, the TCP performance is improved greatly, which means that the TCP performance over OBS can be guaranteed in the case of 0.1% burst loss probabilities. Experimental results verify QoS provision in OBS network which means the TCP throughput of high QoS traffic can be guaranteed. Delay penalty observed in experiment means that the transmission protocols in high layer needs to be modified for a higher transmission bandwidth in OBS network. At last, a joint project of interconnection of OBS and GMPLS network is reported for further investigation.

References:

1. C. Qiao and M. Yoo, "Optical burst switching (OBS) – A new paradigm for an optical internet," J. High Speed Networks, vol. 8, no. 1, pp. 68-84, 1999.

2. Lv Z., Wang H., Zhang M., "Longest Queue First Scheduled Assembly for Optical Burst Switching Networks", Acta Photonica Sinica , 2006 , 35 (1) :74~77

3. Pan Y., Ye PD., "Analysis for the Effect of Optical Packet Assembly Mechanism on Self2similar Traffic Shaping", Acta Photonica Sinica , 2005 , 34 (9) :73~77

4. Guo H., Wu G., Zuo Y., "Edge Router Design for Hybrid Optical Burst Switching and Optical Circuit Switching" , Acta Photonica Sinica , 2006 , 34 (8) :38~42

5. Sunil Gowda , Ramakrishna K Shenai , Krishna MSivalingam, Hakki Candan Cankaya, "Performance Evaluation of TCP over Optical Burst-Switched (OBS) WDM Networks," IEEE ICC 2003

6. Detti and M. Listanti, "Impact of Segments Aggregation on TCP Reno Flows in Optical Burst Switching Networks," Proceedings IEEE, INFOCOM 2002, New York, NY, June 2002

7. Robert Pleich, "Performance of TCP over Optical Burst Switching Networks", ECOC'05.September 2005.

8. M. Yoo, C. Qiao, and S. Dixit. QoS performance of optical burst switching in IP-over-WDM networks. IEEE Journal on Selected Areas in Communications, 18(10):2062–2071, October 2000.

9. V.M. Vokkarane, Q. Zhang, J.P. Jue, and B. Chen. Generalized burst assembly and scheduling techniques for QoS support in optical burst-switched networks. In Proceedings, IEEE Globecom, volume 3, pages 2747 - 2751, November 2002.

10. V. M. Vokkarane and J. P. Jue. Prioritized burst segmentation and composite burst assembly techniques for QoS support in optical burst switched networks. IEEE Journal on Selected Areas in Communications, 21(7):1198 - 1209, September 2003.

11. F. Farahmand and J. P. Jue. A preemptive scheduling technique for OBS networks with service differentiation. In Proceedings, IEEE Globecom, December 2003.

12. Guo H., Lan Z., Wu J., "A Testbed for Optical Burst Switching Network," OFC 2005, March 2005

13. Zhang W., Wu J., Lin JT. "TCP performance experiment on OBS network testbed", OFC2006, March 2006.

14. Guo H., Wu J., Liu X., "Multi-QoS Traffic Transmission Experiments on OBS Network Testbed," ECOC'05.September 2005

15. Y. Xiong, M. Vandenhoute, and H. Cankaya, "Control Architecture in Optical Burst-Switched WDM Networks," IEEE JSAC, vol. 18, Oct. 2000, pp. 1838–51.

16. Guo H., "Joint Research Project on Photonic Network: Interworking of OBS and GMPLS", 2006

TCP traffic analysis for timer-based burstifiers in OBS networks

Kostas Ramantas[1], Kyriakos Vlachos[1], Óscar González de Dios[2] and Carla Raffaelli[3]

[1]Computer Engineering and Informatics Dept., and Research Academic Computer Technology Institute, University of Patras, Rio, Greece (email: _kvlachos@ceid.upatras.gr_)
[2]Telefónica I+D, Emilio Vargas 6, Madrid, Spain
[3]Dep. of Electronics, Computer Science and Systems, University of Bologna, Italy

Abstract. The purpose of this paper is to assess the impact of timer-based burst assembly algorithms for TCP traffic. We present an analysis for short, medium and long assembly times and investigate segment and flow distribution over the assembled bursts. Further, we also analyze their impact on the congestion window evolution and on the effective throughput achieved. It has been found out that short assembly times are ideally suitable for sources with small congestion windows, allowing for a speed up, while large assembly times yield a lower throughput variation among the individual assembled flows. For long assembly times, the transfer of more segments from the same source is trading off the increase of the burstification delay but no throughput gain is obtained. However, large assembly times smooth out individual flow performance and provide a significant lower variation of throughput. To this end, in this paper, we propose a new adaptive burst assembly algorithm that dynamically assigns flows to different burstifiers based on their instant window size.

Keywords: Optical burst switching, transport control protocol, burst assembly

1. Introduction

Optical burst switching (OBS) [1] has been introduced to combine both strengths of packet and circuit switching and is the most promising technology for next generation optical Internet. An OBS network consists of a set of optical core routers, with edge routers at its edges that are responsible for the burst assembly/disassembly function. In OBS networks, an optical burst is constructed at the network edge, from an integer number of variable size packets. Two distinct burst assembly algorithms have been proposed in the literature: the *timer-based* and the *threshold-based*. In the timer-based method, also denoted as T_{MAX} in the literature, [2],[3] a time counter starts any time a packet arrives and when the timer reaches a time threshold (T_{MAX}), a burst is created; the timer is then reset to 0 and it remains so until the next packet arrival at the queue. Hence, the ingress router generates periodically bursts, every T_{MAX} time, independently of the yielding burst size. In the second scheme, [4], a threshold is used to determine the end of the assembly process. In most cases the threshold used is the burst length denoted in the literature as B_{MAX}. In that case, bursts are thought as containers of a fixed size B_{MAX}, and as soon as the container is completely filled with data, the burst is transmitted.

I. Tomkos et al. (Eds.): ONDM 2007, LNCS 4534, pp 176-185, 2007.

The timer-based method limits the delay of packets to a maximum value T_{MAX} but may generate undesirable burst length, while the burst-length based method generates bursts of equal size, but may result in long delays when the traffic load is light. To address the deficiency associated with these assembly algorithms, hybrid (mixed time/threshold based) assembly algorithms were proposed [5], where bursts are created when either the time limit or the burst-size limit is reached, whichever happens first. Apart from the aforementioned assembly schemes, other more complex schemes have been also proposed, which are usually a combination of the timer - based, and the threshold-based methods [6].

The performance of TCP over OBS networks has been studied in previous works [7]-[9] where it has been observed that the burst assembly process at the edge nodes has a significant impact on the end-to-end performance of TCP, mainly because it introduces an unpredictable delay, [10], that challenges the window mechanism used by TCP protocol for congestion control. TCP performance is also challenged by the burst lost ratio that results in multiple segment losses for multiple sources. A useful insight on TCP traffic statistics is given in [11],[12]. In particular, it was found that short assembly times yield a higher throughput to TCP sources primarily because they reduce the total end-to-end delay associated with the round trip-time delay. Long assembly times, are more efficient especially for *fast* TCP flows [11], since they allow the transmission of multiple segments from the same flow per burst. However, this throughput gain may be canceled by the large burstification delay.

In this paper, we present a thorough analysis of TCP traffic over OBS networks. We first analyze how segments and flows are distributed over the assembled bursts for various assembly times and further analyze their effect in the number of transmitted and lost bursts per flow. It is shown that the short assembly times result in a significant increase in the number of bursts needed for a transfer completion, while for larger assembly times, this is relative constant. Furthermore, it is investigated how congestion window increases and what its effect is in average throughput. We argue in this paper that the characterization of a flow as slow, medium or fast depends on its instant congestion window size and we show that a mix burst assembly timer, where burstification delay varies with the size of the congestion window yields a higher throughput together with a smaller variance. For the performance evaluation and traffic analysis, we have developed a comprehensive and detailed TCP over OBS simulator based on ns-2 tool, capable of simulating ~hundreds of active TCP sources per edge node.

The rest of the paper is organized as follows. Section 2 provides an overview of TCP variants and most suitable assembly schemes, whiles Section 3 presents a flow/segment analysis of TCP traffic over OBS. Section 4 discusses the effect of burstification delay on the congestion window expansion and the yielding throughput, and finally Section 5 presents the performance of a new assembly scheme based on the flow congestion window size.

2.   Overview of TCP variants and aggregation schemes for OBS networks

There are a number of TCP versions such as Tahoe, Vegas, Reno, New Reno and SACK, combined with a number of different burst assembly strategies. The most interesting are the three last ones. The main differences among them are the algorithms that they employ when congestion is detected. TCP Reno refers to TCP with *Slow Start, Congestion Avoidance, Fast Retransmit* and *Fast Recovery* algorithms. When Reno starts, it enters the Slow Start phase first with a congestion window of one segment size and an exponential increase, upon the acknowledgement of all the packets transmitted. When the window reaches a certain threshold of w, it enters the Congestion Avoidance phase, according to which the window is now increasing only by one segment after all segments have been acknowledged.

In TCP Reno, there are two kinds of losses identified; the *Time Out* (TO) and *Triple Duplicate* (TD) loss. In the *Triple Duplicate* (TD) case, the sender receives three duplicates ACKs, that acknowledge a new segment, but not the one with the highest sequence number. In that case TCP Reno enters the Fast Retransmit phase, and starts transmitting the lost segments. For every successful transmission of these segments, the sender halves its congestion window and receives a TD ACK message for the next lost segment in the burst. In Reno, the maximum number of recoverable segment losses in a congestion window without timeout is limited to one or two segments in most cases. In the case of a *Time Out* (TO) loss case, no ACK is received in a certain time period, denoted by the expiration of a timer. In that case TCP Reno enters the Slow Start phase, and resets its window back to one segment size. TCP New Reno is a slight modification according to which the sender retransmits one lost segment per round-trip-time upon receiving a partial ACK message, without waiting for a TD ACK and without halving its window until all lost segments are successfully acknowledged.

On the other hand, SACK (Selective Acknowledgment) TCP implements a different ACK message, where the non-contiguous set of data that have been received are stored. To this end, the sender is aware of the lost packets which are transmitted altogether. In that case the congestion window is halved, before linearly increasing again. Detailed SACK performance in OBS networks is clearly superior, as shown in [11], since all the segments that were employed in a burst that was dropped can be identified and subsequently retransmitted at the same round.

TCP's performance (e.g., throughput) depends heavily on burst assembly time due to the extra delay enforced (denoted as burstification delay). Therefore TCP mechanism adjusts its window mechanism upon a burst transmission or reception and thus timer-based assembly schemes may perform better than size-based algorithms. For the timer (T_{MAX}) threshold there could exist an optimal value that maximizes throughput performance in a TCP over OBS network [11]. In [4], it has been shown that optimal performance can also be achieved with an optimal burst length algorithm, while in [6], it is shown that a dynamic assembly algorithm that adjusts the threshold values (e.g., time, burst-length or both) according to traffic statistics can achieve an even better performance.

3.   Segment and Flow distribution

In this section, a segment and flow analysis of TCP traffic is presented when OBS is used as the underlying transport technology. Work presented hereinafter concerns TCP-SACK variant and timer-based assembly schemes that although it is the most promising combination, very few works exist on providing an in-depth analysis of how segments, flows and their parameters vary with burstification delay. We have developed a dedicated TCP-over-OBS simulator using ns-2 platform capable of

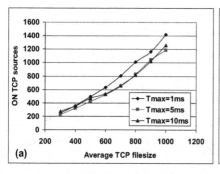
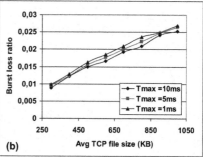

Fig. 1. (a) Number of simultaneous active TCP sources at each edge node versus average file size. (b) Corresponding burst loss ratio.

simulating ~hundreds of active sources per edge node. Such a scenario is close to reality, but requires significant amount of memory and CPU resources. We have modified the raw ns-2 code to efficiently manipulate TCP flows and available CPU resources. The experiments were carried out on the NSF network topology, with 8 edge and 6 core nodes, where each link was employing a single wavelength at 10Gbps. Access rate was set to 100Mbps, equal for all sources. TCP arrivals were modeled with an exponential process with a mean of λ=50 flows/sec, while flow file size was modeled with a Pareto distribution of p load and a minimum ON size of 40KByte. Using this set of metrics, it was possible to vary the TCP arrival rate and/or the mean file size, and obtain measurements for a different number of active sources. Fig. 1(a) displays the number of active sources at each edge node versus the average file size for three different assembly timers namely 1, 5 and 10*msec*, while Fig. 1(b) displays the corresponding burst loss ratio. Results shown hereinafter correspond to the steady state (constant number of active sources) of the experiments, which took place after 200sec of simulation time.

From Fig. 1(a), it can be seen that the number of active sources in the case of 5 and 10*msec* timer is close, while in the case of 1*msec*, it is 200 more for all flow sizes above 600KB. It must be noted here that the number of active sources measured is a dynamic parameter of the simulating experiment that can vary when burst losses occur and thus we argue that this corresponds to a real, instant picture of the network under study. An important issue noticed is that the simulating scenario was entering the steady state much earlier in the case of 5 and 10*msec* timers, than in the case of 1*msec*.

In what follows we have selected a mean file size of 700KB, which corresponds to a burst loss ratio of 2% and 600 or 800 active source respectively for 5, 10 and 1*msec* timers. Using these as reference, we have measured two basic statistics; the distribution of segments and the distribution of flows over the assembled bursts. Fig. 2 displays the cumulative density function (CDF) of (a) the number of segments and (b) the number of different TCP flows per transmitted burst. The results shown correspond to all the bursts transmitted in the network, for all source-destination pairs. Fig. 2 provides a useful insight of how many TCP sources will experience a segment loss from a single burst loss as well as how many segments will be lost. The exact number of lost segments per source is the conjugate probability of Fig. 1(b), Fig. 2(a) and Fig. 2(b).

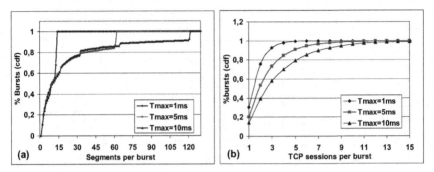

Fig. 2. (a) Segment and (b) flow distribution over the transmitted bursts for 1, 5 and 10*msec* burstifiers.

The sharp increases in Fig. 2(a) are due to the finite access rate that does not allow more segments to be sent within the burstification time delay. To this end, in the case of 1*msec* assembly time up to 13 segments can be loaded onto a single burst, independent of the flow window size. If TCP sources have more segments to send, these are transmitted with a next burst. To this end, it can be easily derived that the congestion window of a high number of sources is incompatible larger than the 1*msec* assembly time. For larger assembly times, this sharp increase is shifted to higher numbers of segments.

Furthermore from Fig. 2(b) and in the case of 1*msec* timer, it can be seen, that the 80% of the transmitted bursts employ segments from only 2 different sources, while for 5 and 10*msec*, this increases to 4 and 6 respectively. To this end, it is clear that large aggregation times may result in the transfer of a higher number of sources and segments per burst that in turns may lead to smaller completion times, fewer bursts generated but with the trade off that more sources will be potentially affected. The latter is the side effect when upon a burst loss a higher percentage of sources (see Fig. 2(b) will face a multiple segment loss. Thus, more sources will halve their window and then either time-out and enter a slow-start phase or try to recover from the loss and linearly increase their window. However, at which point the network will balance is still unknown but it is reflected in the segment and flow distribution, shown in Fig. 2 and the number of active sources, shown in Fig. 1.

We have also measured the actual number of bursts (transmitted and lost) per flow which are necessary to complete a flow transfer. Fig. 3(a) and (b) shows the corresponding cumulative density functions (cdf) for a single (randomly selected) source-destination pair, independently of the flow size. The selection of a single source-destination pair allows for fair conclusions since all bursts are transmitted over the same network path and thus with the same round-trip-time delay and blocking probability.

From Fig. 3, it is clear that small assembly times result in a significant increase in the number of bursts needed to a complete a transfer, while the lost bursts per flow vary much less. In particular, in the case of 1*msec* timer, 80% of the flows need on average up to 80 bursts to complete their transfer, while only 40 and 34 are needed in the case of 5 and 10*msec* respectively. The corresponding lost bursts are 5 and 4. Of course the number of bursts needed to complete a transfer depends heavily on the

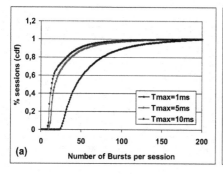
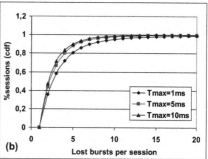

Fig. 3. Distribution of (a) number of bursts needed to complete a TCP session and (b) number of lost bursts per session for 1, 5 and 10*msec* burstifiers for a single source-destination pair.

amount of data to be transferred. Therefore, in Fig. 4, we have further analyzed the transmitted and lost bursts per flow with respect to the corresponding flow size.

From Fig. 4(a), it can be seen that the number of transmitted bursts increase almost linearly with flow size, while 1*msec* curve diverges rapidly. In particular, for the maximum flow size of 2MB, 3.5 times more bursts are needed in the case of 1*msec* timer. This was however expected, since the corresponding 80% of that bursts were carrying less than 13 segments from only 2 flows at most. However, this does not result to smaller throughputs or higher completion times, as shown in the next section. The behavior of the number of lost bursts per flow size is similar (see Fig. 4(b)) but the difference between the curves diverges at a smaller rate.

Based on the above analysis, we may conclude that small assembly times increase the network overhead but potentially constrain individual flow performance due to limited number of segments per flow per burst transmitted. The latter may result to longer file transfer times that in turns leads to more flows remaining active. However, on average burstification delay is by default smaller and thus it is expected that short assembly time to result in higher throughputs as discussed in the next section. Further, they may also result to less *Time-Out* events since it is unlikely a single flow to employ its complete window onto a single (lost) burst. On the other hand, large assembly times service more flows at a time, carrying more segments from each individual flow. This smoothes out any traffic instabilities in the sense that individual flow throughput is absorbed and diluted but however impose such a burstification

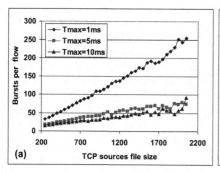
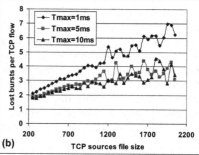

Fig. 4. Number of bursts needed to complete a TCP session and (b) number of lost burst per session versus file size for 1, 5 and 10*msec* burstifiers for a single source-destination pair.

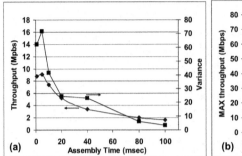 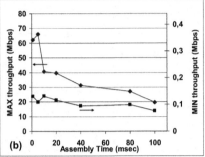

Fig. 5. (a) Average and variance of throughput of all flows of a single source-destination pair versus burst aggregation time. (b) Corresponding maximum and minimum values measured.

4.   Efficient Throughput and Congestion Window expansion

In order to qualitatively investigate the performance of the different assembly timers, we have measured the average throughput achieved along with its variance, its maximum and minimum value. In general, TCP performance depends on the TCP variant used in combination with the number of segments lost and further depends on the flow access rate. A TCP source is characterized as *slow*, *medium* or *fast*, depending on its access rate. From [11], it can be inferred that a flow with a slow access rate (*slow flow*) must be accompanied with a small assembly time while flows with a large access bandwidth (*fast flows*) with relatively larger assembly timers. However, in most cases, access rate for all flows at the edge node is the same and service is differentiated by other means.

Fig. 5(a) displays the average throughput and variance for all flows of a single source-destination for different burstifiers, while Fig. 5(b) shows its maximum and minimum value. From Fig. 5(a), it can be seen that there is no throughput gain for large assembly times albeit *delay first loss* (DFL) gain is maximized [11]. This is also clear by Fig. 5(b), where maximum throughputs higher than >60Mbps were measured only for 1 and 5*msec* timers. As a result, flow transfer time increases with assembly time and specifically, it was found to increase from 1.25*sec* for 1*msec* timer to *1.6*sec and 6,7*sec* for 10*msec* and 100*msec* respectively. However, performance (throughput) variance drops also fast with assembly time (see Fig. 5a) and this is because individual flow performance is diluted via the long burstification delay process. To this end, we may argue that large assembly times can provide a higher notion of fairness among the aggregated flows and smooth out any performance instabilities.

In order to further analyze this, we have measured the evolution of the congestion window of three lossless flows transmitted over the same source-destination route and with a similar file size. Fig. 6(a) displays in detail the rising edge of their congestion window, while Fig. 6(b) displays the sequence number (modulo 132) of the transmitted segments. The results shown correspond to 1*msec* (left column), 5*msec* (middle column) and 10*msec* (right column) burstifiers.

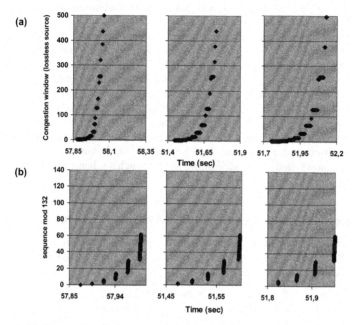

Fig. 6. (a) Congestion window evolution and (b) Sequence number of the transmitted segments under identical timescales of three different, lossless flows with similar files sizes for 1*msec* (left column), 5*msec* (middle column) and 10*msec* (right column) assembly times.

From Fig. 6(a), it can be seen that the congestion window rises faster for 1*msec* timer rather than for 5 or 10*msec*. In this particular case, a 500segment window is reached within 0,171*sec*, while for 5 and 10*msec* timers in 0,290 and 0,38sec. The highest speed up gain is noticed for a window increase from 1 to 100segments. This is the result of a higher number of burst transmissions over the same time period. In particular, in the case of 1*msec*, 6 bursts are transmitted within 0,15*sec* (see counts of segment sequence number in Fig. 6(b), left column), while only 5 and 4 in the rest.

A significant finding is that the maximum instant throughput measured for all three burstifiers was 92Mbps, and which was achieved when window size was 128segments wide for all cases. However this was obtained much faster in the case of 1*msec*, and thus the average yielding throughput was much higher. In particular, average throughput was found to be 23Mbps whereas only 17 and 14Mbps in the cases of 5 and 10*msec* timers. However, it must be clearly noted that the yielding, instant throughput was equal for all three different timers and we may argue that the resulting average throughput actually depends on the flow size and the burst per flow losses.

A second important finding is that after having reached the maximum throughput and before flow completion, instant throughput was constant to its maximum value (92Mbps) only in the case of the 10*msec* burstifier. In the rest and especially for 1*msec* burstifier, this was instantly varying from 50Mbps to 90Mbps. This instability was primarily due to fact that segment distribution over the assembled bursts was not constant due to the short burstification delay. On the contrary, this was varying significantly and thus a steady segment per burst transmission rate could not be obtained albeit the higher number of burst created.

It is therefore clear that measuring averaging throughputs is not indicative for TCP performance. Further, a fixed assembly time does not provide maximum performance but only optimal performance for individual flows with similar characteristics (i.e. flow size, loss ratio, etc.). Large assembly times offer an advantage of carrying more segments but only when flows can send more segments, otherwise they constrain throughput performance. In order to truly enhance TCP performance, the instant congestion window is a metric to be considered for determining the optimum assembly time. For example, short assembly times should be applied to sources with a relatively small congestion windows (<100segments), while larger timers to sources with larger windows. In the next section, we analyze such a scheme and investigate its performance gains.

5.   Congestion Window-based Burst Assembly Scheme

In this section, we propose a new adaptive burst assembly scheme that assigns different burstification delays to flows based on their congestion window size. In particular, we define burst assembly time (T_{BAT}) as follows:

$$T_{BAT} = \begin{cases} 1\,msec & if\ 1 \leq congestion\ window < B\ segments \\ 5\,msec & if\ B \leq congestion\ window < C\ segments \\ 10\,msec & if\ C \leq congestion\ window \end{cases}$$

In our scheme, we propose the characterization of a flow as *slow* or *medium*, when its instant congestion window is less than B or C segments wide and as a *fast* flow when it is even higher. Thus, *slow* TCP flows with a congestion window of less than B segments are aggregated together under 1*msec* delay. When their congestion windows reach the limit of B segments, the flows are upgraded to *medium* rate flows and their segments are assembled under a 5*msec* timer. Similarly, when their congestion windows reach the C segments limit, these flows are upgraded again to *fast flows* and their assembly time is increased to 10*msec*. In this way, each flow is treated separately and thus upon a burst loss only the flows that will suffer from a segment loss will be downgraded to *medium* flows or even to *slow* flows if they time out.

The implementation of the proposed assembly scheme requires three different queues per destination, one for each type of flow, as well as the communication of the window size to the burstifier. Albeit the latter require a modification of the TCP mechanism, we have implemented the scheme in ns-2 platform to particularly measure the yielding throughput and variance for various B, C values. TABLE 1 summarizes our findings, where the performance of the simple cases of 1, 5 and 10*msec* burstifiers is also denoted for comparison. The cases with no B value means that the intermediate transition to 5*msec* assembly is omitted. From TABLE 1, it can be seen that best performance is obtained for B, C values of 32 and 100 segments. In that case, average throughput is increased to 11Mbps, while variance is dropped significantly to 50. This combination merges the optimum operation points of all three cases providing fast transmission times for *slow* sources that are in an early slow-start phase, *medium* rates for flows with up to 100 segments window size and slow transmission rates, (large assembly delays) for the rest.

TABLE 1. Average throughput and variance for various combinations of the proposed congestion-window based, burst assembly scheme

B	C	Average Throughput (Mbps)	Variance
100	200	8,4	60
32	100	11,2	50
-	32	9,1	55
-	200	7,7	59
1msec		8,8	62
5msec		9,0	71
10msec		7,4	41

In the rest cases, performance either approximates the performance of 1*msec* timer, when B is too high ($B = 100$, $C = 200$) or 10*msec*, when C is too high ($B = C = 200$). However, best performance is obtained upon the optimum combination of both values. It must be noted here however, that these depend on the actual flow and segment distribution and thus can be different for different arrival rates and burst loss ratios.

6. Conclusions

In this paper, an analysis of TCP traffic over OBS networks was presented for various timer-based burstifiers. It was found that short assembly times provide a higher average throughput but result to a significant performance variation of the individual aggregated flows. On the other hand, large assembly times are capable of smoothing out performance differences but eventually lead to poor throughputs. However, it was found that the yielding instant throughputs were the same in all cases and large assembly times are beneficial only when there are pending segments to be sent. Otherwise they delay burst transmission unnecessarily. In order to truly enhance individual TCP performance, we proposed a mixed timer-based algorithm that assigns flows to different timer-based burstifiers based on their instant window size. It was found that short assembly times provides a significant performance speed up when windows sizes are less than 100segments, and thus a mixed algorithm can combine the advantages of both a fast throughput increase along with a low performance variation among the individual flows.

Acknowledgments. This work has been supported by EC through the NoE E-Photon/ONe+ project via the Joint Project on *Optical Burst Switching* (JP-B).

References

[1] C. Qiao and M. Yoo, J. High Speed Networks, vol. 8, no. 1, pp. 69–84, 1999.
[2] F. Callegati and L. Tamil, IEEE Commun. Lett, Vol 4, pp. 98-100, Mar. 2000.
[3] M. Düser and P. Bayvel, J. Lightwave Technol., vol. 20, pp. 574–585, Apr. 2002.
[4] V. Vokkarane, K. Haridoss, and J.P. Jue, in Proceeding of Opticomm, pages 125-136, 2002.
[5] X. Yu, Y. Chen, and C. Qiao, in Proc. Opticomm, 2002, pp. 149–159.
[6] X. Cao, J. Li, Y. Chen, and C. Qiao, in Proc. IEEE GLOBECOM, vol. 3, pp. 2808–2812, Nov. 2002,.
[7] M. Casoni, E. Luppi and M. Merani, in Proceedings of Workshop on Optical Burst Switching.
[8] S. Malik and U. Killat, in Proc. of ONDM, 2005.
[9] A. Detti and M. Listanti, in Proc. IEEE, INFOCOM 2002.
[10] M. Izal and J. Aracil, IEEE Globecom 2002, Taipei, Taiwan, November 2002.
[11] Xiang Yu et al. J. of Lightwave Technology, vol. 22, no. 12, pp. 2722 – 2738, Dec. 2004.
[12] Óscar González de Dios, Ignacio de Miguel, Víctor López, in Proceedings of ONDM 2005.

TCP performance experiment on LOBS network testbed

Wei Zhang, Jian Wu, Jintong Lin, Wang Minxue, Shi Jindan

Key Laboratory of Optical Communication & Lightwave Technologies, Ministry of
Education

Beijing University of Posts and Telecommunications, Beijing 100876, P.R.China

covery_zhang@hotmail.com

Abstract: An LOBS testbed is briefly introduced and TCP
performance over OBS is experimentally investigated. Simulation and
experimental results show that TCP performance is displayed
differently with the loss of ACK (TCP Acknowledgement) burst and
data burst.

Keywords: OBS, TCP, Data Burst, ACK Burst, Loss Probability

1. Introduction

Labeled Optical Burst Switching (LOBS) [1], combining the best of multi-protocol
label switching and optical burst switching, has received an increasing amount of
attention from both academia and industry worldwide. As the main part of current
Internet traffics, TCP performance is directly relative to the performance of whole
network. Moreover, it is supposed that TCP will still play a dominant role in the
future IP over WDM network, thus it is very important to investigate the behavior of
TCP flows over OBS network.

Up to now, most researches about TCP performance over OBS concern the
influence of TCP data burst loss on network throughput and the ACK signal link is
always supposed to be lossless, which is still far from the actual situation. So it is
important to study the influence of ACK burst loss on network performance
comparing with data burst loss[2][3][4].

In this paper, we present an LOBS testbed and TCP performance over OBS is
investigated both by simulation and experiment [5][6]. Individual experiments of data
burst loss and ACK burst loss are carried out to test the influence on OBS network
performance. Experimental results show that at the same burst loss probability, the
loss of data burst result in worse network performance comparing with ACK burst
loss. To lowering the bad influence form traffic contention when network is busy,
attaching higher QoS level to data burst traffic is also contributive to the improvement
of TCP performance.

I. Tomkos et al. (Eds.): ONDM 2007, LNCS 4534, pp 186-193, 2007.

2. LOBS Testbed Architecture

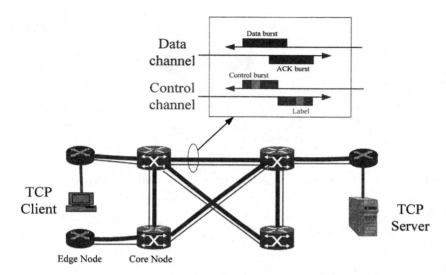

Fig.1. Experimental setup in LOBS testbed.

The LOBS testbed is upgraded based on [1] and enhanced for many aspects. As shown in Fig.1, the testbed is a mesh network comprising three edge nodes (EN) and four core nodes (CN). Each fiber link carries 2 data channels and 1 dedicated control channel, and each channel is 1.25Gbps. At present, each EN supports 4 GE and 8 FE ports to access IP traffic with various QoS demands. In the CN, 4X4 switching matrix, switching speed less than 100ns and processing time less than 10 micro seconds are achieved. Comparing with [1], EN improves the processing capability and applies more flexible algorithms both in assembling and scheduling so that it can combine MPLS mechanism to assembling and support wider range for burst length. In the CN, a much more simple switching architecture is applied to the optical cross-connect providing stable and cost-effective performance. Powerful control plane is constructed to perform routing, signaling and network monitoring function. Both EN and CN is expandable for further research and the total cost of the system is much lower than the first one.

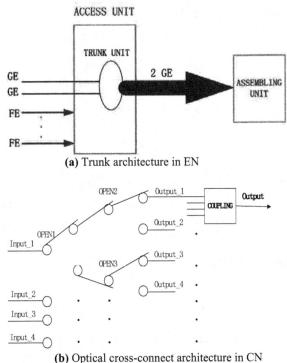

(a) Trunk architecture in EN

(b) Optical cross-connect architecture in CN

Fig.2. Node specialties in LOBS testbed.

3. Experimental Setup

The experiment setup is shown as Fig.1 and the simulation scenario is set similarly. The system is constructed by four core nodes and three edge nodes. Standard interface such as Fast Ethernet and Gigabit Ethernet are provided to access IP-based traffic. And input packets classified according to their ingress node IP and MAC address to form the FEC (Forward Equivalence Class) queue, based on which data burst is assembled. Meanwhile, the corresponding label is inserted to the control burst. Assembled burst is sent into LOBS network following a corresponding control burst for an offset time. In this experimental setup, burst assembly algorithm with fixed time-1ms and length threshold is applied, and for burst scheduling algorithm latest available unused channel with void fill (LAUC-VF) is used, each with a maximal delay of 2.5 us and 10 us.

As shown, a TCP server is connected to one edge node and the client is connected to the other one, by which a TCP connection can be set up through the LOBS network. Because there are two individual links for each TCP data flow, different burst loss probabilities for data burst and ACK burst can be achieved by configuring these links. In the TCP transmission, a maximum congestion control window of 64KBytes and maximum segment size of 512Bytes are employed in the TCP server. No background traffic is added for current experiment and the bursts are dropped deliberately by introducing a specific dropping mechanism to achieve certain burst loss probability.

4. Simulation, experiment and discussions

4.1 Simulation results

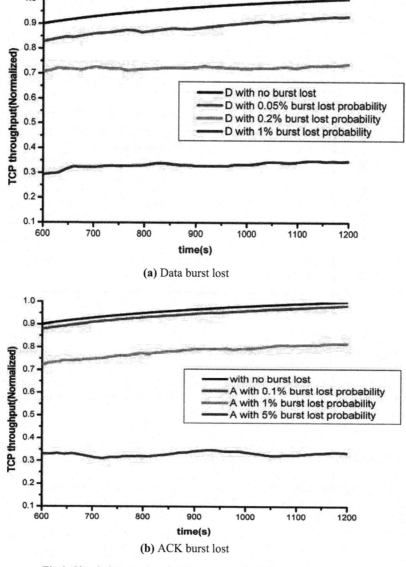

(a) Data burst lost

(b) ACK burst lost

Fig.3. Simulation results of TCP throughput in OBS network

In the simulation scenario, the data burst Dropping Probability (DP) varies from 0 to 1%, as it is shown in Fig[3]a, with increasing dropping probability, TCP throughput decrease accordingly. When the DP is set to 1%, TCP throughput is only 30% of 0 DP, which shows very bad network performance. The ACK burst lost transmission is shown in Fig[3]b, the TCP throughput is also dropping as the DP increase, but as the DP is set to 1%, the TCP throughput is about 73% of 0 DP, which is much higher than the situation of data burst loss. If we want to get the same results of 1% DP from data burst, the DP of ACK burst must set to 5%. It is shown from those results that the influence of ACK burst loss on TCP throughput is much lower compared with data burst loss.

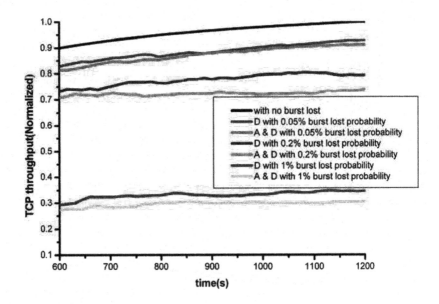

Fig.4. Simulation results of TCP throughput in OBS network, both ACK and data are lost

The case of both ACK and data are lost is shown in Fig[4], it can be investigated from the figure that the loss of ACK burst introduced much lower decrease in TCP throughput.

4.2 Experimental results and discussions

In the experiment scenario, FTP server and client are used to transmit TCP traffic. As shown in Fig[1], ACK burst and data burst can be dropped separately. The results are shown as following:

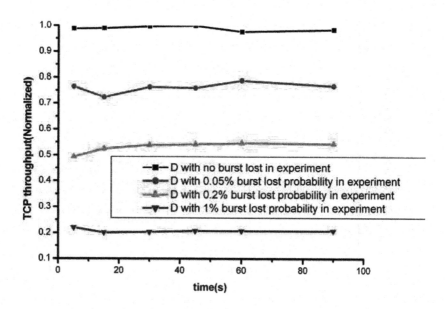

(a) Data burst lost

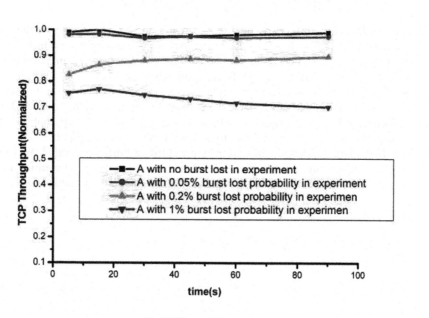

(b) ACK burst lost

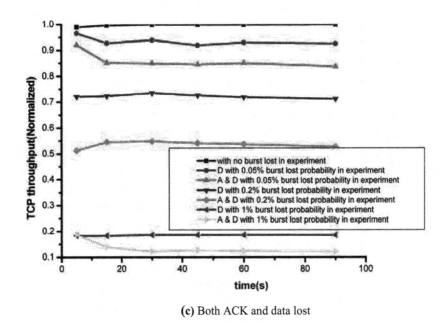

(c) Both ACK and data lost

Fig.5. Experiment results of TCP throughput in OBS network

Similar results from experiments is shown in Fig[5]. The reason for the result that ACK burst loss affect less than data burst loss on TCP throughput performance can be explained by TCP slide Window control mechanism. As we can see from Fig[6], Client A sends data to Client B base on the slide Window control mechanism. Suppose the WIN=400B,Length=100B/segament, initial number is 1.When B has received all the data sent by A, Client A will continue sending data to B, though missed a ACK signal. Therefore, if the data transmit with no loss probability, low lost probability of ACK signal will hardly affect the network throughput.

Considering that QoS provision is provided in OBS network[7], we can apply higher class of QoS to data burst and drop the ACK burst in contention resolution while maintaining the network performance to a comparatively stable level.

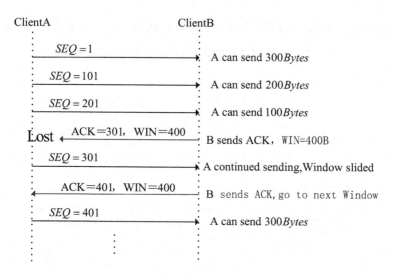

Fig.6. TCP slide Window control mechanism

5. Conclusions

In this paper, the TCP performance over OBS is experimentally investigated on OBS testbed [5][6]. Experimental results show that the performance of TCP traffic is influenced in the case of burst loss, both for ACK burst and data burst. But ACK burst loss affect less than data burst loss on TCP throughput performance, caused by TCP slide Window control mechanism. Moreover, considering that QoS provision is provided in OBS network, contention resolution mechanism of dropping ACK burst when network is busy is proposed in further investigations.

References:

[1]   C. Qiao and M. Yoo, "Optical burst switching (OBS) – A new paradigm for an optical internet," J. High Speed Networks, vol. 8, no. 1, pp. 68-84, 1999.

[2]   Sunil Gowda , Ramakrishna K Shenai , Krishna MSivalingam, Hakki Candan Cankaya, "Performance Evaluation of TCP over Optical Burst-Switched (OBS) WDM Networks," IEEE ICC 2003

[3]   A. Detti and M. Listanti, "Impact of Segments Aggregation on TCP Reno Flows in Optical Burst Switching Networks," Proceedings IEEE, INFOCOM 2002, New York, NY, June 2002

[4]   Robert Pleich, "Performance of TCP over Optical Burst Switching Networks", ECOC'05.September 2005.

[5]   Guo H., "A Testbed for Optical Burst Switching Network," OFC 2005, March 2005.OFC2005

[6]   Guo H., "Multi-QoS Traffic Transmission Experiments on OBS Network Testbed," ECOC'05.September 2005

[7]   Zhang W., Wu J., Xu K. and Lin J.T. "TCP performance experiment on OBS network testbed", OFC2006, March 2006.

Improvement of TCP Performance over Optical Burst Switching Networks

Jun Zhou, Jian Wu, Jintong Lin

Key Laboratory of Optical Communication & Lightwave Technologies, Ministry of Education,
Beijing University of Posts and Telecommunications, Beijing, 100876, China
zhoujuner@gmail.com, {jianwu, ljt}@bupt.edu.cn

Abstract. Transmission Control Protocol (TCP) performance over optical burst switching (OBS) is experimentally investigated on an OBS network testbed. The effect of burst losses on TCP performance over the OBS testbed is studied and the result shows that burst losses will lead to a network wide drop in TCP throughput and there exists an optimal assembly period to maximize the available TCP bandwidth. Then a new assembly mechanism constraining the ratio of acknowledgement (ACK) number to all the segment number in one burst is proposed. Simulation results demonstrate that compared with conventional assembly mechanism, the new one can improve TCP performance over OBS networks to great extent with lower network burst loss probability and higher TCP throughput.

Keywords: optical burst switching, transmission control protocol, assembly, throughput, burst loss probability

1   Introduction

As one of the promising technologies for future Internet Protocol (IP) over wavelength division multiplexing (WDM), optical burst switching (OBS), which combines the best of optical circuit switching (OCS) and optical packet switching (OPS) [1], has attracted considerable research attention recently. Given that Transmission Control Protocol (TCP) is today's prevailing transport protocol and likely to be adopted in future optical networks, so the evaluation and improvement of TCP performance over OBS networks is an important issue [2].

Since in OBS networks data is transmitted and switched in the format of large bursts, which is different from small packets in traditional packet switching networks. Therefore the existing upper-layer transmission protocols such as TCP will exhibit different characteristics, such as delay penalty and correlation benefit, which will influence TCP performance greatly [3].

This paper firstly investigates burst assembly's influence on TCP performance over OBS network testbed experimentally, concluding that there exists an optimal assembly period to maximize the available TCP bandwidth, while burst losses lead to a network wide drop in the bandwidth. Then, a new TCP/ACK-based assembly mechanism is proposed, and simulation results show that the mechanism can improve

I. Tomkos et al. (Eds.): ONDM 2007, LNCS 4534, pp 194-200, 2007.

the performance of OBS networks to great extent, providing lower network burst loss probability and higher available TCP bandwidth.

2   Experiments and Discussion

2.1   Experimental Setup

Fig.1 illustrates the experimental setup and for simplicity only two edge nodes and one core node are utilized. Each edge node is connected with a core node through a fiber link carrying 8×1.25Gb/s channels including a control channel. Both edge and core nodes employ efficient latest available unused channel with void fill (LAUC-VF) scheme to allocate proper output wavelengths for the data bursts, and the maximum processing delays are 2.5us and 10us respectively. Furthermore, the 32×32 optical switching fabric features a switching speed is less than 100ns [4].

In the experiment, we focus on static timer-based algorithm to analyze burst assembly effects on TCP transmission under different burst loss probabilities. Burst losses are obtained through intentional dropping at nodes. As shown in Fig.1, TCP server is connected to one edge node and the client is connected to the other one, by which a TCP connection is set up through the OBS network. In the TCP transmission, a maximum congestion control window of 64KBytes and maximum segment size of 512Bytes are employed in the TCP server.

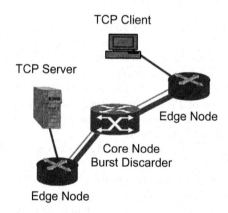

Fig. 1. Experimental setup in OBS testbed.

2.2   Experimental Results and Discussion

In OBS networks, several consecutive packets or segments from different sessions are included in one burst and a burst loss will result in loss of many packets per session.

The continuous packet loss pattern in OBS networks will trigger different retransmission pattern and worsen the network performance [3].

Fig.2 shows that burst losses lead to a network wide drop in throughput. For example, burst loss probability P_b of 0.8% makes the available bandwidth drop by 52%. The figure also shows that higher burst loss probabilities will result in lower TCP bandwidth. So in order to improve the performance of TCP, it is very important to reduce the burst losses in the OBS networks [5].

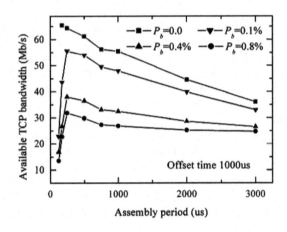

Fig. 2. TCP bandwidth vs. burst assembly period under different burst loss probabilities

Fig.2 depicts that in the case of no burst losses longer assembly period results in lower available TCP bandwidth. Reference [6] has reported the bandwidth is also reduced with the increasing of offset time for the same reason of delay penalty. Therefore, with the increase of assembly period and offset time, the round-trip time (*RTT*) is increased, resulting in delay penalty to decrease the available TCP bandwidth. Here only the instance of P_b=0 is considered because in the case of no burst losses only the delay penalty is in effect, that is to say, the correlation benefit vanishes with extreme value of P_b (i.e., 0 and 1) [7].

As discussed above, the burst assembly process causes delay penalty to decrease the available TCP bandwidth. On the other hand, the assembly process can also introduce a degree of correlation among the loss events for TCP segments interfering with TCP recovery mechanisms [7], and more important, an increase in the assembly period will result in an increase in the TCP bandwidth because of the increase in the number of segments per burst, which is the effect of correlation benefit in OBS networks.

Fig.2 shows that with burst losses introduced, the available TCP bandwidth is increased rapidly and then decreased slowly with the increasing of assembly period because of the combined interaction of the delay penalty and correlation benefit, i.e. there exists an optimal value assembly period 250us to maximize the available TCP bandwidth and the value is independent of burst loss probabilities. The reason is explained in detail as follows. On the one hand, when the assembly period is larger than the optimal value, the available bandwidth is decreased because in this case the delay penalty has greater influence than the correlation benefit. On the other hand, if the assembly period is smaller than the optimal value, similar *RTT* results in similar delay penalty and the correlation benefit becomes the primary factor.

In addition, it can also be seen that the available TCP bandwidth under low burst loss probabilities varies quickly compared with that under high loss probabilities. For example, the bandwidth at 3000us assembly period drops by 40% and 16% for burst loss probabilities of 0.1% and 0.8% respectively, compared to the corresponding bandwidth at 250us assembly period, because for lower burst loss probabilities, the delay penalty effect is the dominant factor to decrease the bandwidth compared with the correlation benefit, while for high loss probabilities the correlation benefit is the primary element.

3   New TCP/ACK-Based Assembly Mechanism

From the above experiment results, it can be seen burst losses lead to a wide drop in TCP throughput. So it is important to reduce burst losses for improving network performance. Furthermore, in the previous studying the ACK loss of TCP connection is either ignored or set to zero using lossless links. However, ACK losses indicate all the correlated data segments have to be retransmitted, causing the network performance to be much worse. Therefore, ACK losses can not be avoided and must be taken into consideration. In this section, we propose a TCP/ACK-based assembly mechanism to constrain the number of ACK packets in one burst by introducing a parameter R to depict the ratio of ACK number to all the segment number in one burst. Therefore, the number of ACK losses due to burst losses can be effectively controlled.

3.1   Principle of the New Mechanism

In the new TCP/ACK-based assembly mechanism, ACKs and other TCP segments are assembled together into one burst, just like the conventional timer-based assembly algorithm. However, the value of R is controlled to be approximately equal to or less than the pre-determined threshold R_{MAX} through the R control algorithm as shown in Fig.3. More details need to be explained about the R control algorithm is that the Extra-Queues are only for ACKs due to the control mechanism, that is to say, if R has never exceeded R_{MAX}, the Extra-Queues are always void and all ACKs are assembled into bursts directly from the Input Queues, as well as other TCP segments. More important, the delayed ACK packets in Extra-Queues are given higher priority than those from the Input Queues to be assembled into bursts, ensuring the delayed ACKs to be transmitted as soon as possible for lower *RTT* and higher TCP throughput [8].

Therefore, with the new TCP/ACK-based assembly mechanism, the number of ACK packets in one burst is constrained to be approximately equal to or less than the threshold R_{MAX}.

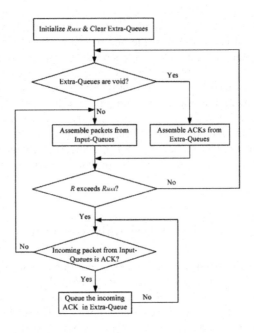

Fig. 3. Schematic description of *R* decision algorithm.

3.2   Simulation Results and Discussion

Simulation is carried out of the new mechanism on the network scenario as in Fig. 4 with two core nodes (CN) and six edge nodes (EN), assuming a TCP segment size is 512byte, RTT=600ms, the sending window W_m=128, the access bandwidth 200Mb/s, the maximum burst length 200Kbyte, the minimum burst length 30Kbyte and the mean burst size is 100Kbyte.

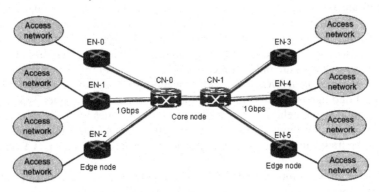

Fig. 4. Network simulation topology with 2 CNs and 6 ENs.

Fig. 5 shows the simulation results compared with the conventional timer-based assembly mechanism that the new mechanism shows better TCP performance than the conventional assembly one with lower burst loss probability and higher TCP throughput.

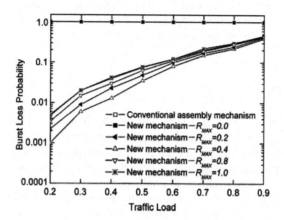

(a) Lower burst loss probability characteristic of the new mechanism

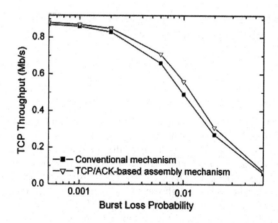

(b) TCP performance improvement of the new mechanism

Fig. 5. Performance comparison of the new mechanism vs. the conventional mechanism..

It can be seen from Fig.5 (a) that the curve of the new mechanism with $R_{MAX}=1$ is approximately coincident with the curve of the conventional one because $R_{MAX}=1$ means that only when all packets in one burst are ACKs (the case seldom happens), the assembly process needs to be adjusted by the R decision algorithm. In other words, the new mechanism in this case works as the conventional assembly scheme. On the other hand, in the case of $R_{MAX}=0$ (i.e. all ACK packets are queued and only data segments are assembled to be transmitted, which means that no TCP connection is setup), the burst loss probability will always be equal to one. Fig.5 (a) also depicts that with the increasing of R_{MAX} to a certain extent, more ACK packets can be assembled to bursts to improve the network performance. However, the value of R_{MAX} can not be set too large, because with too larger R_{MAX} the function of the new mechanism will be more similar to the conventional ones. Fig.5 (b) demonstrates that the new assembly mechanism improve TCP performance with higher throughput compared with the conventional assembly algorithm.

4   Conclusions

The TCP performance over OBS is experimentally investigated on the OBS testbed. Burst losses lead to a network wide drop in throughput. Then the effects of delay penalty and correlation benefit are taken into consideration, concluding that there exists an optimal assembly period independent of the burst loss probabilities to maximize the TCP bandwidth. To improve the performance of TCP over OBS networks, a new TCP/ACK-based assembly mechanism is introduced. Simulation results demonstrate that compared with conventional assembly mechanism, the new mechanism can improve TCP performance to great extent, such as lower network burst loss probability and higher TCP throughput.

Acknowledgments. This work is supported by Program of Ministry of Education of China (Grant No. 105036 and Grant No. NCET-04-0116).

References

1. Qiao, C., Yoo, M.: Optical burst switching (OBS) —a new paradigm for an optical internet. J. High Speed Netw., Vol. 8 (1999) 69–84
2. Detti, A., Listanti, M.: Amplification effects of the send rate of TCP connection through an optical burst switching network. Optical Switching and Networking, Vol. 2 (2005) 49–69
3. Yu, X., Li, J., Cao X., Chen, Y., Qiao, C.: Traffic statistics and performance evaluation in optical burst switched networks. J. lightwave technology, Vol. 22 (2004) 2722–2738
4. Guo, H. X., Lan, Z., Wu, J., Gao, Z., Li, X., Lin, J., Ji, Y., Chen, J., Li, X.: A testbed for optical burst switching network. In: Proc. of Optical Fiber Communication Conference 2005, Anaheim, CA (2005)
5. Bi, F.J., Zhang, M., Ye, P.D.: A novel prioritized scheme for contention resolution in optical burst-switched networks. Acta Photonica Sinica, Vol. 34 (2005) 900–904
6. Zhou, J., Wu, J., Lin, J.: Experimental Study of TCP Performance on an Optical Burst Switching Network Testbed. Accepted to be published in Acta Photonica Sinica
7. Detti, A.,Listanti, M.: Impact of segments aggregation on TCP reno flows in optical burst switching networks. In: Proc. IEEE INFOCOM 2002, New York (2002) 1803–1812
8. Comer, D.E.: Internetworking with TCP/IP: volume I-principles, protocols, and architectures, 4th edition, Prentice-Hall, Upper Saddle River, New Jersey (2000)

Routing Optimization in Optical Burst Switching Networks

M. Klinkowski[1,2], M. Pióro[3], D. Careglio[1], M. Marciniak[2], and J. Solé-Pareta[1]

[1] Universitat Politècnica de Catalunya (UPC),
Advanced Broadband Communication Centre – CCABA
Jordi Girona 1-3, 08034 Barcelona, Spain
[2] National Institute of Telecommunications,
Department of Transmission and Fibre Technology,
1 Szachowa Street, 04-894 Warsaw, Poland
[3] Warsaw University of Technology (WUT),
Institute of Telecommunications,
15/19 Nowowiejska Street, 00-665 Warsaw, Poland
{mklinkow@ac.upc.edu, mpp@tele.pw.edu.pl, careglio@ac.upc.edu,
mmarcin@itl.waw.pl, pareta@ac.upc.edu}

Abstract. This paper addresses the problem of routing optimization in optical burst switching (OBS) networks. We use a simplified analytical model of OBS network with an *overall burst loss probability* as the primary metric of interest. Since the objective function of optimization problem is nonlinear we propose two solutions based on a *non-reduced link load* (NR-LL) model and a *reduced link load* (R-LL) model. In order to find partial derivatives of the cost function we apply a calculation considered previously for circuit-switched networks. We derive exact partial derivatives of NR-LL model and we approximate the partial derivatives of R-LL model. Simulation results demonstrate that our solutions effectively reduce the overall burst loss probability over the shortest path routing. Moreover, in many cases, they over-perform an alternative routing.

Keywords: optical burst switching, optimization, routing.

1 Introduction

Optical burst switching (OBS) [1] is a promising solution for reducing the gap between switching and transmission speeds in future networks. The client packets are aggregated and assembled into optical *burst* units in the edge nodes of OBS network. A burst *control packet* is transmitted in a dedicated control channel and delivered with a small *offset-time* prior to the data burst. In this way the electronic controller of an intermediate node has enough time both to reserve a wavelength in its output link, usually for the duration time of the incoming burst, and to reconfigure dynamically the switching matrix. When the burst transmission is finished in a node the output wavelength is released for other connections. Such a temporary usage of wavelengths allows for higher resource utilization as well as better adaptation to highly variable input traffic in comparison to optical circuit-switching networks.

I. Tomkos et al. (Eds.): ONDM 2007, LNCS 4534, pp 201-210, 2007.

OBS architectures with limited buffering capabilities are susceptible to congestion states. The existence of a few highly congested links may seriously aggravate the network throughput (see e.g. [2]). The congestion can be reduced either by an appropriate network dimensioning or by a proper routing in the network. The dimensioning approach fits the node and link capacities according to the matrix of actual traffic load demands and after such optimization it needs only either a simple shortest path algorithm or a similar mechanism (see e.g. [3]). Some parts of such network, however, may encounter the congestion problem if the traffic demands change. On the contrary, the routing approach introduces some operational complexity since it often needs advanced mechanisms with signalling protocols involved. Nevertheless, the advantage is that it facilely adapts to the changes in the traffic demands.

A great part of the research on routing in OBS networks addresses the problem of alternative, also called deflection, routing (e.g. [2], [4]). In such routing scheme the burst is allowed to be deflected dynamically to an alternative routing path in a node if it contents with another burst on the primary routing path. However, a deflection routing can improve the network performance under low traffic loads it may intensify the burst losses under moderate and high loads [5].

Another approach to the routing problem makes use of the optimization theory and a few works can be found in this area [6], [7], [8]. In OBS network a burst loss probability is the primary metric of interest which adequately represents the congestion state of entire network. An approximated form of the overall burst loss probability, which can be found e.g. in [9], has a nonlinear character which may produce some difficulties in formulating an optimization problem. The solutions presented in works [6]-[8] use a linear programming (LP) formulation which either does not consider the overall burst loss probability as a metric of interest or it takes an approximated form of this metric.

The intention of this work is to fill the gap. Namely, we formulate a nonlinear optimization problem for the routing problem in OBS network. Our objective is to distribute the traffic over a set of pre-established routing paths so as to minimize the overall burst loss probability in the network.

The rest of the paper is structured as follows. In Section 2 we present a routing scenario under the study. In Section 3 we provide a loss model of OBS network. In Section 4 we formulate a nonlinear optimization problem and give the partial derivatives of the cost function. In Section 5 we present simulation results that prove the effectiveness of our solutions. In Section 6 we discuss some implementation issues. Finally we conclude the paper in Section 7.

2   Routing Scenario

Consider an OBS network such as that illustrated in Fig. 1. There are finitely many links, labelled $k = 1, 2, ... K$, and link k comprises, for simplicity, a fixed number of wavelengths C. A subset $p \subset \{1, 2, ... K\}$ identifies a path; we define a matrix $[A_{kp}]$ such that $A_{kp} = 1$ if link k belongs to path p, and $A_{kp} = 0$ otherwise. In the network there is a set P of paths pre-established between source (s) and destination (d) nodes. A subset $P_{sd} \subset P$ identifies all the paths originated in node s and terminated in node d.

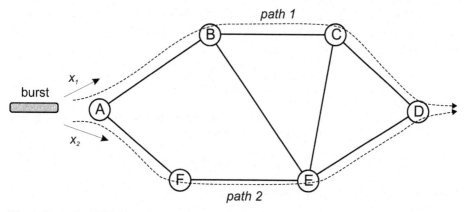

Fig. 1. Example of OBS network with source-based routing; x_1 and x_2 are the splitting factors and $x_1 + x_2 = 1$.

We assume that the network operates with a source-based routing, in that the source node determines the path of a burst that enters the network (see Fig. 1). Moreover, the network uses a multi-path routing where each subset P_{sd} comprises a small number of paths and a burst can take one of those paths. The path selection is performed according to a given splitting factor x_p, such that the sum of x_p of all the paths p belonging to a given subset P_{sd} is equal to 1.

The traffic pattern is described by the matrix $[t_{sd}]$ and bursts destined to a given node d arrive to a node s as a Poisson process of (long-term) rate t_{sd}. Let $t_p = t_{sd}$ for each $p \in P_{sd}$. Thus traffic v_p offered to path p can be calculated as:

$$v_p = x_p t_p. \tag{1}$$

Here the key factor is the vector (x_p) because it determines the distribution of traffic over the network and it may be selected so that to reduce congestion and to improve overall performance (i.e. in our case the overall burst loss probability). In the next two sections we present analytical derivation of this metric and we formulate an optimization problem.

3   Loss Model of OBS Network

A burst going over a path p is blocked and lost if on a given link k that belongs to p there are no free wavelengths. Otherwise a wavelength in the link is reserved for the burst duration and then released immediately after the burst transmission. The reservation holding period is independent of earlier arrival times and holding periods; the holding periods on each link are identically distributed with a mean equal to the mean burst duration, for simplicity we assume that it is equal to 1.

Given the difficulty in obtaining an exact formula for the blocking of a burst we assume that each blocking event occurs independently from link to link along any path inside the network (see e.g. [9]).

The independence assumption implies that the offered load to link k is a Poisson process with rate ρ_k and so the burst loss probability in link k is:

$$E_k = Erl(\rho_k, C) = \frac{\rho_k^C}{C!} \cdot \left[\sum_{i=0}^{C} \frac{\rho_k^i}{i!} \right]^{-1}. \tag{2}$$

For a moment let us assume that we have a vector (E_k) given. Thus the approximate loss probability L_p of bursts offered to path p satisfies:

$$L_p = 1 - \prod_{k=1}^{K}(1 - A_{kp}E_k), \tag{3}$$

and the overall burst loss probability B is equal to:

$$B = \sum_{p \in P} v_p L_p \cdot \left[\sum_{p \in P} v_p \right]^{-1}. \tag{4}$$

In order to calculate (2) we consider two different models that estimate the traffic ρ_k offered to a link, namely a *non-reduced link load* (NR-LL) model and a *reduced link load* (R-LL) model.

In the NR-LL model the traffic offered to link k is calculated as a sum of the traffic offered to all the paths that cross this link:

$$\rho_k = \sum_{p \in P} A_{kp} v_p. \tag{5}$$

This formula roughly approximates the volume of traffic offered to a link; nevertheless, it allows easily to find the solution of equation (2).

In the R-LL the traffic offered to link k is obtained as a sum of the traffic offered to all the paths that cross this link diminished by the traffic lost in preceding links along these paths. Thus:

$$\rho_k = \sum_{p \in P} A_{kp} v_p \prod_{j=1}^{K}(1 - S_{pjk}E_j), \tag{6}$$

where S_{pjk} equals 1 or 0 depending whether or not link j strictly precedes link k along path p, respectively.

In order to solve the formula (6) which incorporates equation (2) we take advantage of the Erlang fixed-point approximation (see e.g. [9]).

4   Formulation of the Optimization Problem

From equations (1) and (4) we define a cost function to be the subject of optimization:

$$B(x) = \sum_{p \in P} x_p t_p L_p \; , \tag{7}$$

The optimization problem is formulated as follows:

$$\min B(x) \tag{8}$$

subject to:

$$\sum_{p \in P_{sd}} x_p = 1 \quad \forall P_{sd} \; , \tag{9}$$

$$0 \le x_p \le 1 \quad \forall p \in P \; . \tag{10}$$

Since the overall burst loss probability is a nonlinear function of vector (x_p) the cost function is nonlinear as well. According to [10] for solving such optimization problem we can use for instance the modified reduced gradient method described in [11]. In order to achieve an optimal solution it is necessary to compute a direction, which leads to a reduction in the total cost. Such direction is indicated by partial derivatives of the cost function.

4.1 Calculation of Partial Derivatives in NR-LL model

The partial derivatives in NR-LL model can be derived directly from (1)-(5). Their computation, however, may be time-consuming in larger networks since $|P|$ partial derivatives have to be recalculated. Instead we use a similar derivation like the one proposed by F. Kelly for circuit-switched (CS) networks [12]. Due to space limitation here we provide only the final form of the solution.

Let for each link k:

$$c_k = \left[E(\rho_k, C-1) - E(\rho_k, C)\right] \cdot \sum_{p:k \in p} x_p t_p \left(1 - L_p\right). \tag{11}$$

Then

$$\frac{d}{dx_i} B(x) = t_i \left[L_i + \sum_{k \in i} c_k\right]. \tag{12}$$

4.2 Calculation of Partial Derivatives in NR-LL model

The application of Kelly's calculation to NR-LL model in OBS network may be quite complex. Therefore we take the derivation initially obtained for CS network model and use it as a rough approximation.

Let $c = (c_1, c_2, \dots, c_K)$ be the (unique) solution to the equation:

$$c_k = \left[E(\rho_k, C-1) - E(\rho_k, C) \right] \cdot (1 - B_k)^{-1} \sum_{p:k \in p} x_p t_p \left(1 - L_p \left(1 - \sum_{j \in p - \{k\}} c_j \right) \right). \tag{13}$$

Then

$$\frac{d}{dx_i} B(x) \approx t_i \left[1 - (1 - L_i)(1 - \sum_{k \in i} c_k) \right]. \tag{14}$$

The formula (14) corresponds strictly to CS network case. Nevertheless, as the numerical examples we used for this study shows it always yields, in OBS network, to the same near-optimal solution of the optimization problem. Moreover, for small networks, where it is feasible to compute direct partial derivatives, the corresponding values of the optimized cost function are identical.

The conformity of results under NR-LL model is kept as well.

We can also observe that low values of the cost function (7) lead to the same near-optimal solutions in both NR-LL and R-LL models.

5 Performance

We evaluate the performance of our routing scheme in an event-driven simulator. In order to find a splitting vector (x_p) that yields to a near-optimal routing in both NR-LL and R-LL models we use a solver *fmincon* for constrained nonlinear multivariable functions available in the Matlab environment. Then we apply this vector in the simulator. The optimized NR-LL and R-LL routing is compared with two other routing strategies, namely with a simple shortest path (SP) routing and a by-pass routing (BP) which is an alternative-like routing (see [2] for more details).

5.1 Evaluation Scenario

Two network topologies are studied, namely a Simple network topology (see Fig. 3a) and the NSFnet network topology (see Fig. 3b). We assume that the links are bi-directional and each link multiplexes $C=32$ wavelengths if not specified otherwise. Each node acts as an edge and a core node. We consider 2 paths per each source-destination pair of nodes. The paths are pre-established according to the Dijkstra algorithm (with regard to the number of hops) and they are not necessarily disjoint. In SP routing only 1 path is available.

The nodes are capable to perform a full wavelength conversion according to the random wavelength-selection algorithm. There is no FDL buffering for the burst contention resolution.

Regarding the traffic modelling, each node is capable to generate a burst destined to any other node with the uniform distribution. Each source node has C input wavelengths with the input traffic per a wavelength equal to ρ (this parameter is specified later). The burst inter-arrival times to the network are exponentially distributed. The mean burst duration is 1ms. We assume that the source nodes do not buffer the bursts after completing their aggregation.

All the simulation results have 99% level of confidence.

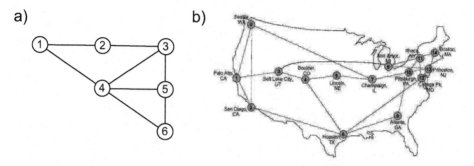

Fig. 2. Network topologies; a) Simple, b) NSFnet.

5.2 Network Performance

In Fig. 3a) we can see the overall burst blocking probability in the function of traffic load (ρ) obtained in the Simple network for different routing strategies. The number of input wavelengths as well as the number of wavelengths in a link is equal to 8. We can observe that both BP and NR-LL and R-LL routing offer a similar performance and all of them outperform SP routing.

Fig. 3b) presents the performance obtained in the Simple network with 32 wavelengths per link. Now the optimized routing outperforms both BP and SP routing, however, the last one is the worst.

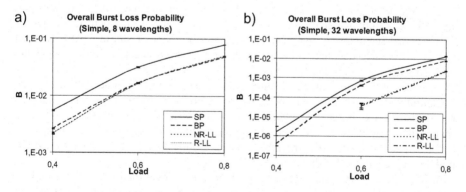

Fig. 3. Comparison of routing schemes in the Simple network; a) $C = 8$, b) $C = 32$.

Finally in Fig. 4) we compare the performance in larger NSFnet network. The number of wavelengths is equal to 32. Similarly as in the previous case the lowest burst loss probabilities are achieved with the optimized routing. The BP routing is only a bit better than the SP routing. Such low performance can be explained by a small number of alternative paths (1 path) available in BP routing.

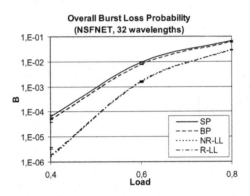

Fig. 4. Comparison of routing schemes in the NSFnet network ($C = 32$).

Note that both NR-LL and R-LL routing offer the same performance without regard to the network and traffic load scenarios. It is because under low link losses (as in the provided examples) the NR-LL model approximates well the traffic offered to a link. As a result both the NR-LL and R-LL models give similar results.

6 Implementation Issues

The proposed optimization framework can be used to calculate a traffic splitting vector that determines the distribution of traffic over the network in a multi-path source based routing scenario. We assume that there is a virtual path topology pre-established that comprise, for instance, a limited number of shortest paths between each pair of source-destination nodes. Such virtual topology can be built in a labelled OBS (LOBS) architecture [13].

Routing optimization can be executed either in centralized or in decentralized way.

Centralized optimization can be used in a static (pre-planed) routing, where the traffic distribution is calculated based on a given (long-term) matrix of demands. Then either a periodic or a threshold-triggered update of the splitting vector can be performed if the matrix of demands changes. Both NR-LL and R-LL optimization models can be applied for this purpose.

A distributed routing should react rapidly to a local disturbance at the point of the disturbance, with slower adjustments in the rest of the network. Similarly like it was proposed for circuit-switched networks [12] the formulas (13) and (14) can be used to design a distributed adaptive routing algorithm for OBS network.

In such distributed adaptive routing we assume that in the network there is the possibility of limited communication between the nodes. The nodes are capable to measure the loads carried through the links and the source nodes are able to measure the loads carried on the paths. Moreover, there is a (limited) arithmetical processing ability for each link and route, which may be distributed over the nodes of the network; for example the processing for routes might be carried out at the sources nodes. Then the measurements of actual loads together with computing (13) and (14) can be used to implement a decentralized hill-climbing search procedure able gradually to vary routing patterns in response to changes in the demands placed on the network (see [12] for details).

7 Conclusions

In this paper the problem of routing optimization in optical burst switched networks with multi-path source-based routing is studied. In this context two different network models are defined, namely with non-reduced link load calculation (NR-LL) and with reduced link load (R-LL) calculation. For both of them a nonlinear optimization problem is formulated and solved.

Simulation results demonstrate that both models equally well find a solution for our routing problem. As a result the traffic is effectively distributed over the network and the network-wide burst loss probability is reduced. The optimized routing always performs better than the shortest path routing. Moreover, it outperforms an alternative routing in a network with high number of wavelengths per link.

Both proposed models can be applied to optimize the distribution of traffic in a centralized way, e.g. in a static routing. Moreover, R-LL model can be used in adaptive distributed routing as it was presented in Kelly's work.

For the purpose of optimization problem we provide the exact partial derivatives in NR-LL model. Moreover, we approximate them in R-LL model by the partial derivatives corresponding to the circuit-switched network model. All the numerical examples we used in this work show that these derivatives lead to a near-optimal solution in OBS network. Nevertheless, the study will be continued in order to prove the correctness of approximation used in R-LL model for any OBS network scenario. The accuracy of OBS network loss models and the properties of the cost function (concavity/convexity or neither of them) will be investigated as well.

The issue not addressed in this paper and related with the multi-path source routing is the problem of out-of-order burst arrival. The burst reordering is common for any multi-path or deflection routing scenario and dedicated mechanisms have to be introduced in order to cope with this problem.

Acknowledgments. The results presented in this work were obtained during a joint Short Term Scientific Mission of the actions COST 293 ("Graphs and algorithms in communication networks") and COST 291 ("Towards digital optical networks"). This work has been partially funded by the MEC (Spanish Ministry of Education and Science) under the CATARO project (Ref. TEC2005-08051-C03-01).

References

1. Qiao, C., Yoo, M.: Optical Burst Switching (OBS) – a New Paradigm for an Optical Internet. Journal of High Speed Networks, Vol. 8. No. 1. (1999) 69-84
2. Klinkowski, M., Herrero, F., Careglio, D., Solé-Pareta, J.: Adaptive Routing Algorithms for Optical Packet Switching Networks. Proceedings of 9th IFIP Working Conference on Optical Network Design and Modelling ONDM2005 (2005)
3. Köhn, M., Gauger, C.M.: Dimensioning of SDH/WDM Multilayer Networks. Beiträge zur 4. ITG-Fachtagung Photonische Netze. (2003) 29-33
4. Cameron, C., Zalesky, A., Zukerman M. : Shortest Path Prioritized Random Deflection Routing (SP-PRDR) in Optical Burst Switched Networks, ICST International Workshop on Optical Burst Switching (WOBS). San Jose (2004)
5. Zalesky, A. Vu, H. L., Rosberg, Z., Wong, E. M., Zukerman, M.: Reduced Load Erlang Fixed Point Analysis of Optical Burst Switched Networks with Deflection Routing and Wavelength Reservation. Proceedings of the First International Workshop on Optical Burst Switching (2003)
6. Zhang, J. et al.: Explicit Routing for Traffic Engineering in Labelled Optical Burst-Switched WDM Networks. International Conference on Computational Science ICCS (2004)
7. Teng, J., Rouskas, G.: Traffic Engineering Approach to Path Selection in Optical Burst Switching Networks. Journal of Optical Networking, Vol. 4. No. 11. (2005)
8. Hyytia, E., Nieminen, L.: Linear Program Formulation for Routing Problem in OBS Networks. Proceedings of the 9th IEEE Symposium on Computers and Communications (ISCC 2004)
9. Zukerman, M. et al: Blocking Probabilities of Optical Burst Switching Networks Based on Reduced Load Fixed Point Approximations. Proceedings of INFOCOM Conference (2003)
10. Pioro, M., Wallstrom, B.: Multihour Optimization of Non-Hierarchical Circuit Switched Communication Networks with Sequential Routing. 11th International Teletraffic Congress ITC-11 (1985)
11. Harris, R.J.: The Modified Reduced Gradient Method for Optimally Dimensioning Telephone Networks. Australian Telecom. Research. Vol. 10. No. 1. (1976) 30-35
12. Kelly, F. P.: Routing in Circuit-Switched Networks: Optimization, Shadow Prices and Decentralization. Advanced Applied Probability, Vol. 20. (1988) 112-144
13. Qiao, C.: Labeled Optical Burst Switching for IP-over-WDM Integration. IEEE Communication Magazine. December (2000)

Performance analysis of routing algorithms for optical burst switching

Óscar González de Dios[1], Miroslaw Klinkowski[2], Carlos García Argos[1],
Davide Careglio[2], Josep Solé-Pareta[2]

[1] Telefonica I+D, c/Emilio Vargas, 6
28043 Madrid, Spain, {ogondio, cgarcia}@tid.es

[2] Universitat Politècnica de Catalunya, Jordi Girona 1-3, Campus Nord,
08034 Barcelona, Spain, {mklinkw, careglio, pareta}@ac.upc.edu

Abstract. In this paper we study the performance of several routing algorithms for optical burst switching network. The main aim is to find ways for performing dynamic load balancing and reduce congestion situation. In our analysis, we consider a common network scenario and, as a performance reference, we use the simple shortest path routing giving results of both analytical and simulation models. Therefore we propose many different routing algorithms based either on adaptive or non-adaptive strategies as well as distributed or isolated path selection. The obtained results highlight that none of the proposed strategies improve significantly the performance over the simple shortest path approach in the considered scenarios.

Keywords: optical burst switching, routing algorithms, performance evaluation.

1 Introduction and motivation

Optical burst switching (OBS) is a photonic network architecture directed towards efficient transport of IP traffic [1]. OBS pretends to be an intermediate solution for optical networks lying between Optical Circuit Switching (OCS) that nowadays in stage of standardization process and Optical Packet Switching considered as a far-term solution for optical networks. The potential advantages of OBS have caused a huge interest in the research in technologies to provide OBS functionality.

OBS architectures with limited (or even without) buffering capabilities are sensitive to traffic overloads. In particular, these traffic overloads cause burst drops, degrading network performance. Either a proper routing strategy with Traffic Engineering (TE) enhancement or adequate network dimensioning can help in the reduction of the congestion on specific links resulting in the increase of network throughput. Both approaches treat the congestion problem otherwise. Teng [2] proposes a TE approach to select the optimal paths for a given traffic matrix. Also, dimensioning the network considering a Shortest Path routing can also be performed [3]. The dimensioning approach fits the node and link capacities according to the matrix of actual traffic load demands and after such optimization it needs only either a

I. Tomkos et al. (Eds.): ONDM 2007, LNCS 4534, pp 211-220, 2007.
© IFIP International Federation for Information Processing 2007

simple Shortest Path algorithm or similar mechanism. However, some parts of such network may encounter congestion problem if the traffic demands change. Another approach for the congestion problem is to use routing strategies that improve network performance by means of online load balancing. Thodime [4] proposes an algorithm that statically computes link-disjoint alternate paths and dynamically selects one of the paths based on the collected congestion information. Thodime also proposes to periodically re-calculate the routes based on several metrics such as the physical distance, hops count, congestion information and link utilization. The authors in [5] present a proactive approach for tackling the problem of a burst contention based on adaptive use of multiple paths between edge nodes.

In this paper we investigate the routing strategies in OBS networks, starting from the Shortest Path approach to adaptive strategies that take into account network state. Moreover, we propose and study novel routing strategies with the aim of reducing burst loss probability. To asses the study, simulations are performed.

The rest of the paper is organized as follows. Section 2 describes the network scenario, the node characteristics and traffic modeling. In Section 3, we introduce a classification of the routing algorithms that are feasible for an OBS network. In Section 4 we consider the application of two static approaches and propose an analytical model to compute the burst loss probability for Shortest Path routing. Several adaptive routing approaches are considered in Section 5. Finally, Section 6 draws the conclusions of this work and outlines the further work.

2 Network scenario

In order to evaluate and compare the routing strategies, a reference scenario has been chosen. The topology is the 15 node NSFNet shown in Fig. 1. Every link has 16 data channels, with 10 Gbps per channel. The nodes are enhanced with full wavelength conversion and 4 feed-forward FDLs. The average burst size is set to 40 kbytes, leading to 32 μs of average burst length. The FDL granularity is the average burst length and up to 4 FDL delays are allowed, leading to a maximum of 128 μs delay per burst. Nodes implement the Just Enough Time (JET) signaling protocol and a First In-First Out resource reservation without preemption or priorities. The scheduling strategies that have been considered are Horizon/LAUC and Minimum starting void filling algorithms [6].

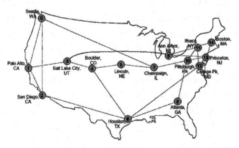

Fig. 1 15-node NSF network topology

Regarding the traffic modeling, each node acts as an edge node, generating bursts that are sent with equal probability to all the other nodes. Both interarrival time and burst length are exponentially distributed. The study is performed varying the overall load introduced by the nodes, which is normalized with respect to the bandwidth of the links in order to allow a fair comparison of the results. The normalized load is defined as follows:

$$\rho = \frac{T_{node}}{W \times B} \tag{1}$$

Where T_{node} is the mean traffic generated by each node, W is the number of data wavelengths per link and B the bandwidth of a single wavelength. Recall that the bandwidth of a link is $W \times B$.

Let N be the number of nodes and L the mean burst length (in bytes), the mean burst inter-arrival time (IAT) between each pair of nodes is

$$IAT_{node-node} = \frac{L \times (N-1)}{W \times B \times \rho} \tag{2}$$

And hence the overall network load is

$$T_{network} = N \times W \times B \times \rho \tag{3}$$

3 Routing algorithms classification

Two routing approaches exist: pure routing strategy like those used in IP networks and labeled routing strategy like MPLS. It is well-known that pure routing is not feasible for high speed networks requiring very fast lookup table processing. In OBS this problem is more evident for the huge amount of control packets to process. Therefore, setting up Labeled Path Switch in the so called Labeled OBS (LOBS) is considered the best approach; both Explicit Routing (ER) and Constraint-Based Routing (CBR) can be extended to provide and engineer the network resources [7]. In this paper we only consider ER solutions for the LOBS network. In particular, we consider that the LSPs between any pair of nodes are computed in advance, off-line, and downloaded to the nodes when the network is booted.

In such an environment, the ER can be set up in different way. Here we give a brief classification.

The path decisions can be grouped into two major classes: non-adaptive and adaptive. Non-adaptive ones do not base their routing decisions on measurements or estimates of the current traffic and topology, whereas adaptive ones do. This is sometimes called static routing and either single LSP or multi LSPs can be set up between any source node to any destination node. In case of multi LSPs, the source node usually balances the traffic among the LSPs.

On the other hand, the adaptive approach needs to set up multi LSPs and attempts to change their path decisions to reflect changes in topology and the current traffic. Three different families of adaptive algorithms exist, namely centralized, isolated and distributed, which differ in the information they use. In the centralized solution, a single entity uses information collected from the entire network in an attempt to make optimal decisions. This solution is clearly unviable in wide area networks where the delays can be excessive. The other two solutions are more feasible. In the isolated approach, a local algorithm runs separately on each node, using local information, such as queue length. Finally, the third class of adaptive algorithms uses a mixture of global and local information.

Usually, the distributed approach is practical if decisions are taken only at the source node. In fact, it can receive congestion information from other nodes and take global optimal decisions selecting only the current best LSP or balance the use of multi LSPs according to a weight function. On the other hand, isolated approach is more beneficial if any node can decide the best route to the next node.

4 Shortest Path and Equal Cost MultiPath analysis

First, we consider two simple static routing techniques: the Shortest Path (SP) and Equal Cost Multi Path (ECMP). In the SP, only one LSP is available at each source node to get any destination node and we assume that the metric is the number of hops.

The SP strategy can be improved by using ECMP routing, which implements a multi LSPs strategy. For each possible source-destination pair, ECMP sets up all possible shortest LSPs (i.e., all those paths that have the same number of hops). Then, the source nodes balance the usage of the LSPs (i.e., each LSP is equally loaded).

4.1 Iterative analytical model for Shortest Path routing

We wish to analytically obtain the overall blocking probability for a given topology and traffic matrix, using Shortest Path routing. Some analytical methods have been already proposed in literature (e.g. [8]); here we are interested to propose a simple and iterative method. Its simplicity will give us, in future works, the possibility to easy enhance it in order to model adaptive routing strategies.

We assume that between every pair of nodes i,j there is an offered traffic T_{ij}. The path from node i to node j is fixed according to a Shortest Path routing strategy in the sense of number of hops. For the sake of simplicity, we will assume that all the links have the same number of wavelengths. Let us consider an OBS meshed network, where every node has several output links. We will consider JET scheduling and constant offset time. We will assume that burst arrivals follow a Poisson process. Thus, the link burst blocking probability B is given by the Erlang-B formula

$$B(W,\rho) = \left(\frac{\rho^W}{W!}\right)\bigg/\sum_{j=0}^{W}\frac{\rho^j}{j!} \tag{4}$$

where W is the number of wavelengths and ρ is the traffic intensity in Erlangs in the link. The end-to-end blocking probability for a traffic flow $T_{i,j}$ is given by

$$B_{i,j} = 1 - \prod_{k \in path(i,j)} (1 - B_k)$$ (5)

where B_k is the blocking probability of the output link of a node in the path from node i to node j subject to a ρ_k traffic intensity:

$$B_k = B(W, \rho_k)$$ (6)

We wish to obtain the overall blocking probability for a given network. The overall blocking probability is given by

$$B_{overall} = \frac{\displaystyle\sum_{i=1}^{N} \sum_{j=1}^{N} B_{i,j} \cdot T_{i,j}}{\displaystyle\sum_{i=1}^{N} \sum_{j=1}^{N} T_{i,j}}$$ (7)

The proposed iterative method to find the overall burst blocking probability $B_{overall}$ is shown below:

1. The first step is to apply the Dijkstra algorithm and calculate the Shortest Path, for every pair of nodes (i, j), where $i = 1,...,N$, and $j = 1,...,N$, with $i \neq j$.
2. For each traffic intensity from node i to node j, calculate the contribution of such traffic to link k

$$T_{k(i,j)}^{n} = T_{i,j} \prod_{k \in path(i,j)} (1 - B_k^n)$$ (8)

At the first iteration ($n = 0$), B_k is set to 0 for every link k. In the next steps, this value will be updated.
3. Once all the traffic intensities at every output link have been calculated, traffic intensities of the paths that cross link k are summed

$$\rho_k^n = \sum_{i=1}^{N} \sum_{j=1}^{N} T_{k(i,j)}^{n}$$ (9)

4. For any k, calculate new values of B_k^n using equation (6) and determine the difference with the B_k^{n-1} calculated in the previous iteration

$$D_k^n = \left| B_k^n - B_k^{n-1} \right|$$ (10)

If $D_k^n < D$ for all k, the iteration process has converged. If the iteration has not converged, update $n=n+1$ and go back to step 2.
5. If the iteration has converged, the overall burst blocking probability is calculated using equation (7).

4.2 Simulation and analytical results

The analytic results have been verified with the simulation. In order to provide more realistic results, the switching and processing times in the nodes are taken into account in the study. The BCP processing time in the intermediate nodes is set to 2.5 μs. The speed of the switch matrix (switching time) is an important parameter in the scheduling process, and has been set to a conservative value of 1 μs. Fig. 2 depicts a comparison between the analytic results and the simulations, considering and not the realistic switching times. Results show that these switching times slightly increase the burst loss probability.

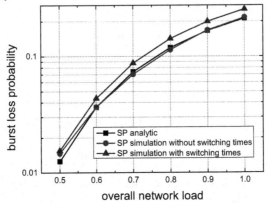

Fig. 2 Analytic vs simulation for SP routing without FDL buffering

Results for the SP and ECMP routing strategies are shown in Fig. 3. Four cases combining lack or presence of FDL buffering and Horizon or Minimum Starting Void scheduling are considered in obtaining these plots. It can be clearly seen that the use of buffers to avoid contention improves the overall blocking probability, even on high load conditions. The difference increases even more when Void Filling algorithms are used.

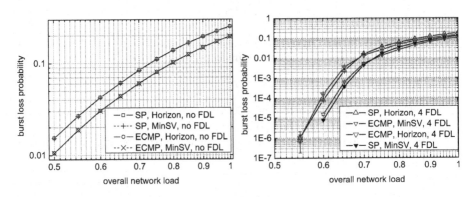

Fig. 3 Comparison between SP and ECMP routing strategies with and without FDL buffers

The comparison between both routing strategies shows a very slight improvement on the burst loss probability parameter when no FDL are used and a performance loss when FDL buffers are used and the network load is low if ECMP routing is used to balance links load. The main reason of the slight difference is that, in this scenario, the network load is relatively better balanced when only one of the Shortest Paths among two nodes is considered then when using several shortest paths with no control over the quantity of traffic deflected. Although the global network load is more balanced with ECMP, alternate routes may get slightly overloaded and therefore increase their blocking probability, which causes a global increase in the network burst loss rate. However, the performance seems to be very dependent on the topology and some scenarios may benefit from the use of the ECMP routing.

5 Adaptive routing

For the following adaptive solutions, we consider that k pre-established shortest LSPs between all source-destination pairs of nodes are available. These multi-LSPs are established only on knowledge of the network topology. Since it is possible to select one path from the set of k available, each node (both source and intermediate) can make per-burst decision according to some parameters. Both isolated and distributed approaches are considered in the following.

5.1 Isolated approach

The isolated adaptive routing approach performs the path selection in consecutive nodes based on local node state information (like congestion conditions, actual link/buffer occupancy), i.e., each node can take a decision according to the state of its own output interfaces. It is a suboptimal solution since it only considers local information but provides good flexibility and no additional signaling is required. From this general concept, several specific algorithms can be inferred. Here we describe 4 different solutions: Path Excluding (PE), Bypass Path (BP), Multipath Routing with Dynamic Variance (MRDV), and Adaptive Multipath OBS Routing (AMOR).

5.1.1 PE description. The behavior of PE [9] algorithm is the following: each node always selects the less congested output queue among all the ports included in the set of available paths. This selection determines the next hop and excludes from the set of available paths all those paths that not include this hop in their route. Hence, from the k original LSPs, each node is removing some paths as long as remains only one path.

5.1.2 BP description [9]**.** In this case, the source node selects one LSP from the k available according to the state of its output queues and the minimum distance. The route can be modified only when traveling burst finds a congested link. In this case, the node tries to 'bypass' it using an intermediate node to reach the next hop.

5.1.3 MRDV description. MRDV [10] is a decentralized and dynamic link-state routing algorithm that balances the load in the network designed for IP Networks. The use of MRDV in OBS networks may help improve network performance. Thus, instead of pre-establishing only the shortest LSPs, longer paths are available to be used when the Shortest Paths are in high load condition. In [11] one of the authors investigated several metrics to balance the load among the possible paths, focusing on the stability of the load distribution.. MRDV assigns dynamic costs to all output links of a node in a traffic path and periodically searches for alternate output links in order to distribute network load among spare links.

5.1.4   AMOR description. This algorithm follows the idea that lies beneath the MRDV algorithm of assigning dynamic costs, but in a more proper way to adapt to OBS nodes particularities. These characteristics of OBS nodes limit their capability of reaction against congestion. While IP routers have finite memory that acts as a buffer, an OBS node may have FDL buffers which act as an optical memory but they are time limited. In the AMOR strategy, available paths are chosen more carefully than in MRDV strategy, as well as the load distribution function, which reacts to traffic profile changes.

5.1.5   Simulation results. From the Fig. 4 we can see that neither the PE nor BP isolated routing protocols help in resolving the congestion, as performance results of both strategies are only slightly better than the ones obtained for the SP algorithm. MRDV provided very similar results as ECMP due to the network topology and traffic profile, so readers are referred to Fig. 3, which show the ECMP behaviour. The AMOR strategy is still under study, and no definitive results are available yet.

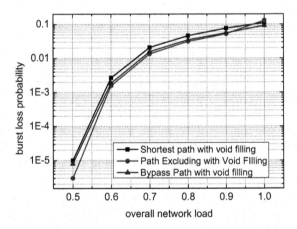

Fig. 4 Comparison among SP, PE and BP routing (nodes use 4 FDLs)

5.2 Distributed approach

The distributed adaptive routing approach, which performs the path selection in consecutive nodes based on global node state information. Each node collects the global state information that the other nodes send by means of flooding and applies a cost function to determine the lowest cost path. This solution introduces the problem of inaccuracy in network state information; in fact, routing decisions performed by such algorithm is optimal as long as this information perfectly represents the actual network state, what is impossible to achieve in real networks. Moreover, distributed routing involves additional signaling complexity so as to exchange the state information inside the network. Here we present one solution: the Distributed Path with n alternatives (DP-n).

5.2.1 DP-n description. For this strategy, each node takes samples of its states (in our implementation the states are represented by the average utilizations of each output link) during a given period T. When this time expires, each node floods its collected data to all other nodes. At the same time, the node recalculates the cost of each multi LSPs using the last update messages received. The cost function f at time t for LSP y between node i and j is

$$f_{i,j}^y(t) = \alpha \sum_{x \in path(i,j)} \overline{UL}_x + (1-\alpha) f_{i,j}^y(t-T) \tag{11}$$

where UL_x is the average utilization of link x during period $(t - T, t)$ and α is the memory factor, which takes into account previous periods.

Once the costs of every LSPs are computed, source nodes balance the traffic among a subset n of the pre-established k LSPs according to their costs.

5.2.2 Simulation results. In the following part we analyze the performances of the algorithm previously described. We have compared DP-1 and DP-4 with the burst loss probability obtained by using the SP algorithm. Four LSPs ($k = 4$) are pre-established for any source-destination pair.

Analyzing Fig. 5 it is really difficult to find a routing algorithm with better performance, as all the curves are really close. Moreover, the results show that DP-4 does not significantly improve the performances of the DP-1 algorithm.

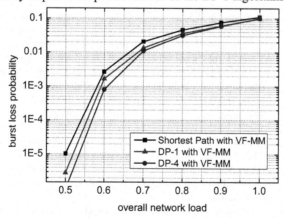

Fig. 5 Comparison among SP, DP-1 and DP-4 routing (nodes use 4 FDLs)

6 Conclusions and further work

In this paper we have presented and compared several routing strategies aimed at improving OBS network performance. By means of simulation the algorithms under study were compared. Furthermore, a method to calculate the burst loss probability analytically with the shortest path strategy was presented and validated against the simulations. The main conclusion is that none of the proposed strategies improve significantly the performance over the simple shortest path approach for this network topology and traffic profile. Thus, it is questioned whether it is worth increasing complexity in the network routing in OBS. However, further scenarios that take into account a realistic network evolution must be studied to have more general results and conclusions. Specifically, one scenario with a dynamic traffic profile may benefit from the adaptive strategies presented in this work, and another scenario with link and node failures that unbalance the network and require a mechanism to automatically route the traffic in the most efficient manner.

Acknowledgments. This work has been partially funded by IST project NOBEL II (FP6-027305) and the MEC (Spanish Ministry of Education and Science) under the CATARO project (TEC2005-08051-C03-01).

References

1. C. Qiao, M. Yoo: Optical burst switching (OBS) - a new paradigm for an optical Internet. Journal of High-Speed Networks, vol. 8, no. 1, pp. 69-84, Mar. 1999
2. J. Teng, G. N. Rouskas: Routing Path Optimization in Optical Burst Switched Networks. IEEE 2005
3. C.M. Gauger,, M. Köhn, J. Zhang, B. Mukherjee: Network performance of optical burst/packet switching: the impact of dimensioning, routing and contention resolution. Beiträge zur 6. ITG Fachtagung Photonic Networks, Leipzig, 2005
4. G. P.V. Thodime, V. M. Vokkarane, and Jason P. Jue: Dynamic Congestion-Based Load Balanced Routing in Optical Burst-Switched Networks. IEEE Globecom 2003
5. S. Ganguly, S. Bhatnagar, R. Izmailov, C. Qiao: Multi-path adaptive optical burst forwarding. Proc. of the IEEE HPSR 2004, Phoenix, AR, Apr. 2004
6. J. P. Jue, V. M. Vokkarane: Optical Burst Switched Networks. Springer Science Optical Networks Series, 2005
7. C. Qiao. Labelel optical burst switching for IP-over-WDM integration. IEEE Com. Mag., vol. 38, no. 9, Sep. 2000
8. Z. Rosberg, H.-L. Vu, M. Zukerman, J. White: Blocking Probabilities of Optical Burst Switching Networks Based on Reduced Load Fixed Point Approximations. Proc. of Infocom 2003, San Francisco, CA, Mar. 2003.
9. M. Klinkowski, F. Herrero, D. Careglio, J. Solé-Pareta: Adaptive routing algorithms for optical packet switching networks. ONDM 2005, Milan, Italy, Feb. 2005
10. F.J. Ramón Salguero, J. Andrés-Colas, J Enríquez-Gabeiras, G. García-de-Blas: Dynamic Routing Strategies to Postpone Network Congestion. Technical Report, COST 279, 2002
11. J. Aracil, O. González, J P. Fernández-Palacios, Routing strategies for OBS networks based on MRDV, ICTON 2005, Barcelona, Spain, July 2005

Transport Plane Resource Discovery Mechanisms for ASON/GMPLS Meshed Transport Networks

Jordi Perelló, Eduard Escalona, Salvatore Spadaro, Jaume Comellas,
and Gabriel Junyent

Optical Communications Group (GCO), Signal Theory and Communications Department
Universitat Politècnica de Catalunya (UPC), Jordi Girona, 1-3, 08034 Barcelona, Spain
{jperello, escalona, spadaro, comellas, junyent}@tsc.upc.edu

Abstract. The reduction of both deployment costs and operating expenses has emerged as a critical goal of next-generation optical networks. In this context, as each node maintains a large number of input/output ports (i.e. wavelengths, fibers, time slots ...), automation is needed to reduce long and tedious manual node configurations, thus easing network management and minimizing resource misconfiguration probability. In this paper, we present two novel mechanisms which permit GMPLS Link Management Protocol to deal with the resource discovery of all-optical meshed transport networks. The subsequent evaluation demonstrates their applicability over large meshed network topologies with considerably high nodal degrees.

Keywords: GMPLS, LMP, Automatic Resource Discovery.

1 Introduction

The proliferation of Internet Protocol (IP) technology, coupled with the large amount of bandwidth offered by recent advances such as Wavelength Division Multiplexing (WDM), is moving towards IP over WDM to be the preferred solution to implement next-generation optical networks. This way, supported by extraordinary technology innovations (i.e. ROADMs and OXCs), a migration from legacy SONET/SDH ring architectures to all-optical meshed ones is expected to be carried out, mainly due to their improved bandwidth efficiency and scalability. Moreover, the advent of a distributed control plane, implemented by means of GMPLS related protocols [1], emerges as a promising candidate to provide next-generation networks with the automation required to fulfill new applications requirements [2].

Automatic resource discovery is understood as the process of a node automatically finding the working links towards its transport plane neighboring nodes. This process avoids long and tedious manual node configurations and minimizes resource misconfiguration probability, which can result in high cost savings for network operators. A first approach to the required discovery procedures for the ASON architecture is introduced in [3], where their applicability over either the control or the transport plane is differentiated. It is noteworthy that their description is there carried out in an abstract way. With the introduction of GMPLS, control plane IP adjacency

I. Tomkos et al. (Eds.): ONDM 2007, LNCS 4534, pp. 221–228, 2007.
© IFIP International Federation for Information Processing 2007

discovery is delegated to legacy OSPF operation [4], whereas the discovery of the transport plane port mappings is intended to be done by the Link Management Protocol (LMP) [5]. However, to achieve its purposes, LMP requires optical nodes to transmit in-band information, so requesting link termination on each incoming port. This issue, in all-optical networks, implies additional hardware, high complexities or even signal losses.

In this paper, we propose and compare two different resource discovery mechanisms, which enable LMP to cope with the automatic discovery of all-optical meshed networks. These mechanisms have been called *Parallel Discovery* (PD) and *Concurrent Discovery* (CD). Specifically, PD extends the mechanism for ring topologies previously presented and experimentally evaluated over the ASON/GMPLS CARISMA network in [6], to be also applicable over meshed network topologies. In turn, CD is introduced to optimize PD, permitting to concurrently verify all neighboring nodes at a time. Their ease of adoption becomes a key issue, since no extensions on standard LMP messages are required.

The rest of this paper is organized as follows. In the following section, we explain the current LMP standardization regarding link connectivity verification and resource discovery. Section 3 introduces an alternative method to resolve transport plane interface mappings, enabling its applicability over all-optical networks. Section 4 depicts both proposed mechanisms, whose performances are evaluated by means of simulation results in Section 5. Finally, Section 6 draws up some conclusions.

2   Verifying Link Connectivity with LMP

Within the GMPLS framework, LMP has been newly introduced to perform several functionalities. Mainly, LMP is responsible for maintaining the connectivity between control plane neighboring nodes (i.e. Optical Connection Controllers in [3], OCCs), for correlating the properties of the existent data links between nodes, for performing resource discovery and even for providing fault isolation capabilities upon lightpath failures. While the implementation of the former two functionalities is mandatory in LMP, the implementation of the remaining is left to be optional. In this section, we provide a brief introduction to standard LMP resource discovery. For further reading and extensive explanation about LMP, the reader can refer to [5].

The TE link concept, understood as an aggregation of data links, is introduced in GMPLS so that scalability of both routing and signaling protocols can be improved. Specifically, LMP provides a mechanism to verify the connectivity of the data links, and to automatically learn TE link and data link interface associations, avoiding in this way long manual configurations. It is worth to point out that, these procedures should be performed when establishing the TE link and afterwards on a periodic basis for all the unallocated data links (i.e. to ensure their liveness). Initially, verification sessions are requested by sending *BeginVerify* messages [5], which can be accepted or rejected by responding with *BeginVerifyAck* or *BeginVerifyNack* messages [5]. Along this exchange, several parameters for the current verification session are set (i.e. verification identifier, verification scope, number of data links to be verified, data link transmission properties …). Then, once the session has been established, *Test* messages [5] are sent in-band over the first data link being verified. These messages contain the local interface identifier of that data link so that, whether a node detects the arrival of a *Test* message on an incoming port, it can resolve its remote interface association (i.e the connectivity of that port), which is subsequently notified to the neighboring node by sending a *TestStatusSuccess* message [5]. Conversely, if no *Test* message is received within an observation period (set during session establishment), a *TestStatusFailure* message [5] is returned. In turn *TestStatusAck* messages [5] are sent back, confirming the receipt of either a *TestStatusSuccess* or a *TestStatusFailure*. This operation is sequentially performed for all the data links previously considered under verification. Finally, verification sessions are ended by exchanging *EndVerify* and *EndVerifyAck* messages [5]. It is worth to recall that, since *Test* messages are sent in-band over the transport plane, electrical-optical-electrical conversion capabilities are needed on every data plane port, thus complicating its adoption by all-optical devices (extra hardware might be needed in most situations).

3   Resource Discovery using Loss-of-Light

The transmission of light on the emitter side of a link and its detection on the receiver side can be also a possibility to achieve transport plane interface mappings. This way, there is no necessity to send control information in-band (i.e. *Test* messages), so permitting control and transport planes to be decoupled. Moreover, this solution can be easily adopted by all-optical devices.

Nevertheless, there are two aspects in this method which deserve special attention. The first one is *time ortogonality versus data traffic*. This means that verification can not be performed while data links are carrying user traffic. However, this is not a big drawback if we take into account that resource discovery is performed before to consider data links ready to carry traffic (before the LMP Link Property Correlation process [5]). The second one is *Loss-of-Light (LoL) verification collision*. Firstly note that a node can only perform verification with one neighbor at a time. In addition, it must be avoided the situation where a node receives light coming from different simultaneous sessions at the same time, since this could suppose an erroneous interface mapping. It is worth to mention that, when a node is acting as an emitter of a verification session, it can send light to any of its neighbors, since it does not know in principle the connectivity of its outgoing data links.

4   Parallel and Concurrent Discovery Mechanisms

In this Section, we present both PD and CD mechanisms. These mechanisms, using the LoL to resolve interface mappings, extend the LMP behavior to enable resource discovery (i.e. link connectivity verification) over all-optical meshed networks.

Concerning PD (Fig. 1), when a node wants to perform verification towards a certain neighbor (for instance, OCC1 desires to start a verification towards OCC2), to avoid any verification collision, it has to ensure that none of its neighbors is also performing verification at the moment. Let us suppose the situation where OCC1 starts verification against OCC2 while OCC4 is performing verification towards OCC3. Since OCC1 does not know in advance the connectivity of its outgoing data links, it can emit light on a data link directly connected to OCC3. This situation would lead to a verification collision on OCC3, thus an erroneous interface mapping could happen on OCC3. To avoid such situations, firstly OCC1 tries to reserve in a sequential way the rest of neighbors by sending *BeginVerify* messages (step 1). The reception of a *BeginVerifyAck* message (step 2) from a neighbor indicates that it was not under verification and also that it will not accept any verification until the present one will be concluded. The same is done until all the other neighbors are reserved (steps 3, 4). It is worth to mention that, since *Test* messages are here sent over the control plane, the parameters needed to be negotiated for the verification sessions are drastically reduced compared to [5].

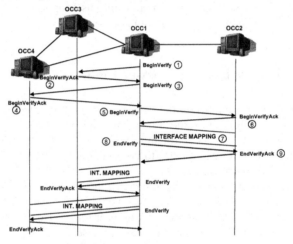

Fig. 1. PD mechanism for meshed networks

In Fig. 1, once OCC3 and OCC4 have been reserved, the session towards OCC2 can be initiated (steps 5, 6). Interface mappings (step 7) are resolved by the exchange of *Test*, *TestStatusSuccess* and *TestStatusFailure* messages like in standard LMP [5], but sending here the *Test* messages over the control plane. This way, in order to test an outgoing interface, the emitter turns its interface related laser on and sends a *Test* message to the neighbor, which contains the local identifier of that interface. When it arrives, the receiver checks if there is any presence of light on any of its incoming interfaces. If so, it resolves the interface mapping and returns a *TestStatusSuccess* containing the local and remote interface association. If no light is detected on any interface, a *TestStatusFailure* is returned. Upon their reception, the emitter confirms it by sending back a *TestStatusAck* message, stores interface association in case of a successful verification and it subsequently turns the laser down. This process is performed for all the emitter outgoing interfaces under verification. Once all of them have been tested, the session is ended by exchanging *EndVerify* and *EndVerifyAck* messages (steps 8, 9). In that moment, if any of the reserved neighbors have not been previously verified, interface mappings are sequentially performed towards each one.

Considering that, in order to initiate verification procedures, a node must reserve all its neighboring nodes so ensuring that none of them will accept any verification request, it is worth noting the possibility to concurrently verify all of them at a time. Unlike PD, where neighboring nodes are verified sequentially, the here proposed CD mechanism (Fig. 2) tries to take advantage of such situation. In the CD mechanism, firstly neighboring nodes are reserved as in PD by exchanging *BeginVerify* and *BeginVerifyAck* messages. Then, interface mappings are performed as follows. In order to test an outgoing interface, first of all the emitter turns its interface related laser on (step 1). Next to, *Test* messages containing the related local interface identifier are sent to all neighboring nodes (step 2). In turn, upon reception of a *Test* message, each node checks the presence of light on any of its incoming interfaces, so that whether the checking has been successful, a *TestStatusSuccess* is responded carrying the interface association (step 3). Otherwise, a *TestStatusFailure* is returned (step 4, 5). Once the emitter has received response from all its neighboring nodes, it stores the interface association (if any) and it subsequently turns the laser down (step 6). This process is done for all the emitter outgoing interfaces. Finally, verification sessions are ended as in PD.

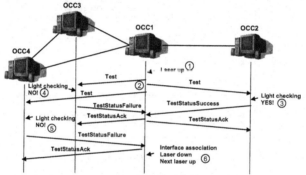

Fig. 2. CD interface mapping

In both PD and CD, if a node, which at a moment is performing verification, receives a *BeginVerify* message, it responds with a *BeginVerifyNack* indicating that it is unwilling to verify. In turn, whether a node receives a *BeginVerifyNack*, it liberates any previously established session by exchanging *EndVerify* and *EndVerifyAck* messages and then waits for a random time (uniformly distributed between 0 and *retry limit*) before re-trying verification. Note that, as in LMP standardization, these mechanisms can be periodically performed for all unallocated data links. In such case, to avoid any traffic disruption, those allocated data links are excluded from verification.

5. Performance Evaluation

In order to quantify the performance of our proposals, simulation studies using the OPNET network simulator [7] over three 28-node different topologies, each one with different average nodal degree (ND), have been carried out. Specifically, the

European Optical Network (EON) *basic reference* (ND = 2.93), *Sparse Ring* (ND = 2.36) and *Densely Meshed Triangular* (ND = 4.36) topologies have been considered [8]. In this way, the obtained results show also how both PD and CD behave as average nodal degree increases in a meshed topology.

The evaluation of the mechanisms has been done taking the discovery time (i.e. the time required to discover the whole transport plane) as the figure of merit. Some assumptions have been initially adopted: (1) eight wavelengths (four bidirectional data links) are available between nodes; (2) the retry values in the PD and CD mechanisms are generated following a uniform distribution between 0 and *retry limit*; (3) 100 μs for laser turn on/off time; (4) 10 ms for the transponder reaction time upon presence/absence of light (i.e. once a laser is turned up, the node waits 10 ms to send the related *Test* message). In this way, we ensure that, upon receiving a *Test* message, the neighbor will have detected the optical power rise on its incoming ports (if connectivity exists between both). Each simulation has been conducted in order to reach steady state results within a 95% confidence interval.

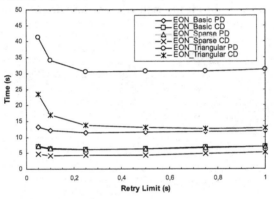

Fig. 3. Transport plane discovery times vs. retry limit

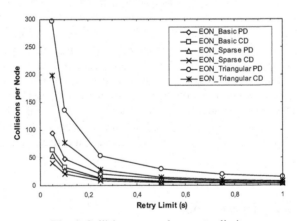

Fig. 4. Collisions per node vs. retry limit

Fig. 3 plots the behavior of PD and CD function of ND and the *retry limit* value used. Although both mechanisms highly depend on ND, it can be noticed their feasibility, even in densely meshed networks. It can be also seen notwithstanding, how CD outperforms PD in every evaluated topology. Furthermore, the difference between both drastically increases as nodal degree gets higher. For instance, when considering the Sparse Ring topology (degree 2.36), the differences are about 2 sec, whereas when considering the Triangular one (degree 4.36), the differences can rise up to 20 sec.

To properly choose the best value for the *retry limit*, we have also quantified the introduced overhead (due to the retrying) for both PD and CD in terms of the number of collisions per node versus the *retry limit* (Fig. 4). First of all, it can be seen how CD introduces less overhead than PD. In addition, it can be concluded that the best compromise between the achieved discovery time and the introduced overhead is obtained for a *retry limit* value equal to 500 ms. Although the lowest discovery times are achieved for a retry limit value of 250 ms in almost all situations (except when using CD over the triangular topology), similar performance is obtained for a 500 ms *retry limit* while introducing less overhead. Therefore, according to these obtained results, we suggest the use of this value to implement both mechanisms. It is noteworthy how, lower *retry limit* values than 250 ms are translated into large undesirable overhead increments.

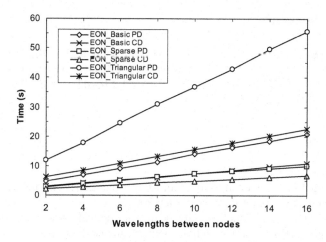

Fig. 5. Transport plane discovery times vs. the number of data links between nodes

So far, both PD and CD have been evaluated taking into account a fixed number of data links between neighbors (8 wavelengths, 4 bidirectional data links). In addition, we have also evaluated how both proposals depend on the number of wavelengths established between nodes. Here, in order to minimize the introduced overhead and according to the previous results, a 500 ms *retry limit* value has been applied. The results are shown in Fig. 5. We can firstly notice the linear dependence of PD and CD on the number of resources between nodes. Moreover, we can observe that this dependence becomes more stressed as ND gets higher, especially for PD. In fact, when considering 16 wavelengths (8 bidirectional data links), we can distinguish differences greater than 30 seconds between PD and CD on the triangular topology.

6. Conclusions

In this paper, we have addressed the automatic resource discovery of all-optical meshed transport networks. To begin with, we have introduced the standard LMP link connectivity verification functionalities, emphasizing its electrical-optical-electrical capability requirement, usually not available in all-optical devices. Alternatively, in order to overcome such requirement, we have also presented a method which resolves data link interface mappings without requiring to send in-band *Test* messages. Then, we have proposed two novel discovery mechanisms for meshed networks, whose performance has been evaluated and further compared by means of simulations over three reference networks with different nodal degrees. Mainly, simulation results show the feasibility of both mechanisms. Moreover, improvements can be seen for CD in front of PD in terms of less achieved total discovery times, less introduced overhead and less dependence on the number of existent resources (i.e. wavelengths) between nodes. Furthermore, such improvements drastically increase as the average node degree gets higher.

Acknowledgements

The work reported in this paper has been partially supported by the Spanish Science Ministry through Project "Red inteligente GMPLS/ASON con integración de nodos reconfigurables (RINGING)", (TEC2005-08051-C03-02), and by the i2CAT Foundation (www.i2cat.cat).

References

1. Mannie, E.: Generalized Multi-Protocol Label Switching (GMPLS) Architecture, IETF RFC 3945, Oct. 2004.
2. Jajszczyk, A.: Automatically switched optical networks: benefits and requirements, IEEE Communications Magazine, Feb. 2005.
3. ITU-T Recommendation G.8080/Y.1304 version 1.0 (and Amendment 2): Architecture for the automatically switched optical network (ASON), Nov. 2001 (and Feb. 2005).
4. Katz,D., Kompella, K., Yeung,D.: Traffic Engineering (TE) Extensions to OSPF Version 2, IETF RFC 3630, Sept. 2003.
5. Lang, J.: Link Management Protocol (LMP), IETF RFC 4204, Oct. 2005.
6. Perelló, J., et al.: Resource Discovery using Link Management Protocol for ASON/GMPLS Ring-based Transport Networks, in Proc. of 32nd European Conference on Optical Communication (ECOC), Sept. 2006.
7. OPNET Technologies Inc., www.opnet.com
8. Inkret, R., Kuchar, A., Mikac, B.: Advanced Infrastructure for Photonic Networks, Extended Final Report of COST Action 266, Sept. 2003.

A Study of Connection Management Approaches for an impairment-aware Optical Control Plane

Elio Salvadori[1], Yabin Ye[1], Andrea Zanardi[1], Hagen Woesner[1], Matteo Carcagnì[1]

Gabriele Galimberti[2], Giovanni Martinelli[2], Alberto Tanzi[2], Domenico La Fauci[2]

[1]Create-Net, Via Solteri 38, Trento, Italy, 38100
{elio.salvadori, yabin.ye, andrea.zanardi, hagen.woesner, matteo.carcagni} @create-net.org
[2] Cisco Optical Networking Group, Via Philips, 12, 20052 - Monza (MI), Italy
{ggalimbe, giomarti, atanzi, dlafauci} @cisco.com

Abstract. Transparent optical networks need novel connection management approaches to take into account the presence of physical impairments in lightpath provisioning. Two main schemes are emerging from literature when considering how to introduce impairment-aware mechanisms in a distributed optical control plane like GMPLS. A well-known approach is based on extending the routing protocol to compute an optically-feasible light-path. Lately, a new approach is emerging which keeps the routing protocol unmodified while leveraging on signaling protocol extensions to find the proper lightpath for the incoming connection request. The aim of this paper is to prove that the *signaling-based* approach has several advantages compared to the *routing-based* one, in term of scalability and robustness especially when link information changes are frequent in the network. Simulation results show that a *signaling-based* approach is much more robust to inaccurate information about network status, therefore it is a suitable approach for considering physical impairments in dynamic optical networks.

Keywords: Generalized Multi-Protocol Label Switching (GMPLS); Optical Control Plane (OCP); Wavelength Division Multiplexing (WDM); Routing and Wavelength Assignment (RWA); Physical Impairments

1   Introduction

Generalized Multi-Protocol Label Switching (GMPLS) has been proposed for managing the control plane in Wavelength Division Multiplexed (WDM) optical networks [1]. The routing protocol used in GMPLS is open shortest path first with traffic engineering extensions (OSPF-TE) [2], which has two main objectives: i) to provide the nodes with a dynamic and more exact view on network status (capacity, load, congestion state and other link attributes) and ii) to enable constrained-based routing (CBR) to be run in each node, in order to perform intelligent path computation (explicit route determination). Resource Reservation with traffic engineering extensions (RSVP-TE) [3] is the signaling protocol often considered to establish the

I. Tomkos et al. (Eds.): ONDM 2007, LNCS 4534, pp. 229–238, 2007.

path whose explicit route has been calculated through some CBR mechanism, by instantiating a label forwarding state along the path itself.

Standard GMPLS does not take into account the physical impairments of real optical networks. That means it assumes an ideal network where optical signals are transmitted from source to destination nodes without degradation. While this assumption is largely correct for optical networks with O/E/O conversion in each node, in upcoming photonic networks employing purely optical crossconnects (OXC), due to transmission impairments (insertion losses, amplified spontaneous emission (ASE) noises, polarization mode dispersion (PMD), chromatic dispersion (CD), crosstalk, etc.) accumulation, the signal significantly degrades when it travels through the lightpath. Most of the currently used route computation algorithms would perform badly, if not fail, in establishing a working lightpath since they are not aware of the specific physical layer constraints and the path selected would possibly lead to unacceptably high bit-error rates (BER) [4-8].

There are two main approaches to introduce impairment-awareness in a GMPLS-based distributed optical control plane [9]. The first approach (*routing-based*) introduces additional physical information into the routing protocol, i.e. OSPF-TE [10-11]. By flooding Link State Advertisements (LSAs), all the nodes populate their Traffic Engineering Database (TED) information which gives them a view of the whole network. Each node can then find the right route for a connection request while standard RSVP-TE signaling is used for lightpath establishment.

The second approach (*signaling-based*) extends the RSVP-TE protocol instead [12]. Each node can select a route based on standard OSPF-TE protocol without knowledge of physical impairments and then the feasibility of the computed optical path is evaluated on a hop-by-hop basis in each node along the path during lightpath set-up. In this case RSVP-TE messages are extended to include physical information.

An initial theoretical assessment of these two approaches is reported in [9], where the authors reflect on the main limitations of the *routing-based* approach, such as TED inconsistency as well as scalability and stability problems when the link information changes are frequent [11]. Another limitation of this approach is the impact on the NE's Control Unit CPU, whose load is heavily stressed by the complexity of the multi-constrained path computation algorithm required for guaranteeing both optimal network performance and sufficient quality of the optical signal. On the other hand, the *signaling-based* approach better handles frequent changes of the physical parameters, and no global flooding of physical information is required, thereby minimizing scalability problems. Furthermore since no complex path computation algorithms are used, the load on the NE's CU is minimized. However, the main drawbacks of this approach are: a bigger set-up delay due to an increased number of set-up attempts as well as a sub-optimal resource allocation due to the impairment-unaware route computation algorithm used.

In this paper, for the first time to our knowledge, an in-depth comparison of these two approaches is provided through an extensive simulation activity. In particular, the increased robustness to imprecise network state information as well as much better scalability of the *signaling-based* approach compared to the *routing-based* is demonstrated, while showing reduced impact on the set-up delay. Notice that the set of impairments considered in this work has been restricted to the linear ones to limit the complexity of the problem we have been studying; however according to the obtained results, including non-linear impairments would have further highlighted the limits of a *routing-based* approach compared to a *signaling-based* one.

Section 2 of this paper provides details of the two optical control plane (OCP) approaches considered. In Section 3 we present and discuss illustrative numerical examples to evaluate the two approaches. Section 4 concludes this paper. The models for the physical impairments used in our simulations are presented in the Appendix A.

2 Impairment-Aware Optical Control Plane Architectural Options

In the following the two approaches previously introduced are described in more detail.

2.1 *Signaling-based* OCP

In this approach, no extensions to the routing protocols to describe transmission impairments and wavelength availability are introduced. Instead, each node has only a local knowledge of the physical parameters related to its adjacent links. In this architectural option the verification of the feasibility of a lightpath is entirely delegated to the signaling phase. The mechanism considered in this paper is very close to the one proposed in [12], but with respect to this a realistic mathematical model for evaluating several linear transmission impairments considered has been used (see Appendix A for more details) instead of a generic additive linear parameter.

Each time a node receives a request to set up a connection, a local path computation element will compute a route to the destination node. Note that this route may not be feasible from the photonic domain point of view. If established, it would result in an unacceptably high BER, violate service level agreements (SLA) while at the same time reserve capacity in the network. In this approach some extensions are therefore introduced in the signaling messages (e.g. RSVP-TE) to collect signal impairments characterizing the traversed links from the ingress to the egress node. In particular, the idea is to include into a **Path** message one or more fields containing information about the optical signal level (e.g. signal power, optical signal to noise ratio OSNR) and accumulated impairments (e.g. CD, Crosstalk, PMD) per wavelength to be set up. This information is updated at every traversed node together with the set of available wavelengths by leveraging the local knowledge of the physical parameters available in the node itself. The egress node will evaluate the feasibility of a suggested wavelength (label) by evaluating the transmission quality according to the client interface terminating the lightpath (in terms of BER, OSNR sensitivity, CD tolerance, etc.). If at least one of the suggested wavelengths is feasible, the egress node will send the **Resv** message back to the ingress selecting the wavelength in order to set up the lightpath. If more than one are feasible, the lower-order one is selected according to a First-Fit policy. If not, a **ResvErr** message must be sent back and a new route to establish the connection request must be calculated.

Compared to [12], which is performing only up to two re-attempts, we have been allowing the ingress node to perform up to K attempts to set-up a lightpath. A K-CSPF path computation algorithm has been implemented in the simulator, which provides the ingress node with up to K possible paths to set up a lightpath to the egress node. While a high K increases the probability to successfully set-up a lightpath request, it also dangerously increases the lightpath set-up delay. The proper value of K should therefore be carefully selected depending on the network topology and on the size of the transparent domain. Anyway, simulation results demonstrate that good performance can be generally obtained even with low values of K.

2.2   *Routing-based* OCP

In this case, each node must have an overall view of the resource utilization and the physical parameters in the network, maintained through a TED available locally. This means that some extensions to the routing protocol (i.e. to LSA messages of OSPF-TE protocol) must be implemented to carry wavelength usage and transmission impairments information in addition to TE attributes. Each time a node receives an incoming connection request, a local path computation element will compute a route through a constrained-based routing algorithm which takes into account the wavelength availability as well as the physical impairments. The ICBR algorithm used in this paper is very close to the one proposed in [8], with two modules, the *network-layer module*[1] which computes a set of candidate lightpaths for the incoming request, among which one of them is selected through a *physical-layer module* responsible for checking its optical feasibility. However, with respect to [8] which was focusing on the performance of the ICBR algorithm only, in this paper the effect of the OSPF protocol extension is studied. Furthermore, more transmission impairments have been considered.

It is important to note that for the mathematical models considered in our simulations the only dynamic information that needs to be updated via LSA messages is the *wavelength availability* per link. All other parameters considered in Appendix A are of static nature or function of the number of wavelengths and can thus be pre-loaded onto each NE during the provisioning phase. Once a node receives the specific *wavelength availability* information per link, it can compute the optical feasibility through its physical-layer module implementing the equations described in Appendix A[2]. The optically feasible computed path would then be set up through standard RSVP-TE selecting one of the available wavelengths according to a First-Fit policy.

In this approach, the TED instance available locally inside each node is supposed to be the same in the all network. However, due to misalignment and routing protocol convergence time in case of network changes (topology, link availability, etc), this database cannot be 100% guaranteed to reflect the real network status. It is worth to note that in the case of an extended LSA any change in the number of available wavelengths in some fiber link will generate new update messages thus frequently misaligning for some time interval the TED instances in the network nodes. Another disadvantage of this approach is that in the case of a heterogeneous multi-vendor

---

[1] Note that for the implementation of the network-layer module we have been re-using the K-CSPF path computation algorithm used in the *Signaling-based* OCP, with a very high value for K.

[2] Note that this is a strong assumption that is reasonable only in the case of a homogeneous network scenario (nodes are provided by the same vendor).

network the TED will potentially need to store information for different kinds of nodes and links. For a change or update in the network, a significant amount of parameters need to be updated, which can cause stability and scalability problems.

3. Numerical Results and Discussion

In this section, the Signaling based-OCP and Routing based-OCP are evaluated through numerical simulations performed with an extended version of the GLASS simulator [14]. In the simulations, typical regular network (Mesh-Torus, Fig. 1(a)), real network (AT&T network [15], Fig. 1(b)) and quasi-regular network (Grid, Fig. (c)) have been studied. Both Mesh-Torus and Grid networks have fiber links 80 km long, so only pre- and booster optical amplifiers are used inside each node, while no in-line optical amplifier are used per link. For simulation purposes, the AT&T topology has been scaled down as well (by a factor of 1:23) to avoid in-line optical amplifiers in all fiber links. In all the networks, nodes are connected through pairs of uni-directional WDM links with 32 wavelengths each; all connections are bi-directional using the same wavelength in both directions.

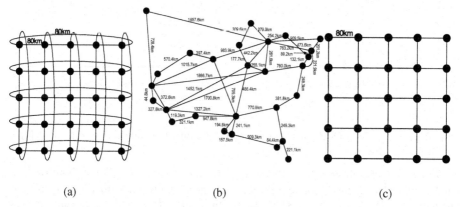

(a)   (b)   (c)

Fig. 1. The network topologies used for simulations (a) 5×5Mesh-Torus network (b) AT&T network (c) 5×5 Grid network

The **traffic scenario** in the simulation consists of a classical traffic model based on lightpath requests with Poisson arrivals average rate ($1/\mu$ per second) and average exponential duration with v seconds. The traffic requests are uniformly distributed among all the nodes. The **traffic load** is defined as the average network resource (link wavelength) usage computed as:

$$\frac{\overline{N}_c \times \overline{L}_c}{M \times W} \times 100\%$$

where $\overline{N}_c$ is the average number of active connections and equals v/μ, $\overline{L}_c$ is the average number of hops in the network (considering only shortest paths between nodes), M is the number of links and W is the number of wavelengths in the network.

In the simulation, route selection is performed in the source node using a modified breadth-first search algorithm that iteratively computes the first K distinct paths satisfying the constraints on available link resources. In *routing-based* OCP only the first selected path satisfying the wavelength continuity constraint and optically feasible is returned (K is virtually infinite).

The purpose of the performed simulations is to compare the *signaling-based* to the *routing-based* OCP approach in GMPLS. We first compare the tolerance of each OCP approach to the inaccurate TED information and then show how strong is the effect of the increased set up time in a *signaling-based* OCP architecture.

In order to study the influence of inaccurate information on both architectures, we artificially introduce an OSPF message processing delay. When each node receives a LSA from its neighbor, it will hold it for a certain number of seconds before flooding it to its neighbors. In this case, all the nodes will have outdated network status.

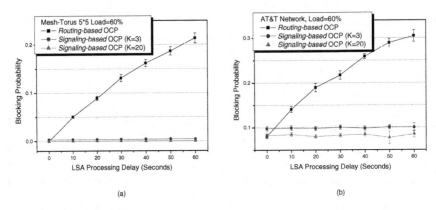

(a)   (b)

Fig. 2 *signaling-based* OCP vs. *routing-based* OCP with increasing LSA processing delay in (a)Mesh-Torus 5×5 network, (b) AT&T network

Fig.2 compares the blocking probability of *signaling-based* OCP (for both K=3,20) and *routing-based* OCP changing with LSA processing delay in Mesh-Tours 5×5 network and AT&T network when the traffic load is 60% and the average request inter-arrival time μ is 2 seconds. This interarrival frequency is supposed to model a dynamic lightpath requests due to a grooming policy requesting new lightpaths each time highly dynamic IP traffic is crossing a 50% occupation threshold per established lightpath, which is a common assumption for most Tier-1 ISPs. The LSA Processing Delay has been overestimated to better show its effect on the performance of both OCP architectures. When there is no LSA processing delay, *routing-based* OCP has similar blocking probability performance to the *signaling-based* OCP with K=20 and it is slightly better than that with K=3 (our simulations proved this result is valid for all traffic loads and on different network topologies, but we do not include them here for space reasons). However, by increasing the LSA processing delay, the blocking probability of *routing-based* OCP architecture increases quickly, while for the *signaling-based* OCP architecture it remains stable for both K=3 and 20. A *signaling-based* OCP is therefore more robust to inaccurate network information than *routing-based* OCP.

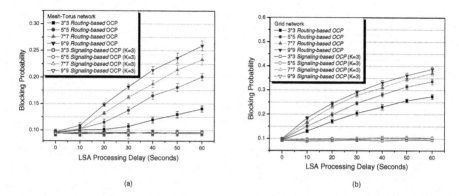

Fig. 3 - Scalability of *signaling-based* OCP and *routing-based* OCP (a) Mesh-Torus Network, (b) Grid Network

In order to check our assumption of a better scalability of the *signaling-based* OCP compared to the *routing-based* OCP architecture we have been running a set of simulations on both a regular (Mesh-Torus) and a quasi-regular (Grid) topology with increasing number of nodes. The curves of the network blocking probability changing with increasing LSA processing delay are shown in Fig. 3. The starting point for different network sizes is obtained by finding the traffic load corresponding to the same network blocking probability when there is no LSA delay. For a *routing-based* OCP architecture, the larger the network, the faster the blocking probability increases with the LSA delay. This is because in large networks, the TED of each node needs more time to be updated and therefore the source node has less accurate information when calculating the K paths. On the other hand, for *signaling-based* OCP architecture, the blocking probability is almost the same for different network sizes and does not change with increasing LSA delay. This means that a *signaling-based* OCP approach is more scalable than a *routing-based* OCP, as deferring in the signaling phase wavelength selection and optical feasibility verification in the signaling phase using local node information greatly reduces the influence of TED misalignments.

It has been proven from Fig.2 and Fig.3 that *signaling-based OCP* is more tolerable to inaccurate TED information and therefore has better scalability than *routing-based OCP*. However, as it is stated in Section 1, its main weakness is the longer lightpath set-up time, which is proportional to the average number of attempts, defined as the number of distinct paths the source node has tried before it receives the **Resv** message.

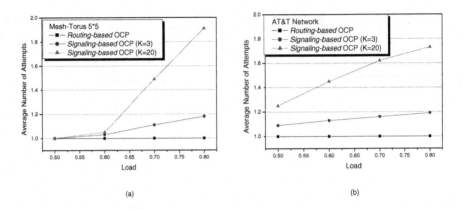

Fig. 4 Comparison of average number of attempts between *signaling-based* OCP and *routing-based* OCP in (a) Mesh-Torus 5×5 network, (b) AT&T network

Fig. 4 shows the average number of attempts changing with the traffic load for *signaling-based* OCP and *routing-based* OCP in Mesh-Torus 5×5 network and AT&T network. The average number of attempts is obtained by averaging number of attempts over all the successfully established connections. While the number of attempts for the *routing-based* OCP is always 1, the number of attempts for *signaling-based* OCP increases with the traffic load. However, it can be found from Fig. 4 that the average number of attempts for *signaling-based* OCP (K=3) is not higher than 20% of the *routing-based* OCP. This indicates that the impact on the network performance of the reattempts performed to discover an optically feasible path is not severe, at least for low K values. Of course for K=20 this impact is stronger as the number of attempts almost doubles for high traffic load.

4   Conclusions

In this paper we have studied the problem of enhancing GMPLS for considering physical impairments in real optical networks. Two approaches have been compared: a *routing-based* OCP architecture which is extending the OSPF-TE protocol to broadcast physical impairments to all nodes and then to find the appropriate route and wavelength based on TED; a *signaling-based* OCP architecture which is instead extending the RSVP-TE protocol to check the feasibility of the lightpath in a hop by hop manner. These two approaches have been implemented in GLASS simulator by considering a set of linear physical impairments (Loss, ASE noise, PMD, CD and Crosstalk, etc). An extensive set of numerical simulation performed on both regular and arbitrary topologies proved that a *signaling-based* OCP architecture is much more robust to inaccurate network information thanks to its distributed nature, with a limited impact on the connection setup time. Furthermore this approach is also more scalable than *routing-based* OCP for networks with high number of nodes.

References

1. RFC 3945 – Generalized Multi-Protocol Label Switching Architecture – October 2004
2. RFC 4203 – OSPF Extensions in Support of Generalized Multi-Protocol Label Switching – October 2005
3. RFC 3473 – Generalized Multi-Protocol Label Switching (GMPLS) Signaling Resource ReserVation Protocol-Traffic Engineering (RSVP-TE) Extensions – January 2003
4. B. Mukherjee: Optical communication Networks, New York: McGraw-Hill, July 1997
5. B. Ramamurthy, D. Datta, H. Feng, J. P. Heritage, B. Mukherjee, Impact of Transmission Impairments on the Teletraffic Performance of Wavelength-routed Optical Networks, IEEE/OSA Journal of Lightwave Technology, vol. 17, no. 10, pp. 1713-1723, Oct. 1999.
6. Teck Yoong Chai, Tee Hiang Cheng, Yabin Ye, Qiang Liu, Inband Crosstalk Analysis of Optical Cross-Connect Architectures, IEEE/OSA Journal of Lightwave Technology, vol. 23, no. 2, pp. 688-702, February 2005
7. R. Cardillo, V. Curri, M. Mellia: Considering Transmission Impairments in Wavelength Routed Networks, Optical Network Design and Models – ONDM 2005, February 7-9
8. Y. Huang, J. P. Heritage, and B. Mukherjee, Connection Provisioning with Transmission Impairment Consideration in Optical WDM Networks with High-Speed Channels, IEEE/OSA Journal of Lightwave Technology, vol. 23, no.3, March 2005
9. R. Martínez, C. Pinart, F. Cugini, N. Andriolli, L. Valcarenghi, P. Castoldi, L. Wosinska, J. Comellas and G. Junyent: Challenges and Requirements for Introducing Impairment-Awareness into the Management and Control Planes of ASON/GMPLS WDM Networks, IEEE Communications Magazine, Vol. 44, No. 12, December 2006, pp76-85.
10. J. Strand, A. Chiu and R. Tkach, Issues for Routing in the Optical Layer, IEEE Communication Magazine, February 2001
11. Shu Shen, Gaoxi Xiao and Tee Hiang Cheng, The Performance of Periodic Link-State Update in Wavelength-Routed Networks, Broadnets 2006
12. F. Cugini, N. Andriolli, L. Valcarenghi, P. Castoldi, A Novel Signaling Approach to Encompass Physical Impairments in GMPLS Networks, IEEE Globecom 2004, pp369-373
13. GMPLS Lightwave Agile Switching Simulator (GLASS) http://snad.ncsl.nist.gov/glass/
14. AT&T network topology: http://www.ssfnet.org/Exchange/gallery/usa/index.html

Appendix A: Transmission Impairments models

The transmission impairments considered in this paper include: Loss, ASE noise, PMD, CD, and Crosstalk. The final Signal Power and optical signal noise ratio (OSNR) will be checked at the end node. If the power and OSNR of signal at the end node can be accepted by the destination transponder, then the lightpath can be established.

By assuming an optical path is composed of M fiber links, $M+1$ optical nodes and N optical amplifiers, for a specific wavelength m,

- The final optical power can be modeled as [4, 5]:

$$P_{out-m} = P_{in-m} + \sum_{j=1}^{N} G_j - \sum_{i=1}^{M} Loss_{fiber-i} \cdot L_i - \sum_{k=1}^{M+1} Loss_{node-k}$$

where: P_{out-m} is the final output power (dBm) of wavelength m; P_{in-m} is the input power (dBm) of wavelength m; G_j is the gain of the jth optical amplifier. $Loss_{fiber-i}$ is the ith fiber attenuation (dB/Km); L_i is the ith fiber length (Km); $Loss_{node-k}$ is the power loss (dB) for the specific switching path (add, drop or express) at the kth node.

- The optical OSNR is degraded by ASE noise in each optical amplifier [4]:

$$OSNR_{out-m} = 10 \cdot \log_{10}\left(\frac{1}{OSNR_{in-m}} + \frac{NF_j \cdot h\nu_m B}{P_{in-m}}\right)^{-1}$$

where: $OSNR_{out-m}$ is output OSNR (dB) of wavelength m due to ASE noise in the jth amplifier; $OSNR_{in-m}$ is the input channel OSNR (dB); h is the Planck constant; ν_m is the frequency of wavelength m; B is the bandwidth of optical filter; NF_j is the noise figure of the jth amplifier.

The optical OSNR is also degraded by accumulated CD, PMD and Cross-talk, which can be modeled as:

- The final CD [4]:

$$CD_{out-m} = CD_{in-m} + \sum_{i}^{M}(CD_{coeff-i} + \Delta\lambda_i \times CD_{slope-i}) \times L_i - \sum_{j=1}^{N}DC_{dcu-j}$$

where: CD_{out-m} is the final output CD (ps/nm) of wavelength m; CD_{in-m} is the input CD (ps/nm); $CD_{coeff-i}$ is the CD coefficient for reference wavelength (ps/nm/Km) in i^{th} fiber; $CD_{slope-i}$ is the CD slope coefficient (ps/nm^2 Km) in i^{th} fiber; $\Delta\lambda_i$ is the lambda deviation of wavelength m from reference wavelength (nm) in i^{th} fiber; DC_{dcu-j} is the dispersion compensation unit (DCU) module compensation value (ps/nm) in j^{th} optical amplifier (we assume that DCU units are only located in the optical amplifiers)

- The final PMD [4]:

$$PMD_{out-m}^2 = PMD_{in-m}^2 + \sum_{i}^{M}PMD_{coeff-i}^2 \times L_i + \sum_{j=1}^{N}(PMD_{DCU-j}^2 + PMD_{AMP-j}^2) + \sum_{k=1}^{M+1}PMD_{node-k}^2$$

where: PMD_{out-m} is the output PMD (ps) of wavelength m; PMD_{f-in} is the input PMD (ps) of wavelength m; $PMD_{coeff-i}$ is the PMD coefficient (ps/√Km) of the ith fiber ; L_i is the ith fiber length (Km) ; PMD_{DCU-j} is the DCU module PMD value (ps) in j^{th} amplifier; PMD_{AMP-j} is the jth Amplifier PMD value (ps) ; PMD_{WXC-k} is the PMD value (ps) of the kth optical node.

- The output Cross-talk after each node [4-6]:

$$Xt_{out-m} = 10\lg(10^{\frac{Xt_{in-m}}{10}} + 10^{\frac{Xt_{node-k}}{10}})$$

where: Xt_{out-m} is the output crosstalk (dBm); Xt_{in-m} is the input crosstalk (dBm); Xt_{node-k} is the crosstalk value associated to k^{th} node (dBm).

In the destination node, the OSNR penalties due to CD, PMD, Crosstalk are calculated. Note that the transmission parameter values used in our simulations have not been included in this paper for space reason mainly; however, as a general indication, we have been assuming that all network links are based on Single Mode Fiber (SMF) and the optical amplifiers considered are based on EDFA (Erbium-Doped Fiber Amplifier) technology and compensated through DCF (Dispersion Compensating Fiber) units.

An Automatic Model-based Reconfiguration and Monitoring Mechanism for Flexible GMPLS-based Optical Networking Testbeds*

Fermín Galán Márquez and Raül Muñoz

Centre Tecnològic de Telecomunicacions de Catalunya (CTTC)
Parc Mediterrani de la Tecnologia, Av. Canal Olímpic s/n, 08860 Castelldefels, Spain
{fermin.galan,raul.munoz}@cttc.es

Abstract. Testbeds play a key role in the evolution of GMPLS-based Intelligent Optical Networks (ION) proving grounds in which new optical networking research ideas (e.g., new constraint-based routing algorithms) can be tested and evaluated. In order to be a productive experimentation environment, a GMPLS optical network testbed should be flexible, allowing the reconfiguration of as many different network topologies and configurations as possible. But usually this flexibility comes at the expense of high management costs when switching from one scenario to another is performed through time-consuming error-prone manual procedures. This paper describes an automatic model-based deployment mechanism that overcomes the limitations of manual reconfigurable testbeds, allowing high flexibility without involving high management costs. The model-based approach is not only suitable for deployment (and undeployment) but also for monitoring. The practical application of the mechanism to ADRENALINE testbed (a GMPLS-based all-optical transport network developed at CTTC) with the ADNETCONF tool is also described.

1 Introduction

The accelerating growth of Internet traffic is motivating the research on dynamic transport networks based upon recent advances in optical networking technologies such as Wavelength Division Multiplexing (WDM), Reconfigurable Optical Add Drop Multiplexers (R-OADM), Optical Cross Connects (OXC) and tunable lasers, capable of providing reconfigurable high-bandwidth, end-to-end optical connections. The introduction of the dynamism or intelligence in future optical networks can be achieved by means of a distributed optical control plane (i.e. routing and signalling). This control plane can be based on Generalized Multiprotocol Label Switching (GMPLS) [1] protocol architecture, an extension of MPLS (Multiprotocol Label Switching) to cover circuit-oriented optical switching technologies such as WDM. One of the major applications of GMPLS is

---

* This work was partially funded by the MEC (Spanish Ministry of Science and Education) through project RESPLANDOR under contract TEC2006-12910/TCM.

I. Tomkos et al. (Eds.): ONDM 2007, LNCS 4534, pp. 239–248, 2007.

constraint-based routing (CBR), which is used to compute paths that satisfy various requirements subject to a set of constraints.

Performance analysis of CBR algorithms in optical networks has been widely studied in the past through simulations. However there is a considerable lack of experimental performance evaluation of real GMPLS-based CBR implemented in optical network testbeds.This experiments require flexible testbed platforms with a high degree of reconfigurability in order to allow the deployment of not only different network topologies (e.g., NSFNet, Pan-European network, metro rings, etc.) but also, different configurations for each topology. (e.g., CBR algorithms to test, available resources per link, etc.).

Manual reconfiguration is the usual procedure to achieve such flexibility, but introducing commands and configurations manually in the different test-bed devices has several important drawbacks. Firstly, is a high time-consuming task due to a lot of time is employed performing tedious and mechanics operations. Secondly, reconfiguration needs specific knowledge not related with the goal of the testbed itself. Third, humans tend to make errors typing commands and writing configurations. Finally, manual reconfiguration is not scalable. The aforementioned problems can be overcome implementing automatic reconfiguration procedures. This paper describes one of such mechanism, based on optical network modelling and the processing of models to perform automatic reconfiguration (deploy and undeploy) actions (in addition, model-based monitoring is also possible) and applies it to ADRENALINE testbed [2], a flexible GMPLS-based all-optical transport network developed at CTTC laboratories.

The rest of the paper is structured as follows. The ADRENALINE testbed is introduced in Sect. 2 as example of flexible GMPLS-based optical network testbed. Then Sect. 3 describes the proposed automatic model-based reconfiguration and monitoring mechanism. After that, Sect. 4 focuses on the application of the proposed approach to ADRENALINE testbed. Finally, Sect. 5 concludes de paper and presents future work lines.

2   ADRENALINE Testbed: An Example of Flexible GMPLS-based Optical Network

The ADRENALINE (All-optical Dynamic REliable Network hAndLINg IP/Ethernet Gigabit traffic with QoS) testbed [2] is a GMPLS-based Intelligent Optical Network (ION) developed at CTTC laboratories (Fig. 1). It is composed by an all-optical transport network constituted by a metropolitan DWDM bidirectional ring with three colourless R-OADM nodes and tuneable lasers, providing reconfigurable (space and frequency) end-to-end lightpaths. Each optical node is equipped with a PC Linux-based Optical Connection Controller (OCC) for implementing the GMPLS-based distributed control plane. These three OCCs are named *optical* OCCs. The control plane is responsible for handling dynamically and in real-time optical node's resources in order to manage automatic provisioning and survivability of lightpaths through signalling and routing protocols. ADRENALINE deploys three optical bidirectional pairs of fiber. The Data Communication Network (DCN) employed for exchange signaling and routing packets between OCCs is based on control channels carried at 1310 nm with a line rate of 100 Mb/s using point-to-point links. Note that the optical transport topology is fixed to a ring network and can not be modified.

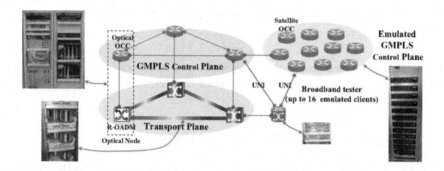

Fig. 1. ADRENALINE functional architecture

Given the fact that one of the focus of ADRENALINE testbed is the performance evaluation of GMPLS-based CBR algorithms and schemes, it was introduced a new set of ten OCCs named *satellite* OCCs. The difference with optical OCCs is that there is no optical hardware associated, that is, the optical hardware is emulated. The satellite OCCs introduce a new degree of flexibility, since there is no restriction neither on the optical network topology nor on the resources per link (e.g., number of available wavelengths, fibers, etc.). Regarding to the DCN, the satellite OCCs can be connected between themselves or with the fix ring of optical OCCs following any topology, through Fast Ethernet control channels. But in this case, the control channel is carried over emulated optical links between any pair of OCCs, allowing QoS constraints configuration (fixed and variable packet delays, packet losses, bandwidth limitations, etc.)[1]. In order to provide a flexible framework for DCN topology reconfiguration, the control channels are implemented using Virtual Local Area Networks (VLAN) 802.1q [3], configured in the layer 2 backbone Ethernet switches (named backbone nodes) and in the OCCs within the testbed. VLAN technology allows performing any layer 2 interconnections between network nodes absolutely decoupled of the physical infrastructure.

All OCCs, both optical and satellite, run the same set of protocols and processes: Resource Reservation Protocol Traffic Engineering (RSVP-TE [4]) for lightpath provisioning, Open Shortest Path First Traffic Engineering (OSPF-TE [5]) for topology and optical resources dissemination, Link Resource Manager (LRM) for management of node's optical resources, Single Network Management Protocol (SNMP [6]) implementing a management agent, and eventually[2], Optical Link Resource Manager (OLRM) for optical hardware control.

---

[1] Thanks to the Netem [7] emulation package installed in the OCCs.
[2] OLRM only runs in optical OCCs, but not in satellites.

Finally, ADRENALINE also includes client devices provided through a broadband tester that emulates a User Network Interface (UNI)-enabled IP router. It generates statistically UNI lightpath requests for Gigabit Ethernet, transmitting and analyzing IP packets once the lightpath is established.

3   Model-based Reconfiguration and Monitoring Mechanism

This section described the proposed model-based mechanism, first introducing scenario modeling concepts (Sect. 3.1), then detailing the model processing in depth (Sect. 3.2).

3.1   Scenarios and Models

From the model-based mechanism point of view, flexible GMPLS testbed is composed of a set of *network nodes* and *backbones nodes* (Fig. 2). The former are OCCs (optical and satellite) and client devices (note the same physical broadband tester emulates up to 16 separated clients), but not the optical hardware in the fixed transport plane. Backbone nodes are the Ethernet switches for DCN interconnection, supporting the 802.1q VLAN technology thus allowing (upon configuration) setting up different logical DCN topologies. Network nodes run processes (backbone nodes don't because their task is only providing VLAN-based interconnection). Processes running in each OCC may be different (e.g., satellite OCCs does not run OLRM, optical OCCs do) and, in different OCCs, the same process running on it can have different configurations (e.g., GMPLS OSPF-TE can be configured with different CBR algorithms in different OCCs).

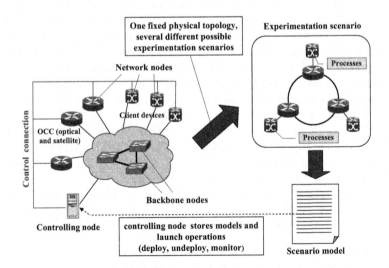

Fig. 2. Experimentation scenarios and models in optical flexible testbed

Experimentation scenarios are conceived by researchers, with their particular investigation goals in mind, obeying to diverse purposes (e.g., validating a theory, get measures, test new algorithm implementations, etc.). An experimentation scenario consists on a particular topology of network nodes (that will be set with the appropriated configuration of interconnection backbone nodes; for example, setting the proper VLAN configuration in switches and network elements operating systems) plus a particular configuration for each one of the processes running in those network nodes. A *scenario model* is a formal specification of an experimentation scenario, written in a particular language (with a particular syntax and semantic) so it can be processed automatically. Models include all the information regarding the network topology (network nodes involved in the scenario, the interconnection links among them, addressing issues, etc.) and the configuration of the processes running in each network node.

In theory any language can be used to write models. However, we will focus on XML-based ones. XML [8] is a W3C (World Wide Web Consortium) recommended general-purpose markup language for creating special-purpose markup languages. Currently, there are many well-known languages based on it (e.g., XHTML, RDF, SOAP, DocBook, etc.). The advantage of using XML is that, being a recognized standard with a wide support in the industry, there are many XML software components available that can be easily reused in order to build model processing tools. As a matter of fact, the ADNETCONF tool (the software that implements the model-based reconfiguration mechanism in ADRENALINE; described in detail in Sect. 4.1) reuses some of such XML modules.

3.2   Model Processing

The processing workflow is illustrated in Fig. 3. Firstly, the user writes the model of the desired scenario, either directly with a general purpose editor (XML is text-based human-readable) or, preferably, using a tool specifically designed for testbed model edition (like ADNETCONF, described in Sect. 4.1). Once the model has been created, it is ready to be processed. The processing engine (implemented with a software program) must run in a node (named *controlling node* in Fig. 2) physically interconnected to the testbed interconnection backbone. In addition, the controlling node must have pre-existing IP connectivity (*control connection* in the Fig. 2 and 3) to all the network and backbone nodes.

1. User launches the processing engine, using as input the model and the desired action to perform with it (deploy, undeploy or monitor). Typically, models are physically stored in the controlling node.
2. The processing of the model produces interaction with the network and backbone nodes through the control connection. There are two possible interactions: issuing commands (always) and installing configuration files (only in some cases during deploy). The nature of the commands depends on the interface provided by the node. In some cases (e.g., PC-based nodes) these commands are executed in the operating system of the node (using a remote shell). In others, they are issued using a proprietary API (Application Programming Interface) (e.g., vendor-specific broadband testers).

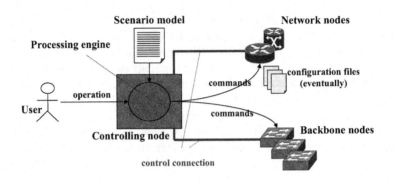

Fig. 3. Model processing

3. Processing engine reports the outcome of the action (i.e., whether the action was performed correctly or there was some error). In the case of monitoring, the result includes information regarding to scenario status.

There are three different processing actions (deploy, undeploy and monitor). The particular interactions performed during each action are described following:

Deploy. Processing of a model in order to set the corresponding scenario up in the testbed. The first commands issued to network and backbone nodes are for establishing the logical topology. It uses to imply enabling VLANs in backbone nodes and assigning VLANs and addresses (e.g., IP addresses) in network nodes. Then, commands are issued in order to start and configure the processes (only for network nodes, due to backbone nodes don't execute processes). The position (first setting the topology, next starting processes) is important, due to some processes need the topology configured in order to start properly (e.g., a routing process starting when no link has been established could fail). In the case some process needs a start-up configuration file, they are generated and installed before executing process starting commands.

Undeploy. Processing of a model in order to set the corresponding scenario down, reverting the testbed to a clear status (i.e., no scenario configured in the testbed). The scenario is supposed to have been previously deployed. Undeploy commands involve stopping of processes in first place, then removing the configuration of the logical topology. In this case configuration files are not required to be generated (they are needed for starting processes but not for stopping).

Monitor. Processing of a model in order to check the corresponding scenario status. The scenario is supposed to have been previously deployed. Monitor actions involve checking the status of links between network nodes and the aliveness of process (i.e., if the process is running or not). For each link and process, the status (up or down) is reported to the user, so he can know the status of the scenario accurately. Although deploy and undeploy are the main actions, monitor is also a very useful auxiliary one that allows knowing if the scenario is working properly or, otherwise, easing the location and fixing of problems.

4 Application to ADRENALINE Testbed

This section presents the practical results of applying the proposed model-based reconfiguration mechanism (described in Sect. 3) to ADRENALINE testbed, first describing the ADNETCONF software tool (Sect. 4.1) and, secondly, showing the experimental results that demonstrates the benefits of the approach compared with manual reconfiguration (Sect. 4.2).

4.1 The ADNETCONF Tool

ADNETCONF (ADrenaline NETwork CONFigurator) is a software tool in charge of scenario model management in ADRENALINE testbed. It has been developed with two main objectives in mind. First, it provides and easy, quick and intuitive way of designing scenario models. Secondly, it implements deploy, undeploy and monitor operations (as described in Sect. 3.2) of such models. The structure of the tool resembles these goals; it is composed of two main modules, a graphic user interface (providing the model editor, in addition to the interface to launch operations) and a processing engine (implementing the actual operations).

As model editor, the GUI (shown in Fig. 4) provides a high-level perspective of the testbed. Models are created just "drawing" the network topology, using a library of items that includes the different network nodes in ADRENALINE testbed (OCCs and client devices) and configuring links among them. In addition, the configuration of the processes is performed just clicking in the particular element and editing a set of dialog boxes (one for each process). Model are stored in files, that can be saved, opened, modified, etc., following the same paradigm that any other conventional editor (e.g., .doc files for Microsoft Word).

Once the model has been completed, the user launches the desired operation clinking in the particular button on the GUI. The model is then serialized to a set of XML files that correspond to the formal representation of the scenario that the processing engine will understand. Up to six different XML files are produced; one describing the logical DCN topology (mandatory), the other five describing the configuration of the different processes (LRM, OSPF-TE, RSVP-TE, SNMP, OLRM), only needed when the particular process has been included in the scenario. This modularization allows a better implementation of the engine, composed of several modules each one specialized in validating and processing one XML file. In fact, XML files can be seen as an internal interface between the GUI (focused in model graphical design but not in processing) and the engine (focused in processing the model but not in designing aspects), hiding the complexity of the XML language to the user by the user-friendly GUI.

The ADNETCONF engine behaves like the one described in Sect. 3.2. The following is used to provide control connections: SSH (Secure Shell) for OCCs, telnet for Ethernet switches in the interconnection backbone and a RPC-based

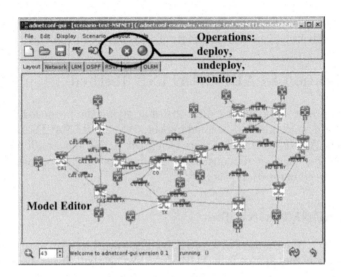

Fig. 4. ADNETCONF GUI

(Remote Procedure Call) vendor-specific proprietary API for the broadband tester implementing client devices. The commands issued in the most complete case (deploy) in ADRENALINE testbed include: VLANs set up for OCCs, client devices and switches; process start-up (LRM, OSPF-TE, RSVP-TE, SNMP and OLRM; two configurations files are generated and installed for OSPF-TE in each OCC before starting the process; one configuration file for LRM) and configuration of client devices in the broadband tester for connection requests, traffic generation and traffic analysis in the configured network topology.

4.2   Results

This section shows some practical results of the model-based automatic deployment method presented in this paper applied to ADRENALINE testbed, with the aim of illustrating its great benefits compared with conventional manual configuration.

The experiment consists in getting and analyzing some parameters when executing deploy and undeploy actions (monitor is left apart for the sake of briefness) with two different scenarios (one simple and the other one complex) using ADNETCONF. The simple one is composed of 3 OCCs (based on a ring topology), 3 client devices (each one attached to one OCC) and 3 links. The complex scenario resembles the mesh topology of the real NFSNet [9], composed by 14 OCCs, 14 client devices (each one attached to one OCC) and 20 links. The parameters analysed are execution time (how long lasts the completion of the operation), number of commands issued to network elements, number of configuration files generated, and their total size (the sum of all the lines in all the generated files). The summary of results is shown in Table 1.

Table 1. Experiment results

	Simple scenario (3 Nodes ring)		Complex scenario (14 Nodes NSFNet)	
	Deploy	Undeploy	Deploy	Undeploy
Commands	168	108	1046	722
Mean execution time	110 s	10 s	370 s	125 s
Commands per second (cmd/s)	1.52	10.8	2.82	5.78
Execution time of equivalent manual operation without errors (0.16 cmd/s)	1008 s (around 17 min)	648 s (around 11 min)	6276 s (around 105 min)	4332 s (around 72 min)
Configuration files	9	Not applicable	42	Not applicable
Configuration files lines	617	Not applicable	6492	Not applicable

One of the conclusions is that ADNETCONF can set up a scenario in the worst case (simple scenario deploy, around 1.52 cmd/s) much faster than any human administrator in the best case (supposing an average never-mistyping user[3], 0.16 cmd/s). Moreover, ADNETCONF ratio can be even higher in the case of undeploy (10.8 cmd/s) and when considering complex (and foreseeably more interesting) scenarios (deployment 2.82 cmd/s; undeployment 5.78 cmd/s).

More important than the improvement in execution time is that ADNET-CONF removes the error debugging and fixing time of manual deployments (due to unnoticed typing errors introducing commands and configurations) since computer-based tools never performs such errors[4]. This time is difficult to estimate but usually is very high (it can take an entire day or even much more to locate and fix a bug) and proportional to the size of the scenario, so the tool is highly-scalable and especially valuable for complex scenarios (like the NFSNet).

Regarding to configuration files, ADNETCONF is able to generate 9 and 42 files respectively (involving 617 and 6492 lines) from a single model. Not only to generate all those files "by hand" can be an overwhelming task, but, even in the case they were created just once and reused after that, keeping the coherence in so many and so long files is difficult. Therefore, the ADNETCONF approach of generating all the needed files "on-the-fly" from the model is much better.

5   Conclusions and Future Work

The paper has described a reconfiguration mechanism for GMPLS-based optical testbeds in order to allow experimental performance evaluation of real GMPLS-based CBR under several topology and resource configurations. The method is based on modeling the desired experimentation scenarios and then processing models with automatic tools that deploy the desired configuration in the testbed

---

[3] Considering 30 wpm (words per minute) [10] and average 5-words commands.

[4] Software-caused errors impact is much smaller than human-caused. The former are deterministic and once fixed in the software they do not happen again. So, after a convenient software debugging time, the tool could be considered error-free.

(involving link connectivity and configuration for the processes running at each optical node) without human interaction. Therefore, the problems of manual reconfigurable testbeds (time consumption, high human error probability, specific knowledge on reconfiguration technologies -like VLAN- and scalability) are overcome, as proved with practical results in ADRENALINE testbed. Moreover, the generation of models does not need to be performed directly (in languages like XML) since specific user-oriented modelling tools can be developed (like AD-NETCONF). Both factors (automatic reconfiguration and tool-oriented model management) lead to a highly productive and flexible experimentation environment for optical networking researches which key elements are *models*.

Furthermore, although this paper is focused in GMPLS-based optical testbeds, the mechanism is general enough to be applied to any other kind of testbed (optical or not) based on IP overlays, maybe using other interconnection technologies apart VLAN (e.g., GRE or IPSec tunnels). Such general model-based deployment method for IP networks is currently being patented by CTTC[5].

Currently, future work is focused in two lines. Firstly, the modelling technology and its performance should be further closely analysed in order to be enhanced. In addition, other fields in computers science (software engineering and knowledge engineering mainly) has long experience in models and meta-modeling and the technologies used there (like ontologies or model driven engineering) should be carefully studied. Secondly, ADNETCONF tool can be improved, increasing its usability (e.g., model coherence checking, real-time monitoring, etc.).

References

1. Mannie, E. (ed.): Generalized Multi-Protocol Label Switching (GMPLS) architecture. RFC 3945, IETF (2004)
2. Muñoz, R., et al.: The ADRENALINE testbed: integrating GMPLS, XML and SNMP in transparent DWDM networks. IEEE Communications Magazine, Vol. 43(8), IEEE (2005), s40–s48
3. IEEE 802.1Q Working Group: 802.1q: Virtual LANs IEEE (2001)
4. Berger, L. (ed.): Generalized Muti-Protocol Label Switching (GMPLS) signalling functional description. RFC 3471, IETF (2003)
5. Kompella, K., Rekhter, Y.: OSPF extensions in support of Generalized Multi-Portocol Label Switching. RFC 3471, IETF (2003)
6. Case, J., Fedor, M., Schoffstall, M., Davin, J.: A Simple Management Network Protocol (SNMP). RFC 1098, IETF (1990)
7. Hemminger, S.: Network Emulation with NetEm. LCA2005 (2005)
8. Bray, T., Paoli, J., Sperberg-McQueen, M., Maler, E.: Extensible Markup Language (XML) 1.0 (Second Edition). W3C Recommendation, W3C (2001)
9. Hülsermann, R., et al.: A set of typical network scenarios for network modeling. Procedings of 5th ITG-Workshop on Photonic Networks (2004)
10. Soukoreff, R., MacKenzie, I.: Theoretical upper and lower bounds on typing speed using a stylus and soft keyboard. Behaviour & Information Technology, Vol. 14, Taylor & Francis (1995), 370–379

---

[5] Method for Logical Deployment, Undeployment and Monitoring of a Target IP Network, PCT/EP2006/009960, October 2006.

Clustering for Hierarchical Traffic Grooming in Large Scale Mesh WDM Networks [*]

Bensong Chen[1], Rudra Dutta[2], and George N. Rouskas[2]

[1] Google Labs
[2] North Carolina State University

Abstract. We present a clustering algorithm for hierarchical traffic grooming in large WDM networks. In hierarchical grooming, the network is decomposed into clusters, and one hub node in each cluster is responsible for grooming traffic from and to the cluster. Hierarchical grooming scales to large network sizes and facilitates the control and management of traffic and network resources. Yet determining the size and composition of clusters so as to yield good grooming solutions is a challenging task. We identify the grooming-specific factors affecting the selection of clusters, and we develop a parameterized clustering algorithm that can achieve a desired tradeoff among various goals.

1 Introduction

Traffic grooming, the area of research concerned with efficient and cost-effective transport of sub-wavelength traffic over multigranular networks, has emerged as an important field of study in optical networks. In *static* grooming [5], the objective is to provision the network to carry a set of long-term traffic demands while minimizing the overall network cost. In *dynamic* grooming [18], on the other hand, the goal is to develop on-line algorithms for efficiently grooming and routing of connections that arrive in real time. The reader is referred to [6] for a comprehensive survey and classification of research on traffic grooming.

Most grooming studies on general topologies [12,15] regard the network as a flat entity for the purposes of lightpath routing, wavelength assignment, and traffic grooming. In general, such approaches do not scale well to networks of realistic size, since the running-time complexity of traffic grooming algorithms increases rapidly with the size of the network and also the operation, management, and control of multigranular networks becomes a challenging issue in large, unstructured topologies. Indeed, in existing networks, resources are typically managed and controlled in a hierarchical manner.

Based on this observation, we have developed a scalable hierarchical approach to traffic grooming [2] modeled after the hub-and-spoke paradigm used within the airline industry. The network is partitioned into clusters, and one node within each cluster serves as the *hub*. Non-hub nodes route their traffic to the hub, where it is groomed and forwarded to the destination cluster; hence, hubs are the only nodes responsible for grooming traffic not originating/terminating locally.

---

[*] This work was supported by the NSF under grant ANI-0322107.

In this work we address an important yet challenging issue in hierarchical grooming, namely, the selection of clusters and hub nodes. Although clustering techniques are used in a wide range of network design problems, there is little work related to traffic grooming. We develop a new parameterized clustering algorithm that is flexible and allows the designer to achieve a balance among a number of conflicting goals. To demonstrate the effectiveness of hierarchical grooming with clustering, we present results for two large networks, including a 128-node, 321-link topology that is approximately an order of magnitude larger than networks that have been considered in previous grooming studies.

We describe the hierarchical grooming approach in Section 2. In Section 3 we present the clustering algorithm, and we obtain lower bounds in Section 4. We present numerical results in Section 5 and we conclude the paper in Section 6.

2   Hierarchical Grooming in Mesh Networks

We consider a network of general topology with N nodes. Each link supports W wavelengths, and the wavelength capacity C is an integer multiple of a basic transmission unit. Traffic demands are provided in matrix $T = [t^{(sd)}]$, where integer $t^{(sd)}$ denotes the amount of traffic to be carried from node s to node d.

The objective of the traffic grooming problem is to determine the lightpaths to be set up so as to carry the traffic matrix T while minimizing the number of electronic ports required. Since each lightpath requires exactly two electronic ports, this objective is equivalent to minimizing the number of lightpaths in the resulting logical topology. This traffic grooming problem is intractable even in simple network topologies, such as paths and stars, for which the routing and wavelength assignment (RWA) subproblem can be solved in polynomial time [13]. Consequently, for networks with more than a few nodes, it is important to develop heuristics which obtain good solutions in polynomial time.

Our framework for hierarchical traffic grooming was inspired by the hub-and-spoke paradigm that is widely used by the airline industry. In our approach, a large network is partitioned into clusters consisting of a contiguous subset of nodes. One node within each cluster is designated as the *hub*, and is the only node responsible for grooming intra- and inter-cluster traffic. Hub nodes are provisioned with more resources (e.g., larger number of electronic ports and higher switching capacity) than non-hub nodes, and are similar in function to airports that serve as major hubs; these airports are typically larger than non-hub airports, in terms of both the number of gates ("electronic ports") and physical space (for "switching" passengers between gates).

Our hierarchical framework consists of three phases [2]:

1. **Clustering of network nodes.** In this phase, the network is partitioned into m clusters and one node in each cluster is designated as the hub. The clustering phase is crucial to the quality of the grooming solution. We describe in detail our clustering algorithm for traffic grooming in Section 3.
2. **Hierarchical logical topology design and traffic routing.** The outcome of this phase is a set of lightpaths for carrying the traffic matrix T, and

a routing of individual traffic components $t^{(sd)}$ over these lightpaths. This phase is further subdivided into three parts: (a) setup of direct lightpaths for large traffic demands; (b) intra-cluster traffic grooming; and (c) inter-cluster traffic grooming. This hierarchical approach is described in detail in [2].

3. **Lightpath routing and wavelength assignment (RWA).** The RWA problem has been studied extensively in the literature. In this work, we adopt the LFAP algorithm [17], which is fast, conceptually simple, and has been shown to use a number of wavelengths that is close to the lower bound for a wide range of problem instances.

3   Clustering for Hierarchical Grooming

Clustering is a function that arises frequently in problems related to network design and organization. Clustering algorithms are classified as either *minimum cut* or *spanning tree*, depending on the underlying methodology [9]. The input to the algorithms generally consists of a set of nodes and edge weights, while the output is a partition of the nodes that optimizes a given objective function. In our case, the goal is to find a clustering that will minimize the number of lightpaths *after* applying the hierarchical grooming (logical design) approach, a fact that adds significant complexity to the problem. Specifically, the input to our problem consists of a traffic demand matrix and several constraints, in addition to the physical topology; also, unlike typical objective functions considered in the literature, ours cannot be easily expressed as a function of the resulting clusters. Therefore, existing clustering techniques such as TPABLO [3] or METIS [16], are not directly applicable to the problem at hand.

One clustering problem that may be relevant to hierarchical traffic grooming is *K-Center* [8,11]. The goal of *K-Center* is to find a set S of K nodes (centers) in the network, so as to minimize the maximum distance from any network node to the nearest center. The set S implicitly defines K clusters with corresponding hub nodes in S. *K-Center* is known to be NP-Complete [8]. We implemented the 2-approximation algorithm in [8], and compare it to our own in Section 5.

More recently, some studies have explored clustering techniques in the context of traffic grooming: a hierarchical design for interconnecting SONET rings with multirate wavelength channels was proposed in [7], and in [4], the "blocking island" paradigm is used to abstract network resources and find groups of bandwidth hierarchies for a restricted version of the traffic grooming problem.

3.1   Important Considerations

We now discuss the tradeoffs involved in selecting the clusters, which set the design principles for the clustering algorithm we present in the next subsection.

To obtain a good clustering, the number of clusters, their composition, and the corresponding hubs must be selected in a way that helps achieve our goal of minimizing the number of lightpaths and wavelengths required to carry the traffic demands. This is a complex task as it depends on both the physical topology and the matrix T. Consider the tradeoffs involved in determining the number m of clusters. If m is small, the amount of inter-cluster traffic will likely be large. Hence, the m hubs may become bottlenecks, resulting in a large number of electronic ports at each hub and a large number of wavelengths to carry the lightpaths over the links to/from each hub. On the other hand, a large m implies small clusters. In this case, the amount of intra-cluster traffic will be small, resulting in inefficient grooming (i.e., a large number of lightpaths); similarly, at the second-level cluster, $O(m^2)$ lightpaths will have to be set up to carry small amounts of inter-cluster traffic. Therefore, one must select the number and size of clusters to strike a balance between capacity utilization and number of lightpaths for both intra- and inter-cluster traffic.

Now consider the composition of each cluster. If the average traffic demand between nodes within a cluster is higher than the average inter-cluster demand, there will tend to be fewer inter-cluster lightpaths, which are typically longer than local lightpaths. Therefore, it is desirable to cluster together nodes with "denser" traffic between each other: doing so reduces the number of longer lightpaths, alleviates hub congestion, and provides more flexibility to the RWA algorithm (since long lightpaths are more likely to collide during the RWA phase).

We also need to consider the cut links that connect different clusters. Each cluster has a number of fibers that link to nodes outside the cluster, and all traffic between a node outside the cluster and one within must traverse these cut links. Since the cut links must have sufficient capacity to carry the inter-cluster traffic, it is important to select clusters so that their cut size is not too small, so as to keep the wavelength requirements low.

Another important consideration arises in physical topologies for which there exists a critical small cut set that partitions the network into two parts. In such a topology, all traffic between the two sides of the bisection will have to go through the cut. Creating clusters that consist of nodes on different sides of the cut is undesirable, because it may generate unnecessary traffic that goes back and forth through the cut. This traffic can be eliminated by forcing nodes on different sides of the bisection to be in different clusters. In Section 3.3, we describe a pre-cutting technique that can be useful in such situations.

The physical shape of each cluster may also affect the wavelength requirements. In particular, it is important to avoid the creation of clusters whose topology resembles that of a path, since in such topologies the links near the hub can become congested. In general, cluster topologies with relatively short diameter are more attractive in terms of RWA.

3.2   The MeshClustering Algorithm

Figure 1 provides a pseudocode description of our MeshClustering algorithm which we use to partition a network of general topology in order to apply our hierarchical traffic grooming framework. The algorithm includes several user-defined parameters that can be used to control the size and composition of clusters, either directly or indirectly. Parameters *MinCS* and *MaxCS* represent the minimum and maximum cluster size, respectively. The algorithm treats these parameters as an *indication* of the desirable range of cluster sizes, rather than as hard thresholds that cannot be violated. Consequently, the final result may contain clusters larger than *MaxCS* (see the discussion below regarding Step 24).

A Clustering Algorithm for Mesh Networks

Input: A mesh network with a set V of $|V| = N$ nodes, capacity C for each wavelength, and reduced traffic matrix $T_r = [t_r^{(sd)}]$.

User-defined parameters: $MinCS, MaxCS$ for the desired minimum and maximum cluster size, respectively, a threshold $0.5 \le \Delta \le 0.8$, a cluster diameter-to-nodes ratio $0 < \delta \le 0.75$, and an intra-to-inter-cluster traffic ratio $0.8 \le \rho \le 1.25$.

Output: A partitioning of the node set V into some number m of clusters, $B_1, \ldots, B_m$, and the selection of node h_i as the hub of cluster B_i.

Procedure **MeshClustering**
begin
1.   **while** $V \ne \phi$ **do**
2. $v \leftarrow$ node in V with max remaining capacity; $V \leftarrow V - \{v\}$
3. $B \leftarrow \{v\}$ // new cluster B with hub v
4.   **while** $V \ne \phi$ and $|B| < MaxCS$ **do** // grow cluster B
5. $Q \leftarrow$ set of nodes $\in V$ adjacent to nodes in B
6.   **foreach** node $q \in Q$ **do**
7. $B' \leftarrow B \cup \{q\}$ // assume q is included in B
8.   HUBTEST: is traffic between B', $\overline{B'} > \Delta \times$ remaining hub capacity?
9.   CUTTEST: is traffic between B', $\overline{B'} > \Delta \times$ remaining cut link capacity?
10.   **if** q passes both tests **then**
11. $x \leftarrow$ total traffic between q and nodes in B
12. $y \leftarrow$ total traffic between q and nodes in $\overline{B'}$
13. $\rho_q \leftarrow x/y$ //intra- to inter-cluster traffic ratio
14. $d \leftarrow$ diameter of induced subgraph B'
15. $\delta_q \leftarrow d/|B'|$ // diameter-to-nodes ratio
16.   **else** $Q \leftarrow Q - \{q\}$
17.   **end for**
18.   **if** $Q = \phi$ **then break**   // cannot grow cluster B
19.   **else**
20. $q_0 \leftarrow$ node $\in Q$ with largest ρ_q and smallest δ_q
21. $B \leftarrow B \cup \{q_0\}$; $V \leftarrow V - \{q_0\}$ // grow cluster B to include q_0
22.   **end while**   // continue until B cannot grow further
23. **end while**
24. Combine clusters of size $< MinCS$ with adjacent clusters
end

Fig. 1. Clustering algorithm for mesh networks

The parameter Δ ($0.5 \le \Delta \le 0.8$, default value $\Delta = 0.8$) is used to test whether there is sufficient capacity at the hub node, as well as the edges connecting the cluster to the rest of the network, to carry the traffic demands. Specifically, we require that the inter-cluster traffic originating from or terminating at a given cluster do not exceed a fraction Δ of the hub capacity (this is the HUBTEST in Step 9); similarly, this intra-cluster traffic must not exceed a fraction Δ of the capacity of the links connecting the cluster to the rest of the network (the CUTTEST in Step 10). The algorithm will consider a node to add to a cluster only if doing so will not violate these two constraints.

The parameter δ controls the ratio of the diameter of a cluster to the number of nodes it contains. To avoid cluster topologies that resemble long paths, we require that $0 < \delta \le 0.75$. We used the value $\delta = 0.75$ which corresponds to a 4-node path, hence restricting the longest path within a cluster to at most three links. The parameter $\rho, 0.8 \le \rho \le 1.25$, specifies the acceptable range for the ratio of intra- to inter-cluster traffic for a cluster. Since it is desirable to cluster together nodes that exchange a substantial amount of traffic among themselves relative to traffic they exchange with the rest of the network, we used $\rho = 1.25$.

The MeshClustering algorithm in Figure 1 generates one cluster during each iteration of the main **while** loop between Steps 1 and 23. In Steps 2-3, the hub of a new cluster B is selected as the node with the maximum remaining capacity among those not yet assigned to a cluster; by "remaining capacity" we mean the capacity remaining on its incident links after subtracting the bandwidth taken up by any direct lightpaths (refer to the second phase of the hierarchical grooming approach in Section 2). The cluster grows by adding one node during each iteration of the **while** loop between Steps 4 and 22. At each iteration, the set Q of candidate nodes for inclusion in cluster B consists of all nodes, not yet assigned to another cluster, which are adjacent to nodes in B. For each node $q \in Q$, we first check whether including q in B would result in a cluster that passes both the HUBTEST (Step 8) and CUTTEST (Step 9); if not, node q is removed from consideration for inclusion into cluster B (Step 16). For all nodes q that pass both tests, we compute the diameter-to-nodes ratio δ_q and intra-to-inter-cluster traffic ratio ρ_q, assuming that q is added to cluster B (Steps 10-15). Let q_0 be a node that passes both tests and has the largest ρ_q value among the candidates; if there are multiple such nodes, we select the one with the smallest δ_q value. We include q_0 to cluster B (Steps 20-21), and the process is repeated. If clusters with fewer than $MinCS$ nodes are created, Step 24 removes them and includes their nodes into adjacent clusters. As a result, at the end some clusters may contain more than $MaxCS$ nodes.

3.3   Pre-Cutting for Imbalanced Topologies

As we mentioned in Section 3.1, when the topology has a bisection of small cut size, the cut links are likely to become congested. Hence, it may be necessary to disallow nodes on different sides of such a bisection from being in the same cluster. However, identifying such a critical bisection in a large, imbalanced topology, is a difficult task. To tackle it, we use the CHACO software [10] which implements the partitioning algorithm in [14]. The software uses the parameter KL-IMBALANCE to control the relative sizes of the node sets on either size of the bisection. We apply CHACO several times, varying the KL-IMBALANCE parameter, and obtain several different bisections of the physical topology. We then select the bisection with the most traffic flowing along the cut links.

Once we identify a critical bisection, we apply the following approach. First, we use the MeshClustering algorithm to determine a clustering that does not take the bisection into consideration. Then, we partition the network into two parts along the bisection, and we apply the MeshClustering algorithm on each part separately; this ensures that no cluster contains nodes from both sides of the bisection. We then select the clustering that requires the fewest lightpaths after the logical topology and RWA phases, unless it requires a significantly larger number (e.g., 10% or more) of wavelengths; in this way, we achieve balance between the lightpath objective and the wavelength requirements.

4   Lower Bounds

A simple lower bound on the number of lightpaths (our main objective) can be calculated based on the observation that each node must source and terminate a sufficient number of lightpaths to carry the traffic demands from and to this node, respectively. This bound can be determined directly from the traffic matrix T. However, we obtain a better lower bound based on the following observations. Let b_{sd} denote the number of direct lightpaths set up from s to d. Since all traffic originating (respectively, terminating) at node s (respectively, node d) must be carried on some lightpath also originating (respectively, terminating) at s (respectively, d), the following constraints must be observed:

$$\sum_d b_{sd}C \ge \sum_d t^{(sd)} \quad \forall s \qquad \sum_s b_{sd}C \ge \sum_s t^{(sd)} \quad \forall d \qquad (1)$$

We can obtain a lower bound on the number of lightpaths by solving the ILP:
    Minimize $\sum_{s,d} b_{sd}$ subject to constraints (1).

We emphasize that this ILP will not necessarily yield a meaningful solution to the original grooming problem, only a lower bound. By configuring CPLEX to use dual steepest-edge pricing, we are able to compute this bound within a few seconds even for the 128-node topology we consider in the next section. Although this bound is better than the bound above, we believe that it is somewhat loose.

For a lower bound on the number of wavelengths, consider the bisection cut of the network we identify using the approach described in Section 3.3. Let t be the maximum amount of traffic that needs to be carried on either direction of the links in the cut set. Also, let x be the number of links in the cut set, and C the capacity of each wavelength. Then, the quantity $\lceil t/xC \rceil$ is a lower bound on the number of wavelengths for carrying the given traffic matrix.

5   Numerical Results

We now present experimental results for two network topologies: a 47-node, 96-link network [1] with a balanced topology (i.e., there is no bisection with a small cut size that can be bottleneck in traffic grooming), and a 128-node, 321-link network which corresponds to the worldwide backbone of a large service provider (http://www.caida.org). The latter topology is imbalanced, as there exists a bisection with a cut size of 5 links that divides it into two parts of 114 and 14 nodes, respectively, hence we use the approach we described in Section 3.3 to create clusters that contain nodes on one side of the cut only.

The traffic matrix of each problem instance is generated by drawing $N(N-1)$ random numbers (rounded to the nearest integer) from a Gaussian distribution whose mean and standard deviation depend on the traffic pattern. We consider three patterns: *random*, which is challenging since the matrix does not have any structure that can be exploited by a grooming algorithm; *falling*, which is such that the amount of traffic between two nodes decreases with the distance between them; and *rising*, which is the opposite of the falling pattern.

For a given topology and traffic pattern, we generate thirty problem instances and we compare our MeshClustering algorithm to the *K-Center* algorithm [8]. We consider two performance metrics: the *normalized lightpath count* and the *normalized wavelength count*. The former is the ratio of the number of lightpaths required for hierarchical traffic grooming with one of the clustering algorithms, to the lightpath lower bound of Section 4; the latter is the ratio of the number of wavelengths required to the wavelength lower bound of Section 4.

Figure 2 plots the normalized lightpath and wavelength count for each of thirty problem instances with a falling traffic pattern. For each problem instance, four values are shown, corresponding to four different clusterings. The first two are from the *K-Center* algorithm, with the number of clusters K equal to 4 and 6, respectively. The other two are from our MeshClustering algorithm. Recall that our algorithm does not take the number of clusters as input, rather, it tries to optimize it. Consequently, the algorithm may produce different clusters for two different problem instances. To make the comparison against *K-Center* as fair as possible, we selected two sets of values for the user-defined parameters of MeshClustering so that the average number of clusters over all thirty instances is 3.52 and 5.45, respectively.

From the figure, we observe that the number of lightpaths required for hierarchical grooming is about 40% higher than the lower bound, regardless of the clustering algorithm used. As we mentioned earlier, we believe that this lower bound is rather loose, hence the performance of hierarchical grooming is better than the curves imply. Also, except for a couple of instances, the curves corresponding to the MeshClustering algorithm lie below those corresponding to the *K-Center* algorithm. Although the difference is not high, note that a 1% reduction in the number of lightpaths in this network would result in about 40 fewer electronic ports, a substantial savings in cost.

We also observe that the MeshClustering algorithm requires significantly fewer wavelengths than the *K-Center* algorithm. This result is due to the fact that our algorithm is designed to take the wavelength requirements into account. In absolute terms, the difference in the number of wavelengths for these problem instances is in the order of 10-12. We also note that the large values of the normalized wavelength count are due to the fact that the lower bound is loose in this case. Recall that a good bound depends on finding a cut of small size and large cross-cut traffic, but this 47-node network does not have such a bottleneck cut. Furthermore, the falling pattern makes it unlikely that a large amount of traffic will cross any cut.

Figure 3 is similar to Figure 2 but shows results for the rising pattern. Due to the nature of this pattern, relatively large amounts of traffic will cross any network cut, resulting in the much tighter wavelength bounds observed. Again, except for a few instances, our clustering algorithm outperforms the *K-Center* algorithm. We also observe that hierarchical grooming provides good solutions regardless of the clustering algorithm.

Finally, Figure 4 presents results for the 128-node network and the random traffic pattern. For the *K-Center* algorithm, we let the number K of clusters be either 9 or 10, and we selected the parameters of the MeshClustering algorithm so that it also produces either 9 or 10 clusters (the average over all instances is 9.33). Our clustering algorithm slightly outperforms *K-Center* in terms of the number of lightpaths, and both are relatively close to the lower bound. However, in terms of wavelengths, our algorithm produces results that are within 5% of the lower bound, whereas *K-Center* requires twice that number of wavelengths.

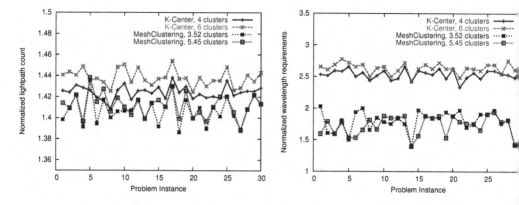

Fig. 2. Lightpath (left) and wavelength (right) comparison, falling pattern, 47-node network

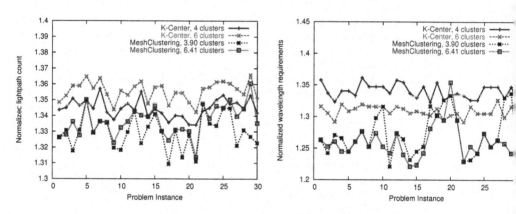

Fig. 3. Lightpath (left) and wavelength (right) comparison, rising pattern, 47-node network

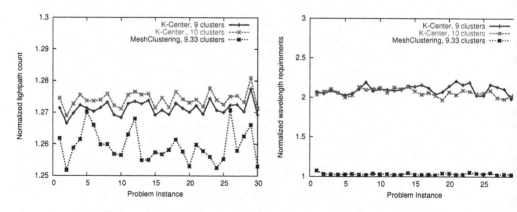

Fig. 4. Lightpath (left) and wavelength (right) comparison, random pattern, 128-node network

6   Concluding Remarks

Hierarchical traffic grooming is a new approach for efficient and cost-effective design of large-scale optical networks with multigranular traffic. The hierarchical grooming framework must be coupled with clustering techniques that follow grooming-specific design principles, and we have presented such a clustering algorithm that is flexible in balancing various conflicting goals via user-defined parameters. The experimental results demonstrate that (1) our hierarchical approach scales to large networks; (2) our clustering algorithm outperforms an algorithm that was not developed with traffic grooming in mind; and (3) hierarchical grooming combined with specially designed clustering techniques produce logical topologies that perform well across a variety of traffic patterns.

References

1. P. Baran. On distributed communications networks. *IEEE Trans. Commun.*, 12(1):1-9, Mar. 1964.
2. B. Chen, G. N. Rouskas, and R. Dutta. A framework for hierarchical grooming in WDM networks of general topology. *BROADNETS 2005*, pp. 167-176, Oct. 2005.
3. H. Choi and D. B. Szyld. Application of threshold partitioning of sparse matrices to markov chains. In *IPDS*, 1996.
4. Z. Ding and M. Hamdi. Clustering techniques for traffic grooming in optical WDM mesh networks. In *GLOBECOM 2002*, vol. 3, pp. 2711-2715, Nov. 2002.
5. R. Dutta and G. N. Rouskas. On optimal traffic grooming in WDM rings. *IEEE Journal on Selected Areas in Communications*, 20(1):110–121, January 2002.
6. R. Dutta and G. N. Rouskas. Traffic grooming in WDM networks: Past and future. *IEEE Network*, 16(6):46–56, November/December 2002.
7. M. Esfandiari, *et al.* Improved metro network design by grooming and aggregation of STS-1 demands into OC-192/OC-48 lightpaths. In *NFOEC 2003*, Sep. 2003.
8. T. Gonzalez. Clustering to minimize the maximum inter-cluster distance. *Theoret. Comput. Sci.*, 38:293–306, 1985.
9. John A. Hartigan. *Clustering Algorithms.* Wiley, 1975.
10. B. Hendrickson and R. Leland. The Chaco user's guide. Technical Report SAND95-2344, Sandia National Laboratories, Albuquerque, NM 87185-1110, July 1995.
11. D. Hochbaum and D.B. Shmoys. A best possible heuristic for the k-center problem. *Math. Oper. Res.*, 10:180–184, 1985.
12. J.Q. Hu and B. Leida. Traffic grooming, routing, and wavelength assignment in optical WDM mesh networks. *INFOCOM 2004*, 23(1):495-501, Mar. 2004.
13. S. Huang, R. Dutta, and G. Rouskas. Traffic grooming in path, star, and tree networks: Complexity, bounds, and algorithms. *JSAC*, 24(4):66-82, Apr. 2006.
14. B. Kernighan and S. Lin. An efficient heuristic procedure for partitioning graphs. *Bell System Technical Journal*, 29:291–307, 1970.
15. V. R. Konda and T. Y. Chow. Algorithm for traffic grooming in optical networks to minimize the number of transceivers. In *HPSR Workshop*, pp. 218-221, 2001.
16. K. Schloegel, G. Karypis, and V. Kumar. A new algorithm for multi-objective graph partitioning. TR99-003, Dept. of CS, Univ. of Minnesota, Sep. 1999.
17. H. Siregar *et al.* Efficient routing and wavelength assignment in wavelength-routed optical networks. *7th Asia-Pacific NOMS*, pp. 116-127, Oct. 2003.
18. H. Zhu *et al.* A novel generic graph model for traffic grooming in heterogeneous WDM mesh networks. *IEEE/ACM Trans. Netw.*, 11(2):285-299, Apr. 2003.

Grooming-Enhanced Multicast in Multilayer Networks[†]

Péter Soproni, Marcell Perényi and Tibor Cinkler

Department of Telecommunications and Media Informatics, Budapest University of
Technology and Economics
H-1117, Magyar tudósok körútja 2, Budapest, Hungary
{soproni, perenyim, cinkler}@tmit.bme.hu

Abstract. In this paper we investigate and evaluate the performance of
multicast routing in grooming capable multi-layer optical wavelength division
multiplexing (WDM) networks. New wavelength-graph models are proposed
for network equipments capable of optical-layer branching of light-paths. We
evaluate the cost-effectiveness of electronic and optical-layer multicast and
unicast as well. The high scalability of optical multicast against unicast is also
showed. All routing and technical constraints are formulated in ILP and realized
in our versatile simulator.

Keywords: optical multicast, WDM, multilayer network, ILP formulation

1 Introduction

In the recent years the traffic load of transport networks has increased significantly
due to the rapid growth of Internet and network based applications. There comes a
time when the network providers cannot satisfy the traffic demands by merely
enlarging network capacities. Many types of demands, responsible for the heavy
traffic load, can be regarded as multicast (point to multipoint) demand instead of
ordinary unicast (point to point) demand. Applications include digital media
broadcasting (e.g. IP-TV, IP-Radio, etc.) or digital media distribution and streaming.
Multicast in a transport network is especially useful when the content has high
bandwidth requirements, e.g. distribution of one or several digital TV channels from
the distributor of the content to the local providers. Other possible application of
multicast is described in [1].

There are quite a few papers in the field of optimizing the cost of multicast routing
in transport networks. Madhyastha et al. proposed in [2] a heuristic method for the
problem of multicast routing and wavelength assignment in WDM ring networks.
They proposed new node architectures with electronic or optical Drop-and-Continue
(D&C) feature. The authors of [3] presented an analytical model of grooming problem
represented as non-linear programming formulation. They compare the results with
heuristic approaches. Several heuristic tree formation algorithms are proposed in [4].
The study assumes a network with sparse splitting capabilities, i.e. only some of the

[†] This work was supported by the EC within the IST FP6 NoE e-Photon/ONe+ research framework.

I. Tomkos et al. (Eds.): ONDM 2007, LNCS 4534, pp. 259–268, 2007.

nodes are able to perform splitting of lightpaths in the network. Huang et al. [5] introduce a multicast dynamic light-tree grooming algorithm for solving the problem of formation and maintenance of a multicast tree. The authors of [6] use an ILP formulation to solve the optimal routing and wavelength assignment problem, and show that a network with only a few splitters and wavelength converters can efficiently transfer multicast demands. Two heuristic optimization algorithms are proposed in [7] minimizing the number of allocated wavelengths in the network. The authors aim to perform optimal QoS (Quality of Service) routing and optimal wavelength assignment together.

2   Problem formulation

A two-layer network is assumed, where the upper, electronic layer is time switching capable while the lower, optical layer is a wavelength (space) switching capable one. The electronic layer can perform traffic grooming, i.e. multiplexing low bandwidth demands into a single WL channel. The two layers are assumed to be interconnected according to the peer model or vertically integrated according to the multi-region network node framework, i.e., while routing, the control plane has information on both layers and both layers take part in accommodating a demand. Note, that the result is applicable to overlay or augmented interconnection models as well.

The network topology and the number of fibers are assumed given as well as the description of traffic demands. The capacity of WL channels and the cost of routing, (e.g. space switching, optical to electronic conversion, WL branching, etc.) can also be given in advance. We assume static traffic consisting of unicast and multicast traffic demands. A unicast demand has one source, one destination node and a given bandwidth, while a multicast demand has more than one target node. The objective is to reach all destination nodes from the source, while observing all routing and technical constraints.

3   Network model

We aimed to use a general network model for routing in two layer networks with grooming and with different types of nodes and arbitrary topologies assuming peer-model that allows optimal routing using the resources of both layers jointly. The model must be able to handle any regular mesh topology. For all these reasons we chose wavelength graph model to represent the network. The WL graph corresponding to the logical network is derived from the physical network considering the topology and capabilities of physical devices.

A simpler version of the model has been first proposed in [8]. ILP formulation of the static RWA problem with grooming and protection has been given in [9].

The types of nodes can also be quite different: Optical Add-and-Drop Multiplexers (OADM), Optical Cross-Connects (OXC: optical core) with full or limited, optical or opto-electrical WL conversion or even an Opto-Electrical Cross-Connect (OEXC: electrical core). Furthermore, some of these nodes support grooming, typically with limited number of optical ports. All these properties can be considered in the WL graph model, together with different protection techniques of traffic demands.

The network consists of nodes and links connecting the nodes. Both ends of an optical link (fiber) are attached to an interface (IF) of a physical device. A physical device contains an internal switching fabric and some IFs. The number of available WLs in a fiber is the minimum of the WLs supported by the end IFs. Every link and every physical device has a specific logical representation in the WL graph.

A physical link is derived to as many logical edges as the number of available WLs in the link. The logical sub-graph of a physical device depends on the capabilities of the device. Every edge in the graph has a capacity and a cost of usage. The capacity of the edge usually equals to the WL capacity, which depends on the used carrier (typically 2.5 Gbps – which was assumed in our simulations – or 10 Gbps). The cost of the edge is determined by its functionality (WL edge, O/E conversion, etc.).

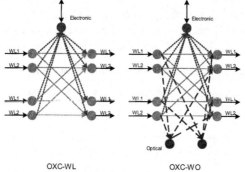

Fig. 1. Sub-graph of an OXC-WL device in the wavelength graph (*left*).
Sub-graph of an OXC-WO device in the wavelength graph (optical splitting capable) (*right*).

A physical device is modeled by a sub-graph. It represents all IFs of the device, and the capabilities of its internal switching fabric. The WL graph model (together with our ILP framework) can support devices with different capabilities appearing in the network at the same time. The model is easily extendable; type of devices can be changed later, if new internal models are introduced.

A sub-graph of a versatile physical device is depicted in Fig. 1. The equipment is a combination of an OXC with WL-conversion and an OADM: it can originate and terminate traffic demands, as well as perform space-switching. WL-conversion is possible only through the electronic layer. This is illustrated by an electronic node in the sub-graph, while other (pair of) logical nodes correspond to interfaces. Fig. 1 assumes two fibers connected to the device, and two WLs per fiber, which results in two input and two output interfaces – because all edges are directed. We will use this complex node and its extension in the simulations.

4.1   New node models

The OXC-WO (OXC with WL-conversion and Optical splitting) devices should be considered as an extension of the OXC-WL type. It introduces a new functionality:

optical splitting of light-paths. The branching function is represented by dashed edges in the sub-graph. By adding these specific logical edges into the sub-graph, we can accurately determine the cost of optical branching. The in-degree of splitting nodes is always less than or equal to one, while the out-degree is more than or equal to two.

Branching in this device is, however, possible in the electronic layer as well. To accomplish electronic branching, the demand must be first routed up to the electronic layer. The returned (branched) lightpaths should not necessarily use the same WLs. Thus WL-conversion can also be performed in addition to branching (and 3R processing) of the signal.

5   ILP formulations

We used the following ILP formulation to route multiple multicast trees in the network (Note, optical branching requires a few extra constraints in addition to these):

$z_{ij}^{or} \in \{0, 1\}$ indicates whether sub-demand o of multicast tree r uses edge (i, j) or not.

$x_{ij}^{r} \in \{0, 1\}$ indicates whether (end-to-end or multicast demand) r uses edge (i, j).

$y_{ij} \in \{0, 1\}$ indicates if edge (i, j) is used by any of the demands.

$$\sum_{\forall j \in V_i^+} z_{ji}^{or} - \sum_{\forall k \in V_i^-} z_{ik}^{or} = \begin{cases} -1 & \text{if } i = s^r \\ 0 & \text{if } i \notin \{s^r, t^{or}\} \\ +1 & \text{if } i = t^{or} \end{cases} \quad (0.1)$$

for every (logical) node $i \in V$, tree r and o sub-demand.

V_i^+ denotes the set of nodes that can reach node i by the use of one directed edge, while V_i^- is the set of nodes reachable from i by one directed edge. Capitals A, V, V_E, O, R denote the set of edges (arcs), nodes (vertices), electronic nodes, sub-demands and the set of multicast trees respectively. The source of tree r is denoted by s^r, while targets are denoted by t^{or}, where o is the corresponding sub-demand.

$$z_{ij}^{or} \leq x_{ij}^{r}, \ \forall (i, j) \in A, \ \forall o \in O, \ \forall r \in R \quad (0.2)$$

$$x_{ij}^{r} \leq \sum_{\forall o \in O} z_{ij}^{or}, \ \forall (i, j) \in A, \ \forall r \in R \quad (0.3)$$

$$\sum_{\forall j \in V_i^+} x_{ji}^{r} = \sum_{\forall k \in V_i^-} x_{ik}^{r} \leq 1, \ \forall i \notin V_E, \ \forall r \in R \quad (0.4)$$

$$\sum_{\forall j \in V_i^+} x_{ji}^{r} \leq \begin{cases} 0 & \text{if } i = s^r \\ 1 & \text{if } i \neq s^r \end{cases}, \ \forall i \in V_E, \ \forall r \in R \quad (0.5)$$

$$\sum_{\forall r \in R} x_{ij}^{r} \cdot b^r \leq B_{ij}, \ \forall (i, j) \in A \quad (0.6)$$

$$x_{ij}^{r} \leq y_{ij}, \ \forall (i, j) \in A, \ \forall r \in R \quad (0.7)$$

$$y_{ij} \le \sum_{\forall r \in R} x_{ij}^r, \ \forall (i, j) \in A \qquad (0.8)$$

$$\sum_{\forall j \in V_i^-} y_{ji} = \sum_{\forall k \in V_i^-} y_{ik} \le 1, \ \forall i \notin V_E \qquad (0.9)$$

Variables:

$$z_{ij}^{ro} \in \{0, \ 1\}, \ \forall (i, j) \in A, \ \forall o \in O, \ \forall r \in R \qquad (0.10)$$

$$x_{ij}^r \in \{0, \ 1\}, \ \forall (i, j) \in A, \ \forall r \in R \qquad (0.11)$$

$$y_{ij} \in \{0, \ 1\}, \ \forall (i, j) \in A \qquad (0.12)$$

Objective function:

$$\text{Minimize} \sum_{\forall (i, j) \in A} c_{ij} y_{ij} \qquad (0.13)$$

(0.1) expresses flow-conservation of each sub-demand of each tree in all nodes. (0.2) tells that edge *(i, j)* must be allocated if any sub-demand of tree *r* wants to use it, while (0.3) ensures that edge *(i, j)* is not allocated in vain. (0.4) ensures that light-paths cannot branch (or disappear) in non-electronic nodes. Constraint (0.6) says that the aggregate bandwidth of demands using edge *(i, j)* should be less than the capacity of one WL. (0.7) expresses that a tree can only use an edge, if it is allocated for the routing. In addition (0.8) ensures that edges used by none of the trees are not allocated. (0.8) can be omitted, however, since it is implicitly included in the minimization objective. Constraint (0.9) is similar to (0.4), but on the highest hierarchy-level. It expresses that the number of allocated input and output edges should be equal in non-electronic nodes. The objective function (0.13) expresses that the total cost of allocated edges should be minimized, i.e. we are looking for a minimal-cost routing solution in the network.

The above mentioned formulation can be applied for a mix of unicast and multicast demands as well. The routing problem of solely unicast demands has equivalent ILP formulation using only a two-level hierarchy of variables. Although that formulation contains less variables and inequalities, the solution time does not differ significantly from the solution time of the general formulation. The ILP solver likely recognizes the simplifications and shortcuts, which does not cause significant (more than an order of magnitude) deviation in the solution times.

6   Technical constraints

We applied several technical constraints to influence certain properties of the routing. These constraints reflect real-world technical restrictions. Two types of constraints are considered: branching limitations and restrictions concerning size of multicast trees.

6.1   Branching limitations

Most of the current switching devices are not able to perform neither optical nor electronic layer branching of the signal due to technical or software restrictions. Thus it is necessary to limit the splitting of the signal in both layers. The electronic layer branching can be constrained by the following simple inequality:

$$\sum_{\forall j \in V_i^-} x_{ij}^r \le \alpha_i, \forall i \in V_E, \forall r \in R \qquad (1.1)$$

The branching limit in node i is denoted by constant α_i.

The all-optical branching of the signal has further implications. The power of the signal decreases by 3 dB when it is split in two. In case of splitting in more than two the attenuation is more significant, which impairs correct detection in a receiver.

The number of times the signal is split can be constrained for each sub-demand, because it determines the quality of the signal in the receiver. The number of optical splits can be constrained by the following formula:

$$\sum_{\forall i \in V_{osplit}} z_{ij}^{or} \le L^{or}, \ \forall o^r \in O, \forall r \in R, \qquad (1.2)$$

where L^{or} means the upper limit for sub-demand o of tree r.

Constraint (1.2) sets an upper limit for the number of splits along the path of each sub-demand of each tree. The number of optical splits along the path can be calculated by counting the number of edges originated from optical splitting nodes. Note, that we only calculate the total number of splits along the path, not the maximum number of consecutive splits along the path. Although that property is more important, it can be formulated only by too complex linear constraints.

6.2   Limitation on the size of multicast trees

Sometimes the optimal solution of a multicast tree produces too long paths between the source and some of the targets. It also implies higher delay, which is unacceptable in some applications or harms the QoS (Quality of Service) agreement. The length (measured in hops in this case) of the path can be limited by the following formula, referred to as *depth-limit constraint*:

$$\sum_{\forall (i,j) \in A} z_{ij}^{or} \le \beta^{or}, \ \forall o \in O, \forall r \in R \qquad (1.3)$$

Different requirements can be specified for each tree r and sub-demand o by assigning different β^{or} constant values. However, inadequate selection of β^{or} values can render the whole routing problem infeasible. In order to measure the length of the path in some metric such weight factors should be introduced that represent the distance.

It is also possible that size of the whole tree should be limited. The number of links contained by the tree can be restricted by the following formula (*tree-size limit*):

$$\sum_{\forall (i,j) \in A} x_{ij}^r \le \mu^r, \ \forall r \in R, \qquad (1.4)$$

μ_r is a constant value meaning the tree-size limit.

A switching device is called a leaf-node, if it does not branch or relay the multicast tree, but terminates it. It implies that the out-degree of such a node equals to zero. The width of the tree is defined by the number of leaf nodes in it. The *breadth-limit* of the tree can be set up by the following inequalities:

$$\sum_{\forall j \in V_i^-} x_{ij}^r \leq \kappa_i \cdot (1 - v_i^r), \ \forall i \in t^r, \forall r \in R \qquad (1.5)$$

$$(1 - v_i^r) \leq \sum_{\forall j \in V_i^-} x_{ij}^r, \forall i \in t^r, \forall r \in R \qquad (1.6)$$

$$\sum_{\forall i \in t^r} v_i^r \leq \eta^r, \forall r \in R \qquad (1.7)$$

Variables:

$$v_i^r \in \{0, \ 1\}, \forall i \in V_{dr}, \forall r \in R, \qquad (1.8)$$

Variable $v_i^r \in \{0, \ 1\}$ expresses whether tree r is relayed (0) or terminated (1) in node i, while κ_i is a properly chosen constant of a large value. Constant η^r expresses the breadth-limit. The sum in (1.7) should only include the electronic nodes of the target switching devices, since these are the only candidate leaf nodes of the tree. Because of the other routing constraints other leaf nodes (other than the target nodes) are not possible.

6.3   Soft constraints

All technical constraints and several of the routing constraints can be reformulated as soft-constraints. It means that a technical or routing recommendation is no longer a member of the ILP constraints; rather it is included in the optimization objective. If a solution does not observe a recommendation, it is not rejected, but penalized. The necessity of a recommendation can be controlled by adjusting the value of the penalty. The advantage of this method is that we can sooner obtain a feasible solution, instead of waiting for a solution observing all the limitations. This technique is especially useful when the whole problem together is infeasible, but lessening the requirements can lead to a feasible solution. For example the soft depth-limit constraint can be expressed by the following formula:

$$\sum_{\forall (i,j) \in A} z_{ij}^{or} \leq \beta^{or} + q^{or}, \ \forall o \in O, \forall r \in R \qquad (1.9)$$

$$q_i^{ro} \in R^+, \forall o \in O, \forall r \in R \qquad (1.10)$$

$$\text{Minimize} \sum_{\forall (i,\ j) \in A} c_{ij} y_{ij} + \sum_{\substack{\forall o \in O, \\ \forall r \in R}} c^{or} q^{or} \qquad (1.11)$$

q^{or} is a positive variable representing the deviation from the recommended value of the limit, while c^{or} is a constant meaning the weight of the penalty. Other constraints can be reformulated as soft constraints in the same way.

7   Results

In all simulations NRS core network topology [10] was used with 5 WLs per link.

Fig. 2 shows that multicast routing scales well with the increasing number of both source (trees) and target nodes. The cost curve does not rise beyond a limit, it has a saturation section. It means inserting a new light-tree or further nodes into the current trees has steadily decreasing cost.

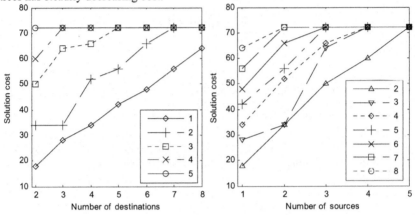

Fig. 2. Cost of routing as a function of the increasing number of targets for different number of sources *(left)*. Cost of routing as a function of the increasing number of sources for different number of targets *(right)*.

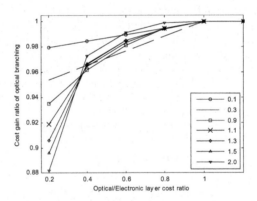

Fig. 3. Cost gain ratio of optical branching versus electronic layer only branching as a function of optical-electronic cost ratio. Different curves assume different WL costs.

Fig. 3 tells that significant cost can be saved if optical layer branching capability is introduced into the network. The figure depicts the cost of routing in an optical branching capable network compared to an only electronic branching capable network. It shows that the lower the price of the optical layer is, the more costs can be saved. This suggests that optical layer branching is particularly worth, if the electronic layer has high cost, which is in accordance with the real-world conditions. The cost gain depends, however, on several properties of the network and the set of demands.

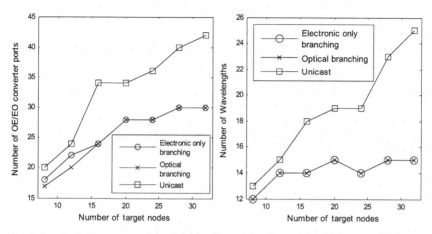

Fig. 4. Required number of converter ports *(left)* and wavelengths *(right)* as a function of the number of target nodes for unicast and multicast routing (with and without optical branching).

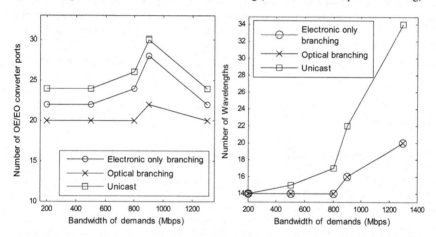

Fig. 5. Required number of converter ports *(left)* and wavelengths *(right)* as a function of the bandwidth of demands for unicast and multicast routing (with and without optical branching).

Fig. 4 shows that the required number of converter ports and wavelengths rises with the increasing number of target nodes. Optical branching outperforms electronic branching only if the number of nodes participating in the trees is low – compared to the total number of nodes in the network. As more and more nodes are involved in the trees the advantage of optical branching disappears. It is likely due to the fact that in member nodes of the trees the electronic layer must be reached anyway, thus there is less necessity for optical layer branching in these nodes. Both the required number of converter ports and WLs scale well with the increasing number of demands, unlike in the case of unicast routing, where the required number of resources increases more rapidly. In case of unicast a set of end-to-end demands equivalent to the multicast demands were routed.

Fig. 5 also shows the required number of converter ports and WLs, but as a function of the increasing bandwidth of the demands. Four trees were routed with three target nodes each. In such a situation optical branching clearly uses less converter ports than electronic branching. However, optical branching does not decrease the required number of WLs. Unicast – naturally – performs far worse in all measurements. Grooming is less and less applicable if the bandwidth of the demands is close to the WL capacity. The sudden fall of the number of OE/EO ports can be explained by this. In case of tiny bandwidths multicast routing does not really show its power, since grooming compensates some of the disadvantages: a huge amount of WLs can be spared by multiplexing tiny demands into a single WL channel.

8   Conclusion

In the paper we proposed WL graph models and a new ILP formulation to route unicast and multicast demands in WDM networks. We evaluated the cost and the resource usage of multicast routing with and without optical layer branching of light-trees and unicast routing as well. We showed that optical branching usually performs better than branching in the electronic layer only. However, if many nodes are involved in the trees then electronic layer is used anyway, and optical branching loses its gain. In case of tiny bandwidths, grooming can well compensate the drawback of unicast over multicast.

References

1. B. Quinn and K. Almeroth, "IP multicast applications: Challenges and solutions", IETF RFC 3170, Sep. 2001
2. Madhyastha et al., "Grooming of multicast sessions in WDM ring networks", OptiComm 2003: Optical Networking and Communications, Nov. 2003
3. G. V. Chowdhary and C. S. R. Murthy, "Grooming of Multicast Sessions in WDM Mesh Networks", Workshop on Traffic Grooming, 2004
4. X. Zhang et al., "Constrained Multicast Routing in WDM Networks with Sparse Light Splitting", Journal of Lightwave Technology, vol. 18, issue 12, p. 1917, Dec. 2000
5. X. Huang et al., "Multicast Traffic Grooming in Wavelength-Routed WDM Mesh Networks Using Dynamically Changing Light-Trees", Journal of Lightwave Technology, vol. 23, no. 10, Oct. 2005
6. D. Yang and W. Liao, "Design of light-tree based logical topologies for multicast streams in wavelength routed optical networks," in Proc. IEEE Information Communications (INFOCOM), San Francisco, CA, Apr. 2003
7. X. H. Jia et al., "Optimization of Wavelength Assignment for QoS Multicast in WDM Networks", IEEE Transactions on Communications, vol. 49, no. 2, Feb. 2001
8. T. Cinkler et al., "Configuration and Re-Configuration of WDM networks", NOC'98, European Conference on Networks and Optical Communications, Manchester, UK, 1998
9. T. Cinkler, "ILP formulation of Grooming over Wavelength Routing with Protection", ONDM 2001, 5th Conference on Optical Network Design and Modeling, Wien, Feb. 2001
10. NRS core network topology (16 nodes, 22 links)
    http://www.ibcn.intec.ugent.be/css_design/research/projects/INTERNAL/NRS/index.html

MUPBED - Interworking Challenges in a Multi-Domain and Multi-Technology Network Environment

Hans-Martin Foisel[1], Jan Spaeth[2], Carlo Cavazzoni[3],
Henrik Wessing[4], Mikhail Popov[5]

[1] Hans-Martin Foisel, T-Systems/Deutsche Telekom, Goslarer Ufer 35, D-10589 Berlin,
Germany, Hans-Martin.Foisel@t-systems.com
[2] Jan Späth, Ericsson, Gerberstr. 33, 71522 Backnang, Germany
[3] Carlo Cavazzoni, Telecom Italia, Via G. Reiss Romoli 274, I-10148 Torino, Italy
[4] Henrik Wessing, Technical University of Denmark, Bldg. 343, Oersteds Plads, 2800
Kgs. Lyngby, Denmark
[5] Mikhail Popov, Acreo AB, Electrum 239, SE-164 40 Kista, Sweden

Abstract. Today's data transport networks are evolving continuously towards customer oriented and application aware networks. This evolution happens in Europe in a highly diverse network environment, covering multiple network domains, layers, technologies, control and management approaches. In this paper, the issues, challenges and the solutions developed in the IST project MUPBED („Multi-Partner European Test Beds for Research Networking"; www.ist-mupbed.eu) for seamless interworking in a typical European heterogeneous network environment are described, addressing horizontal, inter-domain, and vertical, inter-layer topics related to data plane, control plane and applications.

Keywords: ASON/GMPLS, Test Network, Multi-Domain, Multi-Layer, Interworking, Intelligent Optical Networks

I. Introduction

Seamless interworking in a heterogeneous and fast evolving multi-domain, multi-layer, multi-technology network environment is one of the top challenges network operators are facing today, indifferent of their background, e.g. incumbent, newcomer, local or global player, commercial or research network service provider. Additionally, the customer and application awareness of the networks have to be increased, the network flexibility enhanced leading to provisioning of so-called "On-Demand Network Services". This network scenario fits very well to the current European network environment and will most likely also be applicable to the foreseeable future.

Solutions to cope with these network environments are based on the data plane level on standards of ITU-T (International Telecommunication Union – Telecommunication Standardization Sector) and IE   and Electronics

I. Tomkos et al. (Eds.): ONDM 2007, LNCS 4534, pp. 269–278, 2007.
© IFIP International Federation for Information Processing 2007

Engineers – Standards Association) and on the control plane level on ASON/GMPLS (Automatically Switched Optical Network, Generalised Multi-Protocol Label Switching) standards of ITU-T, IETF (Internet Engineering Task Force) and Implementation Agreements of the OIF (Optical Internetworking Forum). Several ASON/GMPLS test network activities were carried out worldwide [1 - 8] aiming at gaining first experiences from these new network functions and providing valuable feedback to the standardisation bodies and forums.

The IST-Project MUPBED, "Multi-Partner European Test Bed for Research Networking" [9, 10] investigates and evaluates many of these interworking issues conceptually and experimentally. The practical evaluations have been carried out in a European scale multi-domain, multi-layer ASON/GMPLS test network, in a joint effort by network operators, industry partners, NRENs (National Research and Education Networks) and research institutes/universities. The test network includes multiple domains, each with a set of different technologies, e.g. IP/MPLS, Ethernet, SDH, all optical cross connects and different control and management approaches.

This paper reports on interworking solutions developed in the MUPBED project, applicable to these heterogeneous network environments on the European as well as on the global scale.

The paper is structured in 4 sections. In section II the European network environment is described. Section III and IV cover the data and control plane interworking solutions, which have been developed, implemented and evaluated in the MUPBED test network.

II. European Networks – a Heterogeneous Environment

Networks have been portioned, layered or deployed with different technologies because of many reasons related to, e.g. investments, operation, services, or they simply have become heterogeneous because of acquisitions. **Fig. 1** depicts schematically such a wide variety of different network domains with respect to the data plane technology implemented, whereas **Fig. 2** illustrates the mix of different solutions at the control plane level. These scenarios are applicable for an intra- as well as inter-carrier network environment. Nevertheless, seamless interworking among all these different approaches is strongly needed and highly desirable for future–proof flexible and efficient solutions. Mandatory to achieve this goal is the integration of data and control plane functions to provide powerful new network functionalities. This integration results in challenges regarding to horizontal interworking, e.g. among network domains and to vertical interworking, e.g. among the different network layers and applications. These twofold interworking directions are schematically depicted in **Fig. 3**. Most of them were tackled and solved in the MUPBED test network, which is based on a similar divergent set of implementations at the five local test beds at TI/Torino, TID/Madrid, Acreo/Stockholm, PSNC/Poznan, and DT/Berlin, and at the European research network's implementation of their interconnections.

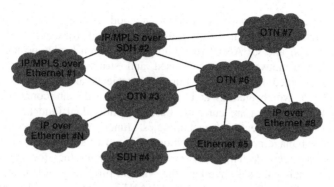

Fig. 1. Data plane: Multi-domain, multi-layer, multi-technology network scenario

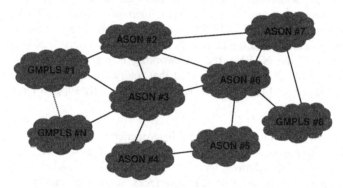

Fig. 2. Control plane: Multi-domain ASON/GMPLS network scenario

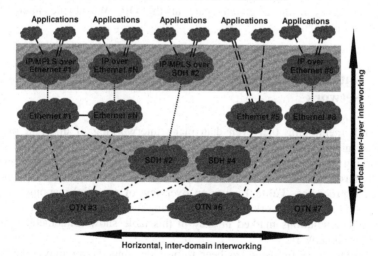

Fig. 3. Horizontal and vertical interworking areas in a multi-domain, multi-layer, multi-technology network environment, most of them tackled and solved in the MUPBED test network

The European research network situation is a good example for a heterogeneous multi-carrier network scenario, with multiple National Research and Education Networks (NRENs) and GÉANT2 (the European research backbone network). The European research networks are organised and structured as follows: Campus networks and selected national and European projects are interconnected to the respective NREN, which enables and supports nationwide interconnectivity. These national research networks are currently based on individual (and therefore largely different) network architectures, technologies, functions, vendor equipment and network control/management implementations. It is most likely that this heterogeneity and individuality per domain will be maintained in the foreseeable future. Additionally to their national coverage, the NRENs provide interconnectivity to the European research backbone network GÉANT2 enabling a European scale connectivity and, additionally, long haul connections to non-European research networks. To enable such pan-European interconnections, GÉANT2 has to interoperate with all the individual NREN networks and is always involved in a multi-domain network scenario, including mainly the connection configuration and control. Today, in most of the NRENs, the interworking is realised at the IP layer only. However, in the future it can be expected that interworking is required for various other network layers and technologies as well, to efficiently support large bandwidth and high QoS services.

Within the MUPBED test network the heterogeneity of the implementation per domain or test bed has been maintained. No alignment of these different data, control and management plane solutions has been postulated, but standard-based solutions have been required at the inter-domain interfaces, to enable seamless interworking within the MUPBED network as well as with other European and national project's test beds and even on a global scale at the OIF Worldwide Interoperability Demonstration 2005.

III. Data Plane Interworking Areas in the MUPBED Test Network

It is worth mentioning that even data plane interworking of different multi-layer networks might raise several challenges, although standards are available. This interworking is especially challenging because the involved network domains are generally based on different technologies, network platforms, vendor equipment and operational processes. **Table 1** gives an overview of potential mapping procedures and solutions among network layers, highlighting the manifold solutions and options available. The solutions marked in grey in this table show implementations in the MUPBED test network.

In MUPBED, a very first step has been to provide connectivity among the local test beds based on transparent data links. With today's technologies, this could be best achieved with SDH/OTN interconnections between the five MUPBED test bed sites. However, the MUPBED project had to find a solution, which is supported at the inter-domain interfaces by all local test beds and the European research networks. This results into a provisioned – statically configured – full-mesh topology of "transparent Layer 2" inter-connections, based on Ethernet over IP/MPLS transport (**Fig. 4**). It shows the five MUPBED test beds, the different technologies they are based on, reflecting very well the divergent European network environment, as well as the involved NRENs and GÉANT2 networks. All switching functions of the MUPBED network were allocated within the different MUPBED test bed sites. In such a way a multi-domain network in a truly multi-partner environment with participants from industry and research community has been setup.

Table 1. Data plane standard based mapping solution overview in multi-domain, multi-layer, multi-technology networks; marked options were implemented in the MUPBED test network

N\M (M over N)	IP/MPLS	Ethernet	SDH	OTN
IP/MPLS	LSPx – LSPy interworking	EoMPLS or L2TP		
Ethernet	LSP ⟷ VLAN mapping or plain transport	VLANx - VLANy interworking or stitching	Emulation (restricted usability)	
SDH	GFP/VCAT/LCAS	GFP/VCAT/LCAS	Basic feature	
OTN	GFP/VCAT/LCAS	GFP/VCAT/LCAS	Basic OTN feature	Basic feature

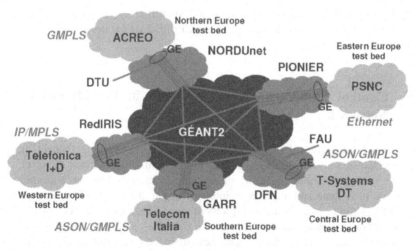

Fig. 4. Data plane implementation in the MUPBED test network – all five local test beds are interconnected via Ethernet over IP/MPLS links over the NREN and GÉANT2 networks

As an example how the end-to-end interworking of these different data plane implementations has been achieved within the MUPBED test network, the detailed implementation and interworking challenges and solutions at DT/Berlin are depicted in **Fig. 5**. The LSPs in the GÉANT2 IP/MPLS platform have been stitched to the LSPs in the IP/MPLS platform of the NRENs, here the X-WIN network of DFN. At the NRENs PoP at the location of the MUPBED test bed, these LSPs have been mapped into VLANs. In DT's local test bed, an Ethernet metro network performs the VLAN resolution and translation functions and makes those VLANs available on physically separated Gigabit-Ethernet (GE) ports, which in turn are connected to GE ports of the Ericsson NG-SDH domain. By setting up Ethernet Soft-Permanent-Connections (Eth-SPCs) with a variable bandwidth between VC-4 to VC-4-7v, these ports can be interconnected to any desired other MUPBED test bed or application location. All these inter-domain and inter-layer mapping schemes are standard-based and interoperable solutions, resulting in the highest possible degree of interworking capabilities.

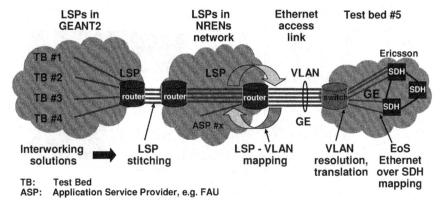

Fig. 5. Data plane interworking solutions in the MUPBED test network - Allocation of functions among GÉANT2, NREN networks and MUPBED local test beds

IV. Control Plane Interworking Areas in the MUPBED Test Net-work

Interworking at the control plane level is even more challenging than at the data plane, because the standards and specifications are still in progress, therefore early prototype implementations need to be chosen to accomplish an automatic and seamless multi-domain and multi-layer interworking in the MUPBED test network.

The ASON/GMPLS inter-domain implementations (**Fig. 6**) have been following a network architecture which is suitable to the current – and most likely also for the near future European networks – very heterogeneous network environment, with many separated network domains largely consisting of different technologies, network architecture, and operational mechanisms. Therefore, solutions for seamless (automatic) inter-working among these network domains are a key issue to be solved for the next generation of research networks. The expected result is an increase in the multi-domain connection dynamics, the possibility for on-demand customer-oriented services, and a significant reduction of manual configurations.

This network architecture and the concept to maintain the individual architecture and technology approaches in each of the five MUPBED test beds while enabling automatic interworking among the domains has been applied to the MUPBED network scenario, resulting into a network topology as depicted in **Fig. 7**. Furthermore, this concept allows using OIF inter-domain interfaces, the interoperable implementations of which were proven and demonstrated at the OIF Worldwide Interoperability Demonstration 2005 [8].

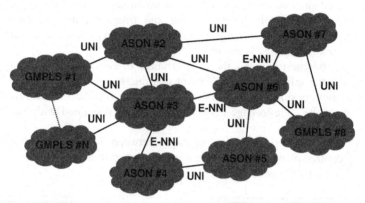

Fig. 6. Control plane: ASON/GMPLS multi-domain and multi-layer solution for a heterogeneous multi-domain network, using OIF inter-domain interfaces.

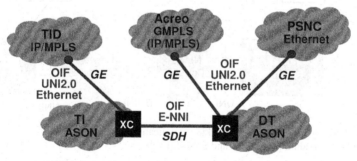

● **UNI-C 2.0 Ethernet proxy server implementations**

Fig. 7. Control plane: ASON/GMPLS network architecture implementation in the MUPBED test network

Additionally to the interworking of network domains and layers, the MUPBED project has addressed the application–network interface issues, aiming to make the MUPBED ASON/GMPLS test network application-aware. As depicted in **Fig. 8**, a chain of functions has been chosen to interface the applications with the OIF UNI-C (client side), comprising Application Programming Interfaces (API), the adaptation function and finally the ASON/GMPLS control plane-aware UNI-C.
The main objective of the API is to provide a uniform access to the adaptation function with a simple communication for the applications, completely decoupled from any ASON/GMPLS signaling or knowledge of the underlying transport

technology. The API implementation chosen in MUPBED is based on Web Service technologies, which enables easy communication on port 80 (HTTP), thus avoiding most of the firewall problems. The functional range of the API comprises the following three main communication messages: resource request, resource release and status request.

The Adaptation Function (AF) is introduced as responsible for interfacing with the network control plane and for deciding, when new network resources should be allocated. The adaptation function receives resource requests from the application via the API, and is responsible for translating and triggering these requests to the network. In this way, decoupling between applications and the currently used transport network technology is ensured.

The adaptation function controls the establishment of connections by interfacing to the UNI-C and the information available at this interface. Therefore, the adaptation function does not consider the network topology of the ASON/GMPLS layer, as it is simply aware of the edge-to-edge connections that are associated with the UNI-C it controls.

The OIF UNI specifies the signaling messages between the adaptation function and the network control plane. At the adaptation function side a UNI-C proxy server or RSVP agent provides the needed control plane functions.

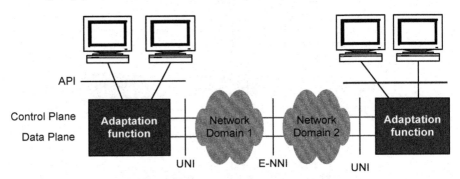

Fig. 8. Vertical and horizontal interworking: application-network interworking principle via API – Adaptation module – OIF UNI and E-NNI functions

In the MUPBED project, the following three application-network interface solutions have been developed (**Fig. 9**) and implemented [9]:

- The Network Provider Stack implementation implements an advanced resource allocation scheme in a multi-layer IP/MPLS over ASON/GMPLS network. As test application it uses a storage and backup software
- The Standalone or GUI implementation includes the resource allocation and is not integrated directly with specific applications. This makes it suitable as a separate, manually controlled tool for applications where integration of these functions could not be easily made
- In the Socket Stack Solution the applications do not communicate through an API and adaptation function to the UNI-C interface. Instead they directly execute socket calls in the UNI-C proxy server

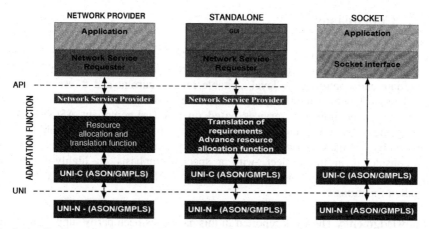

Fig. 9. Application-network interworking solutions implemented in the MUPBED project

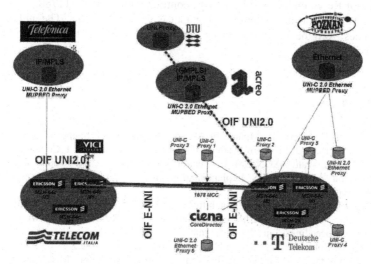

Fig. 10. Control plane: Example of the application-network interworking based on Standalone Application Vertical Integration Stack, depicting the topology of an Ethernet switched connection DTU/Copenhagen – DT/Berlin – TI/Torino initiated at DTU

The most flexible standalone or GUI implementation has been implemented, tested and demonstrated in the MUPBED test network. **Fig. 10** shows a screen shot of the MUPBED network control plane level topology display, highlighting the Ethernet switched connection setup from DTU/Copenhagen via multiple ASON/GMPLS network domains at DT and TI to the UNI-C instance (Avici router) at TI, clearly illustrating the smooth interworking capabilities of applications with transport and data network domains and elements.

V. Conclusions

The MUPBED project has implemented and tested solutions for seamless interworking in a heterogeneous multi-domain and multi-layer network environment, on the data and control plane level, suitable for the current and foreseeable future network environment of the European research networks and carriers. These solutions enable on-demand transport services based on standard-compliant ASON/GMPLS control plane technologies, ensuring the highest possible level of interoperability.

Additionally, seamless interworking with a wide variety of Ethernet-based clients has been considered in the project, with a special emphasis on highly-demanding applications, which could interwork with ASON/GMPLS networks via API and adaptation functions.

Acknowledgments. The work reported in this paper has been partly supported by the European Commission within the FP-6 project MUPBED (Multi-Partner European Test Beds for Research Networking) under contract number IST-511780.

The responsibility for the content of this paper is with the authors. The authors thank all colleagues from the MUPBED project for their work, which has been the basis for this paper.

References

1. C. Cavazzoni, R. Morro, R. Muñoz, R. Martínez, F. Galán, S. Szuppa, H. M. Foisel, H. Dentler, M. Herpers, J. Liebenow, Juan P. Fdez-Palacios, S. Spadaro, J. Comellas, "NOBEL phase 2 - Large scale ASON/GMPLS network demonstrator", ECOC2006 Workshop on Workshop on ASON/GMPLS Implementations in Field Trials and Carrier Networks, www.ist-mupbed.eu
2. PHOSPHORUS overview presentation, http://www.ist-phosphorus.eu/documents.php
3. P. Kaufmann, "VIOLA-Testbed: Current State and First Results", TERENA Networking Conference Proceedings, June 2006, Catania
4. T. Lehman, J. Sobieski, B. Jabbari, „DRAGON: A Framework for Service Provisioning in Heterogeneous Grid Networks", IEEE Communications Magazine, March 2006, Volume 44, No 3
5. HOPI white paper: http://networks.internet2.edu/hopi/hopi-documentation.html
6. Cheetah: X. Zhu, X. Zheng, M. Veeraraghavan Z. Li, Q. Song, I. Habib N. S. V. Rao, "Implementation of a GMPLS-based Network with End Host Initiated Signaling", IEEE ICC 2006 conference proceedings, paper OS11-1
7. Y. Sameshima, S. Okamoto, W. Imajuku, T. Otani, Y. Okano, "JGN II Testbed Demonstration of GMPLS Inter-Carrier Network Control with Actual Operational Consideration, ECOC 2006 Proceedings, paper We 4.1.5
8. J.D. Jones, L. Ong, M. Lazer, Interoperability update – dynamic Ethernet services via intelligent optical networks, IEEE Optical Communication, Nov 2005, Vol3, No4, pp. 2-11
9. MUPBED project web page, presentations and Deliverables
10. C. Gerlach, H.-M. Foisel, S. Szuppa, A. Weber, „MUPBED: A multi-domain ASON/GMPLS test network", NOC 2006 Proceedings, pp. 395-402

Rule-Based Advertisement and Maintenance of Network State Information in Optical-Beared Heterogeneous Networks

János Szigeti, Tibor Cinkler

High-Speed Networks Laboratory
Department of Telecommunications and Media Informatics
Budapest University of Technology and Economics **
{szigeti,cinkler}@tmit.bme.hu

Abstract. A more flexible routing of better performance can be achieved in multilayer networks when the controllers of different network layers are "peering" instead of "overlaying", i.e., instead of the simple overlay model the *peer interconnection* or the *vertically integrated* model is used, and Label Switched Paths are searched in a Wavelength Graph that represents the network state very accurately. However, routing in "Peer network" does not only require large databases for storing the Wavelength Graph and complex path-search algorithms, it also increases the load of control channels, as more link state changes must be advertised. Much of the link state information is, however, redundant as the state changes are not independent. In this paper we propose a method for topology information advertisement and maintenance to significantly reduce the amount of control messages without deteriorating the quality of routing.
Key words: path computation, multilayer, vertically integrated, topology advertisement

1   Introduction

The current core & metro communication networks are based on optical transmission and optical switching equipment. They are able to carry a large amount of data, however, optical packet level forwarding of data is still problematic. The new paradigms are OBS (Optical Burst Switching), and in longer term optical label switching and stripping (OLS) and optical packet switching (OPS) but OCS (Optical Circuit Switching) is still viable as in core networks the volume of aggregated traffic between node pairs does not fluctuate too dramatically, and relatively long-term connections must be set up to serve them. The connection provisioning is the task of the *Control Plane* (CP) and of the *Management Plane* (MP).

Recently, two mayor standardization bodies have made efforts towards the standardization of the Control Plane of Optical Networks: ITU-T has defined

** This work has been supported also by the EC within the IST FP6 NoE e-Photon/ONe+ (www.e-photon-one.org) research framework.

I. Tomkos et al. (Eds.): ONDM 2007, LNCS 4534, pp. 279–288, 2007.
© IFIP International Federation for Information Processing 2007

ASON [1] while IETF has created the GMPLS framework [2]. The latter one has even split the tasks of the controller (*Optical Crossconnect Controller*, OCC in ASON terminology) by separating MPLS and GMPLS Traffic Engineered LSP calculation from other control functions and delegating it to the *Path Computation Element* (PCE) [3].

These PCEs must cope with the increasing number of devices, fibers, wavelengths, connections, etc. in the network. One way is to expand their storage and calculation capacity by square or higher polynomial as the network grows, while the other, the preferred way is to reduce the information to be processed.

2 Routing in Multilayer Networks

Multilayer networks consist of multiple networking technologies and techniques stacked one over the other, e.g., IP/MPLS/OTN or IP/Ethernet/ngSDH/OTN.

Nowadays, in a multilayer network all the lower layers are statically configured either manually or via the MP, while the uppermost layer is switched via the CP. This is typical for both, IP networks and PSTNs today. However, to reduce the OPEX and to speed up provisioning of new services a CP is needed at lower layers as well.

In such networks the simplest case is when all the layers have their own CP, and the upper (client) layer adapts to the changes of the lower (server) layer. This is referred to as *Overlay Model*.

2.1 Vertically Integrated (Peer) Network Model

If enough information is exchanged between the CPs of the layers and the CPs are interfaced well to each other, then they can take routing decision jointly. This is referred to as *Peer Model* since the layers are equal (peer).

Beside the Overlay and Peer Interconnection models discussed before, there is the so-called vertically integrated model where the network layers are typically run by the same operator, and instead of one CP per layer there is a single integrated unified CP for all the layers.

Now, all the information necessary for routing is spread to all network components that are responsible for making routing decisions. Using the IETF terminology, these components are the PCEs. They maintain a *Traffic Engineering Database* (TED), and based on the TED, they calculate Traffic Engineering *Label Switched Paths* (TE LSPs) and provide them to their clients (*Path Computation Client*, PCC), whenever a traffic demand arrives in the network and the ingress node requests a TE LSP.

The information necessary for routing is the state of the network components. Spreading the information is referred to as *Link State Advertisement* (LSA), that is performed by the *Interior Gateway Protocol* (IGP).

Clearly, in the case of the Peer Model, not only the node connectivity, but all the wavelength information is to be spread unless full wavelength conversion capability (e.g., electronic switch core) is assumed in all nodes (*Wavelength Interchangeable* (WI) network). That is not a typical case.

Practically, a *Wavelength Graph* (WG) is used to represent node connectivity, wavelength usability and conversion/grooming capability. In this case a node is modeled as a directed subgraph [4]. However, flooding all this information of all layers to the PCEs loads the Control Channel significantly.

Although the vertically integrated model offers more detailed topology information to the PCE and there are no interaction limitations between the optical and the electronic (MPLS) layers, there are some problems with the conventional WG which need to be solved.

- *Edge state-dependency.* Switching an edge in the WG affects other edges as well, making them unreachable (see the dotted edges in Fig. 2.a). The background of this behaviour is that the optical crossconnect may switch the λ-channel coming at a given input port to *any* output port at the same λ wavelength. Each switching possibility must be advertised. Having N output ports means N advertised links. Whenever the input port is switched to a specified output port, the other $N-1$ output ports become unreachable from the given input port, and vice versa.

  The goal of the proposed model is to overcome this problem and reduce or eliminate the advertisement of affected edges. Of course, there are also devices which can duplicate, split and merge λ-channels in the optical layer. The model should also support these advanced OXCs, however, primarily we focus on simpler optical devices.

- *Electronic port assignment based on stalled information.* Usually, the OXC has either no or just a limited opto/electronic conversion capability [5], i.e., not all of the wavelength channels can be simultaneously converted into electronic signal and back. The O/E converters are relatively large and expensive, and if the routing algorithms tend to avoid O/E conversions, there is no need for full conversion capability. An OXC with limited O/E/O capability is realized as depicted in Fig. 1.a.

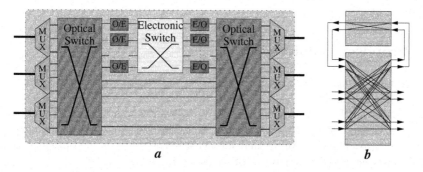

Fig. 1. OXC with limited O/E/O conversion capability: a) physical, b) logical view

The corresponding directed subgraph in the WG is shown in Fig. 1.b.

When we route a connection into the electronic layer, we do not care about which O/E port is assigned to the conversion, as each O/E converter has the same attributes and the traffic in the electronic layer can be re-arranged (groomed). However, the allocation must name a definite O/E port. If the LSA is delayed by any information update strategy [7][8], it may occur, that the next connection – as the decision is based on stalled information – is routed via an already used O/E port (Fig. 2.b) and its SETUP will fail (Fig 2.c) even though there are plenty of other available O/E ports in the OXC. This is the second problem in the conventional WG we want to solve.

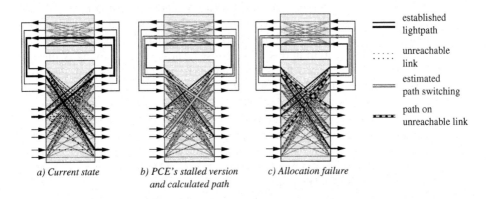

a) Current state   b) PCE's stalled version   c) Allocation failure
                             and calculated path

	established lightpath
	unreachable link
	estimated path switching
	path on unreachable link

Fig. 2. Failed allocation due to inaccurate link state information

Although efforts were made to reduce the amount of advertised LSA messages while keeping the same level of accuracy in the network view [8], these intra-OXC aggregations are still redundant and do not provide a general solution for any type of OXC.

3   Proposed Solution

Assuming a dynamically configured network, unlike the overlay networks, where established lightpaths appear as direct links between non-adjacent nodes in the upper layer's CP, in the peer model the topology of an optical crossconnect and the corresponding subgraph in the WG does not change due to connection setup/release. The only thing that changes, is the state of the optical switches and, consequently, the attributes of edges in the WG.

Whenever a connection is set up or released, the controllers of the physical devices along the connection path are notified about it by the signaling protocol. These notifications trigger a reconfiguration of switching units inside the OXC and also a *Link State Advertisement* procedure to let the PCEs know about the new state of the device.

Knowing the current state of the OXC and the received signaling message, the new state of the OXC (meaning the new attributes of each link advertised about the OXC) is predictable. If the logic, deciding which advertised edges of the OXC become unusable or available after a SETUP or RELEASE call, is given to the PCE, the whole LSA procedure can be replaced by a simple notification which must be sent to the PCE(s). For compatibility reason, this notification can be encapsulated into a single LSA message.

To maintain an accurate TED in the PCE the following information is needed:

1. Logical topology about the physical devices and the fiber cable connections between them
2. Resource allocation/deallocation information
3. Switching rules to decide link reachability

In this enumeration 1 and 3 is invariant information, determined by the type of the physical device, while 2 is network event-dependent.

3.1   Description of the Messages

Static invariant data, i.e., the topology and the switching rules – as they are coherent – can be advertised together and *initially*. The basic block of the static data describes a port, represented by a node in the WG, and its attributes, that are beside the TE attributes the identifiers of incoming and outgoing links. For both groups of incoming and outgoing links a discrete value n in the range $0 < n \leq N$ – where N is the size of the set – denotes how many links of the group can be used simultaneously (e.g., $n = 1$ for simple crossconnects). In this message specification we exploit the fact that the TE metrics of a link are determined by the ports at its endings. Defining the switching rules that way, the topology of the network is also given, as each link is assigned to an input (head) and an output (the link's tail) group.

The simplest realization of the dynamic data is an LSA message naming the concerned link and the allocated bandwidth. Whenever it becomes 0 the PCE knows that the link is not switched.

3.2   Examples

In the following examples we show how the switching rules of the OXC are described beside its topology. All the figures show OXCs with 3 ports and 2 wavelengths on each port.

Figure 3 shows a simple OXC (OXC#0) on the left, where each port behaves the same way allowing 1 – 1 link selection (as shown under the magnified node-box) in both directions. On the right, an extended device (OXC#1) can be seen with 1 λ-converter per wavelength. Figure 4 depicts OXCs with grooming capability: OXC#2 has unlimited grooming capability (all of the 6 output ports may be switched simultaneously from the input port, as denoted under the magnified input node-box), while OXC#3 has 3 grooming ports, which is resolved by a single "grooming" node allowing 3 input and 3 output ports to be simultaneously switched.

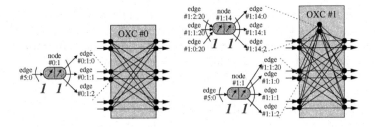

Fig. 3. Topology and switching rules of OXCs without grooming

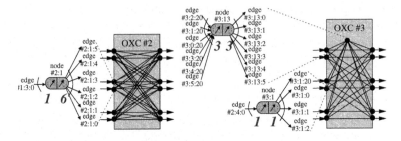

Fig. 4. Topology and switching rules of OXCs with grooming

3.3 Exceptions

In the examples of 3.2 the common attribute of the presented physical devices is that they are *passive* in the sense that their state does not change unless they receive control command from the CP. Their behaviour due to the command is also determined.

There are, however, also *active* physical components, whose state change is not predictable. E.g., the OXC with the so-called *lambda fragmentation/ de-fragmentation* capability[9] can decide on its own whether existing lightpaths should be converted into electronic signal and back for achieving better network performance (throughput) or not.

Another network components considered as *active* are the foreign domains in an inter-domain environment. For scalability and strategic reasons, domains will refuse advertising their detailed inner topology. Besides, a domain is also active in the sense, that it can initiate either intra- or inter-domain connection, which changes its link states without external influence.

4   Performance analysis

We examine the proposed method and compare it to the conventional method from three aspects:

– amount of required topology information messages,

- data storage capacity and
- path computation complexity.

Assuming standard topology update strategy (LSA sending is not delayed), the required number of messages does not depend on how the PCE stores the data and how it calculates the paths; it is implementation-independent. In turn, both data storage and path calculation are implementation-dependent.

Let us assume an implementation where the TED stores raw topology and TE information and the path computation is based on Dijkstra's algorithm where the unreachable links are forbidden by assigning very large (numerical equivalent of infinite) weight to them.

4.1   LSA Message Density Reduction

The analysis of the required number of LSA messages is split into two: first, we show that the advertisement of invariant data does not load the control network more than the conventional initial topology advertisement, next, we compare the LSA message requirements of the proposed method to the conventional solution. Both analysis are carried out on two types of OXCs:

- sOXC meaning *simple OXC* without electronic ports and
- OXC_K an OXC with K grooming ports.

In the analysis we use the following notation: N is the number of fiber ports, each carrying L wavelengths, K is the number of O/E ports and there are also K E/O ports. W_e, W_n, and W_a are the size of *link label*, *node id*, and *TE attributes*, respectively.

Invariant Data The amount of the invariant data (T^*) can be set against the number of initial LSA messages(T).

In the rule-based model the topology and rule initialization of a simple OXC requires for each input and output optical port a basic block and each block is $W_n + 2 + (1 + N) * W_e + W_a$ units long.

OXC type	basic blocks in T^* (new model)	LSAs in T (old model)
sOXC	$2 * N * L$	$N^2 * L$
OXC_K	$2 * N * L + 1$	$N * (N + 2 * K) * L + K^2$

In case of an OXC with K O/E ports, one additional basic block, denoting the electronic conversion and grooming capability, is required which is $W_n + 2 + 2 * N * W_e + W_a$ long. The blocks corresponding to optical ports are $W_n + 2 + (2 + N) * W_e + W_a$ long.

On the contrary, in the old model, on each λ each input port was connected to each output and electronic input port, and the electronic layer had its all-to-all switching capability. This results in far more initial LSA messages, however, these were much shorter, $2 * W_n + W_l + W_a$, denoting the two endings and the TE attributes of an identified link. Figure 5.a compares the total transmitted data size of T and T^*.

Dynamic Information In this section we investigate the impact of a single event (a single resource allocation or deallocation) and the total number of required LSA messages is the sum of LSA messages triggered by single events (M).

The amount of LSA depends on many environmental factors in the conventional model. The most important factor is whether links become unreachable or reachable due to an event or no change occurs at all. If there is no change, then the scope of the change is limited to the affected links. It is the case, when routing is done over an already configured lightpath. Let the ratio of events concerning a link along an already configured lightpaph be γ.

In other cases, the number of affected links is N in a mere OXC and $N+K$ in an OXC with K grooming ports. However, the more lightpaths are established, the more links became unreachable in advance, and their state does not change through a new lightpath setup. Let us denote the average input/output port usage ratio with η. Now the number of messages is:

OXC type	M^* (new)	M (old model)
sOXC	1	$\gamma * 1 + (1 - \gamma) * (1 - \eta) * 2 * N$
OXC_K	$1 + \delta$	$\delta + (\gamma * 1 + (1 - \gamma) * (1 - \eta) * 2 * (N + \delta * K * L))$

In the case of sOXC, the number of affected links is 1 if the event aims an already configured lightpath (with γ ratio), and 2-times N (links to input and output ports) otherwise, but η rate of these affected links is already unreachable.

In OXC_K, this number is increased by the amount of links that a grooming event additionally affects.

In the rule-based topology advertisement these numbers are 1 and $\delta * 2 + (1 - \delta) * 1 = 1 + \delta$ since grooming requires 2 link allocations whereas no grooming only 1 (see Fig. 4).

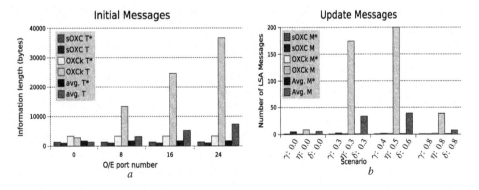

Fig. 5. Numerical results on COST266 Reference network

The amount of invariant and dynamic data is demonstrated in Fig. 5 on the COST266 European Reference Network consisting of 28 nodes, with 24 λs and with grooming capability at the 5 most connected nodes. The figures show 6 columns for each scenario comparing the values in non-grooming nodes, grooming nodes and network-wide average. Fig. 5.a illustrates that the old model was unable to cope with the increasing number of grooming ports, whilst the new model scales well. Fig. 5.b shows 4 scenarios with different network load from empty to near fully loaded (high γ, η, δ values) state. One can see that in our model the number of LSA messages is low and less state-dependent.

4.2 Data Storage Capacity

The data storage requirement is defined by the initial topology advertisement. The dynamic state updates do not influence the storage capacity requirement. The old and the new models require roughly the same amount of storage space:

D^*_{sOXC}	$2 * N * L * (W_n + 2 + (1 + N) * W_e + W_a)$
D_{sOXC}	$N^2 * L * (2 * W_n + W_l + W_a)$

In case of OXC_{K} the storage capacity can be calculated in the same way from T^* and T.

4.3 Path Computation Complexity

The path computation algorithm checks each examined link whether it fits into the TE requirement of the connection and whether it is reachable. TE requirements can be checked in both models the same way. The difference is that in the proposed implementation, the reachability information is not stored but calculated, whenever the algorithm requires it. That means at most $n_i + n_o$ additional comparisons in each examined node, depending on the size of the incoming (n_i) and outgoing (n_o) groups.

Generally, it can be stated that the proposed model requires more computational steps. If the original algorithm performed in each node $d * X$ steps proportional to the nodal degree $(d = n_i + n_o)$, these additional comparisons mean a *constant-times* computation complexity increment.

4.4 Fault tolerance

Finally, it must be also mentioned, that the presented model is less tolerant against network failures (basically lost LSA messages).

In the conventional model the lost LSA messages do not result in unrecoverable network failures, as most information update strategies advertise link states periodically even if no change occurs, secondly, whenever a controller receives an unfeasible resource allocation request, it knows, that the improper request is due to inaccurate information, and marks the link for advertisement.

Contrarily, in the proposed model the controller can notice the improper request, however, it cannot resolve, which lost event has caused the inaccuracy in the TED of the PCE. General protection against information loss, periodical TED synchronization or other supplementary methods are required which may be the subject of further research.

5 Conclusions

Vertically integrated multilayer networks provide far more flexible traffic engineering solutions than overlay networks do. However, there are also more difficulties we have to face with. In this paper we focused on topology advertisement. We pointed out that, in multilayer optical networks, traditional LSA may be applied, however, it increases the control traffic significantly, and the most link state changes are predictable. We presented our solution to reduce the control traffic (carrying topology information), which is a *rule-based state change advertisement*. After discussing the migration possibilities we have compared the proposed LSA method to the existing one, and showed, that the storage and computation requirements of the new method do not increase much while the LSA traffic is reduced significantly.

References

1. ITU-T *Architecture for the Automatic Switched Optical Network (ASON)*, G.8080/Y.1304, 2001
2. L. Berger *Generalized Multi-Protocol Label Switching (GMPLS) Signaling Functional Description*, RFC 3471, 2003
3. A. Farrel et al. *A Path Computation Element (PCE)-Based Architecture*, RFC 4655, 2006
4. T. Cinkler, D. Marx, C.P. Larsen, D. Fogaras: "Heuristic Algorithms for Joint Configuration of the Optical and Electrical Layer in Multi-Hop Wavelength Routing Networks", IEEE INFOCOM 2000, Tel Aviv, March 2000
5. T. Cinkler: "Traffic and λ Grooming", IEEE Networks, March/April 2003, Vol. 17., No.2, pp. 16-21
6. M. Perényi, J. Breuer, T. Cinkler, Cs. Gáspár: "Grooming Node Placement in Multilayer Networks", ONDM 2005, 9th Conference on Optical Network Design and Modelling, pp. 413-420, Milano, Italy, February 7-9, 2005
7. G. Apostolopoulos, R. Guerin, S. Kamat, S. Tripathi: "Quality of Service Based Routing: A Performance Perspective", ACM SIGCOMM '98, Vancouver, Canada, Aug. 1998
8. J. Szigeti, I. Ballók, T. Cinkler: "Efficiency of Information Update Strategies for Automatically Switched Multi-domain Optical Networks", IEEE ICTON 2005, 7th International Conference on Transparent Optical Network, Barcelona, Catalonia, Spain, July 3-7, 2005
9. T. Cinkler, G. Geleji, M. Asztalos, P. Hegyi, A. Kern, J. Szigeti: "Lambda-path Fragmentation and De-Fragmentation through Dynamic Grooming", IEEE ICTON 2005, 7th International Conference on Transparent Optical Networks, Barcelona, Catalonia, Spain, July 3-7, 2005

Enhanced Parallel Iterative Schedulers for IBWR Optical Packet Switches

M. Rodelgo-Lacruz[*], P. Pavón-Mariño[+], F. J. González-Castaño[*], J. García-Haro[+], C. López-Bravo[*] and J. Veiga-Gontán[+]

[*]University of Vigo, Spain, {mrodelgo,javier,clbravo}@det.uvigo.es
[+]Polytechnic University of Cartagena, Spain, {pablo.pavon,joang.haro}@upct.es, javg@alu.upct.es

Abstract. In this paper we propose an enhanced parallel iterative scheduler for IBWR synchronous slotted OPS switches in SCWP mode. It obtains a maximal matching of packet demands without resource conflicts. The analytical and numerical results are highly competitive regarding previous work.

Keywords: OPS, IBWR, Scheduling Algorithms.

1 Introduction

In the Optical Packet Switching (OPS) paradigm of Wavelength Division Multiplexing (WDM), packet payloads stay in the optical domain. OPS offers high bandwidth efficiency due to statistical multiplexing, but it is well-known that packet granularity and optical buffering impose extreme constraints to photonic switching, incurring in unacceptable hardware costs with state-of-the-art technology.

In this paper, we focus on synchronous slotted OPS in Scattered Wavelength Path (SCWP) operational mode [1]. This mode specifies a fixed packet size (slot length) and packet alignment with slot boundaries at the input ports (and thus optical synchronizing stages, which increases cost). However, the performance improvement due to the better contention behavior has encouraged the study of this alternative. Packet length in OPS networks is a current topic of discussion. The European DAVID project [2] selected synchronous slotted OPS with slot lengths of ~1 μs for the WDM backbone network. In WDM OPS networks, there is a mapping of permanent end-to-end connections to link wavelengths. In the SCWP operational mode, optical packet paths (OPP) univocally determine a fixed sequence of transmission fibers, but the transmission wavelength may change in each hop. This provides extra freedom to switch schedulers in packet wavelength selection, boosting the statistical multiplexing effect. Therefore, SCWP achieves a high throughput with a low packet delay in OPS networks, also lowering optical buffering requirements [3][4].

In SCWP it is possible to *simultaneously* transmit several packets of the same OPP through a fiber, in different wavelengths. In this paper we adopt the round-robin packet ordering criterion in [5] that avoids the performance degradation due to unbalanced wavelength assignments. The wavelengths are assigned cyclically. Each

I. Tomkos et al. (Eds.): ONDM 2007, LNCS 4534, pp. 289–298, 2007.
© IFIP International Federation for Information Processing 2007

node uses two sets of round-robin pointers to track packet sequence: one round-robin pointer per input fiber, tracking the wavelength of the next packet in the input traffic sequence, and one round-robin pointer per output fiber, determining the output wavelength of the next packet to be transmitted. Figure 1 shows an example.

This paper focuses on the Input-Buffered Wavelength-Routed (IBWR) switch architecture, for its scalability. Figure 2 shows the WDM adaptation of this architecture [6]. The switch has N input/output fibers, and n wavelengths per fiber. It has a buffering section followed by a non-blocking switching section. The buffering section consists of $n \cdot N$ Tunable Wavelength Converters (TWC) with a tuning range $\lambda_0...\lambda_{K-1}$, K=max $(n \cdot N, M)$ and two $K \times K$ Arrayed Waveguide Gratings (AWGs), which are interconnected by M delay lines of 0 to $M-1$ slots. Due to AWG symmetry, a packet arriving at port i leaves the buffering section at port i, regardless of the selected delay. The wavelength conversion determines the delay line for the packet. The switching section is composed of $n \cdot N$ TWCs followed by a $nN \times nN$ AWG. The switching AWG routes each packet to the proper output fiber/wavelength.

The IBWR switch scheduler assigns packet delays and packet output wavelengths. These two tasks are independent.

- *Packet delay assignment.* Current optical switches employ Fiber Delay Lines (FDLs) due to the lack of optical RAMs. In IBWR switches, delays are assigned at packet arrivals. The scheduler discards a packet if it cannot assign a delay fulfilling two contention conditions: *(i) output fiber contention*: at most n packets can reach any output fiber in a given slot, *(ii) input port contention*: the packets that arrive to the same i-th input port (same fiber and wavelength) in different time slots cannot leave the switch in the same time slot. Otherwise they would collide at the i-th TWC of the switching section, which can only handle one packet at a time.

- *Output wavelength assignment.* The scheduler assigns output wavelengths to the packets when they leave the switch, according to the round-robin criterion.

Remark: Other OPS architectures, with higher hardware costs and less scalable than IBWR, emulate output buffering (OB) [6][7] (the only factor limiting packet delay assignment is output fiber contention).

Previous work has characterized IBWR delay assignment as a matching in bipartite graphs [4]. At every slot, the scheduler seeks a feasible assignment maximizing the number of packet delay assignments (i.e. minimizing packet losses). If there are several alternatives, it minimizes average packet delay. The *sequential* IBWR scheduler for the SCWP mode in [8] is unfeasible in practice (for ~1 μs slots). Conversely, our proposal is parallel, as Virtual Output Queuing (VOQ) schedulers.

The rest of this paper is organized as follows: in section 2 we describe the Parallel Desynchronized Block Matching Scheduler (PDBM), which is the basis for this proposal. In section 3 we present the Insistent PDBM (I-PDBM) algorithm. In section 4 we discuss simulation results. Section 5 concludes the paper.

2   PDBM scheduler

PDBM [9][11] was the first parallel iterative matching scheduler for IBWR switches. We reproduce it here since it is the basis for the enhancements in this paper.

Figure 3 shows an electronic PDBM implementation. The nN input modules (one per input fiber wavelength) are interconnected with NM output modules (one per output fiber and delay line) by three types of signals. The main ones are the *request* signals, from input to output modules, and the *grant* signals, from output to input modules.

Input module i, $i=0,...,nN-1$, keeps an input TWC availability state vector $\overline{x}_i(t)$, $t=0,...,M-1$. Component $\overline{x}_i(t)$ equals 1 if a packet is scheduled to leave the buffering section at the i-th port in t slots (0 otherwise). At every slot the state vector is shifted: $\overline{x}_i(t-1)= \overline{x}_i(t)$ and $\overline{x}_i(M-1)=0$, to reflect FDL propagation after each slot.

Output module (j,t), $j=0,...,N-1$, $t=0,...,M-1$, keeps: (a) a value $n- y_{jt}$ (delay availability) of $log_2(n)$ bits. Variable y_{jt} denotes the number of packets for output fiber j that will leave the switch in t time slots; (b) a *grant pointer* FG_{jt}, of $log_2(N)$ bits. It indicates the first input fiber in the scan; (c) an alternating bit CW_{jt} indicating the search direction. Note that, at every slot, the delay availability register in module (j,t) must be transferred to module $(j,t-1)$, $j=0,...,N-1$, $t=1,...,M-1$. Also, modules $(j,M-1)$, $j=0,...,N-1$, reset the availability registers to n.

At each input fiber controller, a round-robin grant pointer WG_f, $f=0,...,N-1$, indicates the first wavelength to scan in input fiber f.

PDBM Algorithm

At system initialization, $\overline{x}_i(t)$, y_{jt}, WG_f and CW_{jt} are set to 0. All FG_{jt} grant pointers associated to the same output fibers are initialized by maximizing the minimum distance between pointer positions:

$$FG(f,0) = 0$$

$$FG(f,t) = FG(f,t-1) + \min\left(1, \left\lfloor \frac{N}{M} \right\rfloor\right) \quad \begin{array}{l} \forall f = 0...N-1 \\ \forall t = 1...M-1 \end{array}$$

Algorithm iterations consist of three steps (*request, grant,* and *accept*):

Step 1. Request: Each input module i with a packet for output fiber j sends a request signal to every output module in fiber j whose associated delay satisfies the input contention constraint. That is, output modules (j,t) such that $x_i(t)=0$.

Step 2. Grant: Each (j,t) output module scans the request signals from the input modules, starting by the input module indicated by grant pointers FG_{jt} and WG_f. The scans from other input modules proceed in a clockwise or counter-clockwise sense, according to the alternating bit CW_{jt}. The first $n- y_{jt}$ scanned request signals are acknowledged, and a grant signal is sent to the associated input module.

Step 3. Accept: Each input module i receives at most M grants, from the M delays associated to the destination output fiber. The shortest granted delay t is accepted and assigned to the packet that is present at input i. If the input does not receive any grants during algorithm execution, the packet is discarded. Otherwise, an accept signal is sent to the accepted output module and the $\overline{x}_i(t)$ and y_{jt} state vectors are updated to reflect packet allocation. When a packet is granted, its input port does not participate in subsequent algorithm iterations.

Wavelength of the next packet in the sequence

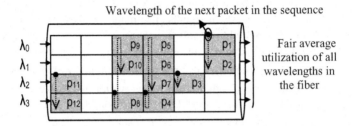

Fig. 1. Round-robin wavelength sequence criterion, fiber with four wavelengths $\lambda_0,...,\lambda_3$.

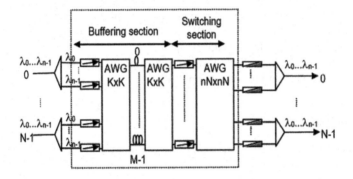

Fig. 2. Adaptation of the Input-Buffered Wavelength-Routed switch (IBWR) to WDM.

At each time slot, after the last iteration, $\overline{x_i}(t)$ and y_{jt} are updated and shifted as described above to consider the allocation and the propagation of the packets in the delay lines. The CW_{jt} bits are negated to alternate request scanning directions each time slot. The FG_{jt} grant pointers are incremented by one (module N), every *two* time slots and the WG_f round-robin grant pointers are incremented by the number of received packets at fiber f in the current slot (modulo n).

Algorithm justification

PDBM converges in $min(M,nN)$ iterations at most [9]. Thus, convergence speed is independent from switch size.

The initialization of the pointers and their evolution are inspired by the desynchronizing scheme of the RDSRR [10] algorithm to minimize the grant block overlapping effect: if an output module (j,t) receives more requests than available delays, it only acknowledges signals from the modules whose indexes are closest to grant pointer FG_{jt}. If the grant pointers take the same value, "close" input modules receive several grants, and "far" input modules receive no grants at all. In PDBM, all grant pointers of a given output fiber get initial values that maximize the minimum distance (modulo N) between two input nodes. The scheduler keeps the desynchronization by increasing (modulo N) all pointers every two time slots. The scanned direction is inverted at each time slot to enforce fairness in case of non-uniform packet arrivals.

Although PDBM does not guarantee packet sequence, input modules are scanned following the round-robin criterion to mitigate mis-sequencing.

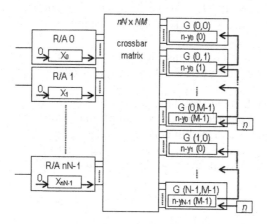

Fig. 3. Electronic implementation scheme for the PDBM scheduler.

3   Insistent PDBM (I-PDBM) scheduler

The basic PDBM scheduler may assign longer delay lines than strictly necessary, ignoring shortest ones even in absence of contention. Specifically, it converges to a maximal size match (no more connections can be established without replacing any existing connections) with suboptimal aggregated delay, i.e., some connections could be individually removed and reassigned to a shorter delay output port. We call this effect "PDBM impatience". We will illustrate it with an example:

Let us assume a switch with two inputs (outputs), two wavelengths per fiber and three delay lines ($N=2$, $n=2$, $M=3$). Two packets arrive at input fiber 0 requesting output fiber 1. The state of the switch is:

- FG_{ft} and CW_{lt} are indifferent because there are no packets in fiber 1.
- $WG_0 = 0$. The round-robin grant pointer of input fiber 0 points to input module 0. The first input wavelength of fiber 0 to be scanned is 0 for all iterations.
- $x_0(t)=x_1(t)=0 \; \forall t$. No input contention.
- $y_{10}=1$, $y_{11}=1$, $y_{12}=0$. There is a free delay line for $t=\{0,1\}$ and two delay lines for $t=2$.

Figure 4 summarizes the state of the node. From that state, the algorithm iteration evolves as follows (figure 5):

Request: input modules 0 and 1 send request signals to all output modules *(1,t)*, since there is no input contention at them.

Grant:

- $t=0$. Output module $(1,0)$ scans the signal of input module 0 ($WG_0=0$) and acknowledges it. The request signal of input module 1 is not acknowledged because there is a single available wavelength, $n-y_{jt}=1$ ($2-y_{10}=1$).
- $t=1$. Output module $(1,1)$ scans the request signal of input module 0 ($WG_0 = 0$) and acknowledges it. The request signal of input module 1 is not acknowledged because there is a single available wavelength, $n-y_{jt}=1$ ($2-y_{11}=1$).
- $t=2$. Output module $(1,2)$ scans the signals of input modules 0 and 1, and acknowledges them both because there are two available wavelengths, $n-y_{jt}=2$ ($2-y_{12}=2$).

Accept. Input module 0 receives three grant proposals ($t = 0,1,2$) and accepts the best one (delay 0). Input module 1 receives a single grant proposal ($t=2$) and accepts it. The packet at input 1 is assigned to delay line 2. However, there is room in delay line 1, which has no input contention. Thus, the assignment is suboptimal. To solve the impatience problem we propose a new algorithm: **Insistent PDBM or I-PDBM**.

I-PDBM algorithm

In the PDBM accept step, a granted input module confirms the received grant to update the state vector y_{jt} and deactivates the other request signals to allow input ports with lower priorities to be granted. It is possible to simplify the algorithm to execute a single accept step after the last iteration. It suffices to change input modules to keep the request signal active for the "accepted" grant and to deactivate all others. Since the number of wavelengths does not decrease and the pointers do not change until the accept step at the end of the slot, each granted input module that keeps an active request signal is granted again, whereas the unrequested granted delays are released and reassigned to other input modules.

The previous simplified scheme easily solves PDBM impatience if each granted input module stops requesting higher delays but it keeps the request signal active for better ones. By stopping higher delay requests, it releases some wavelengths that can be granted to other modules. Subsequent iterations may increase the number of packet assignments and further reduce the delay of previously assigned packets. Thus, grants are *provisional*, until the very last iteration when the accept stage takes place.

Therefore, the differences with PDBM are:

Step 1. *Request*: each input module i with a packet destined to output fiber j sends a request signal to every output module of fiber j whose associated delay satisfies that the input contention constraint is not worse than any granted delay to the same input module in the previous iteration, i.e. the input module sends request signals to output modules (j,t) such that $\overline{x_i}(t)=0$ and $j \le p$, where p is the shortest granted delay.

End of the algorithm. *Accept*: the accept step takes place after the last iteration. So, state vectors and pointers are not updated until the end of the time slot and the granted input modules participate in subsequent algorithm iterations.

Destination of packets that arrive in the current time slot				Delay lines of output fiber 0	
Input fiber 0		Input fiber 1		Delay length (time slots)	Delay availability (packets)
Wavelength 0	Wavelength 1	Wavelength 0	Wavelength 1	0	1
Output fiber 1	Output fiber 1	No packet	No packet	1	1
				2	2

Fig. 4. Node state information.

Signals from input fiber 0 to output fiber 1	Request		Grant		Accept	
Delay	λ_0	λ_1	λ_0	λ_1	λ_0	λ_1
0	yes	yes	yes		yes	
1	yes	yes	yes			
2	yes	yes	yes	yes		yes

Fig. 5. Algorithm stages.

Algorithm justification

I-PDBM converges when the signals get stabilized, i.e. there are no new packet allocations nor assignments of better delays to granted packets. As PDBM does, I-PDBM converges in $min(M,nN)$ iterations at most to a maximal size matching. *Proof*: i) an output port only changes a grant signal if a previous input port (according to the grant pointers) releases a request signal. An input port only releases a request signal if it received a grant in the previous iteration from an output port that is associated to a shorter delay. Since the grants from delay 0 do not change after the first iteration, the algorithm converges in M iterations at most. ii) An input port is granted a shorter delay only if another input port was granted a shorter delay in the previous iteration. Since there are nN input ports, the algorithm converges in nN iterations at most.

I-PDBM avoids PDBM impatience and it is simpler to implement. It has two steps (request and grant), whereas PDBM needs three (request, grant and accept).

4   Results

In this section we present simulation results to compare I-PDBM (in terms of average delay, buffer requirements and practical convergence) with OB architectures and the previous PDBM algorithm, under benign or bursty traffic conditions.

Figures 6(a) and 6(b) show the average delay of I-PDBM and PDBM under n-SCWP Bernoulli traffic (I-PDBM: continuous line, PDBM: dotted line). Switch sizes were $N=\{2,4\}$, $n=\{2,8,32,64\}$ (OPS switches operate in the core network, with a high aggregate bandwidth but few ports). Buffer sizes were adequate for OB architectures (packet loss probability below 10^{-9} under 90% load): $M = \{35,10,3,2\}$ for $n = \{2,8,32,64\}$, respectively. To illustrate the effect of traffic burstiness, figures 6(c) and 6(d) show the average delay of I-PDBM and PDBM under an n-SCWP arrival

Markov-modulated ON-OFF Poisson process (MMPP), for burst lengths of $\beta = 16$ (Figure 6(c)) and $\beta = 64$ (Figure 6(d)). Switch sizes were $N = 4$, $n = \{2,8,32,64\}$, and buffer sizes were the same as above. For ON/OFF input traffic, the average delay of both algorithms is very similar. Bursty traffic affects I-PDBM performance as in the case of PDBM and OB architectures [11]. However, for Bernoulli traffic, the average delay of I-PDBM is lower in all configurations. We conclude that packet delay decreases by avoiding PDBM impatience and thus I-PDBM outperforms PDBM.

Table 1 shows buffer requirements for a packet loss probability of 10^{-7} under Bernoulli traffic (simulations with 10^9 packets). This is a good feasibility metric for OPS nodes, because FDL length is a serious bottleneck nowadays. As we would expect, reducing packet delay leads to lower buffer requirements. We observe that I-PDBM buffer length is very small, and it is close to the ideal OBS case.

Tables 2 and 3 compare the theoretical convergence bound with the number of iterations K to converge with a probability above $1-10^{-6}$ (90% input load). PDBM and I-PDBM behave similarly. Under Bernoulli traffic, they only need extra iterations for few wavelengths ($n=2$). However, in all cases the number of iterations is quite low.

5   Conclusions

In this paper we have proposed the enhanced I-PDBM parallel iterative matching scheduler for IBWR optical packet switches [6], which is significantly advantageous over PDBM [9] in terms of performance and hardware complexity.

Acknowledgements
This research has been supported by project grants TEC 2004-05622-C04-01/TCM CAPITAL/TEC 2004-05622-C04-02/TCM ARPAq (MEC, Spain) and PGIDIT04TIC322003PR/PGIDIT05XIC32201PN (Xunta de Galicia, Spain). The authors participate in the COST 291 action and the e-Photon/ONe+ network of excellence.

Switch size	$\rho=0.1$	$\rho=0.2$	$\rho=0.3$	$\rho=0.4$	$\rho=0.5$	$\rho=0.6$	$\rho=0.7$	$\rho=0.8$	$\rho=0.9$
$N=2,n=2$	2/4/2	3/4/3	3/4/3	4/5/4	5/6/5	5/7/6	7/8/8	10/11/10	18/20/20
$N=2,n=8$	1/1/1	2/3/2	2/3/2	2/4/2	2/4/2	2/5/2	3/6/3	3/7/4	6/9/8
$N=2,n=32$	1/1/1	1/1/1	1/1/1	1/1/1	2/3/2	2/3/2	2/4/2	2/4/2	2/5/3
$N=2,n=64$	1/1/1	1/1/1	1/1/1	1/1/1	1/1/1	1/1/1	2/3/2	2/3/2	2/4/2
$N=4,n=2$	3/5/3	3/5/4	4/6/4	5/7/5	6/8/6	7/10/8	9/13/11	14/19/16	26/30/30
$N=4,n=8$	1/1/1	2/3/2	2/3/2	2/3/2	2/4/2	3/4/3	3/5/3	4/8/5	8/13/10
$N=4,n=32$	1/1/1	1/1/1	1/1/1	1/1/1	2/3/2	2/3/2	2/4/2	2/4/2	3/5/3
$N=4,n=64$	1/1/1	1/1/1	1/1/1	1/1/1	1/1/1	2/3/2	2/4/2	2/4/2	2/5/2

Table 1. Buffer requirements (OB/PDBM/I-PDBM). Bernoulli input traffic, 10^{-7} packet loss probability.

Bernoulli ρ=0.9	n=2	n=8	n=32	n=64
N=2	1/ 3 **4**	2/ 3 **9**	2/ 2 **5**	2/ 2 **4**
N=4	2/ 5 **8**	3/ 3 **13**	2/ 2 **5**	2/ 2 **5**

Table 2. Practical number of iterations for PDBM/I-PDBM convergence vs. theoretical convergence bound (bold), Bernoulli traffic.

MMPP ρ=0.9, N=4	n=2	n=8	n=32	n=64
β=16	5/ 5 **8**	6/ 6 **10**	3/ 3 **3**	2/ 2 **2**
β=64	4/5 **8**	6/ 6 **10**	3/ 3 **3**	2/ 2 **2**

Table 3. Practical number of iterations for PDBM/I-PDBM convergence vs. theoretical convergence bound (bold), MMPP traffic.

References

1. Hunter D. et al. "WASPNET: A Wavelength Switched Packet Network". *IEEE Communications Magazine* 1999; 37(3):120-129.
2. Dittman L. et al. "The European IST Project DAVID: A Viable Approach Toward Optical Packet Switching". *IEEE Journal of Selected Areas in Communications* 2003; 21(7): 1026-1040.
3. Pavon-Mariño P., García-Haro J., Malgosa-Sanahuja J., Cerdán F. "Scattered Versus Shared Wavelength Path Operation, Application to Output Buffered Optical Packet Switches. A Comparative Study". *SPIE/Kluwer Optical Networks Magazine* 2003; 4(6):134-145.
4. Pavon-Mariño P., García-Haro J., Malgosa-Sanahuja J., Cerdán F. "Maximal Matching Characterization of Optical Packet Input-Buffered Wavelength Routed Switches". In *Proc. of 2003 IEEE Workshop on High Performance Switching and Routing (HPSR 2003)*, Torino, Italy, June 2003, pp. 55-60.
5. Pavon-Mariño P., Gonzalez-Castaño F.J., Garcia-Haro J. "Round-Robin wavelength assignment: A new packet sequence criterion in Optical Packet Switching SCWP networks". *European Transactions on Telecommunications* 2006; 17(4): 451-459.
6. Zhong W.D., Tucker R. S. "Wavelength routing-based photonic packet buffers and their applications in photonic packet switching systems". *IEEE J. Lightwave Technol.* 1998; 16(10): 1737-1745.
7. Guillemot C et al. "Transparent optical packet switching: the European ACTS KEOPS project approach". *IEEE J. Lightwave Technol.* 1998; 16(12): 2117–2134.
8. Chia M.C. et al. "Packet loss and delay performance of feedback and feed-forward arrayed-waveguide gratings-based optical packet switches with WDM inputs-outputs". *IEEE J. Lightwave Technol.* 2001; 19(9): 1241-1254.
9. Pavón-Mariño, P. *Contribution to Optical Packet Switching: Architectures, Performance Evaluation and Comparative Analyses* (in Spanish). PhD Thesis. Dep. de Tecnologías de la Información y las Comunicaciones, Univ. Politécnica de Cartagena, Spain.
10. Jiang Y., Hamdi M. "A Fully Desynchronized Round-Robin Matching Scheduler for a VOQ Packet Switch Architecture". In *Proc. 2001 Workshop on High Performance Switching and Routing*, May 2001, Dallas, USA, pp. 407-411.

11. Pavon-Mariño P., García-Haro J., Jajszczyk A. "Parallel Desynchronized Block Matching: A Feasible Scheduling Algorithm for the Input-Buffered Wavelength-Routed Switch". Submitted to *Computer Networks* for publication.

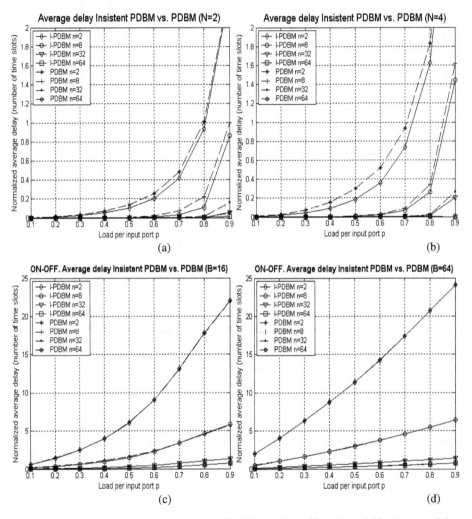

Fig. 6. (a) and (b) Average delay under SCWP Bernoulli traffic; (c) and (d) Average delay under SCWP MMPP traffic; (c) β=16; (d) β=64.

A New Algorithm for the Distributed RWA Problem in WDM Networks Using Ant Colony Optimization

Víctor M. Aragón, Ignacio de Miguel, Ramón J. Durán, Noemí Merayo, Juan Carlos Aguado, Patricia Fernández, Rubén M. Lorenzo, and Evaristo J. Abril

Dpt. Signal Theory, Communications and Telematic Engineering, University of Valladolid, Campus Miguel Delibes, 47011 Valladolid, Spain
Tel: +34 983 423000 ext. 5574, Fax: +34 983 423667
varafer@coit.es, ignacio.miguel@tel.uva.es

Abstract. We present a new algorithm based on ant colony optimization to solve the dynamic routing and wavelength assignment problem in a distributed manner. The algorithm uses the ant colony optimization metaheuristic to obtain updated information about the network state, which is then used to find the routes and wavelengths to establish new connections through an adaptive process able to deal with dynamic changes in network state. By means of simulation we show that the algorithm leads to lower blocking probability and lower mean setup time than other distributed algorithms.

Keywords: Optical networks, Wavelength Division Multiplexing (WDM), Dynamic Routing and Wavelength Assignment (RWA), distributed systems, Ant Colony Optimization (ACO).

1 Introduction

Ants are social insects that despite not having a remarkable individual intelligence are able to coordinate themselves to make complex tasks using chemical signals that modify the behavior of the individuals. The most interesting feature, from the point of view of telecommunications, is the method that they follow to find the shortest path between two points. Each individual deposits a chemical sign, called pheromone, as it moves; and each ant has a certain tendency to follow those trails presenting higher intensity of such a substance. Hence, due to the interaction of many individuals, the pheromone tends to concentrate in the shortest paths. This method can be adapted for the resolution of multiple problems represented through graphs, leading to a metaheuristic known as Ant Colony Optimization, ACO [1].

ACO has been used to solve the problem of routing in telecommunication networks based on circuit switching and on packet switching. In the last years, ACO has also been used to solve the routing and wavelength assignment (RWA) problem in optical networks. In particular, for the dynamic problem, where all-optical connections (lightpaths) are established and released on user demand, there are several proposals [2–4], all of them using centralized control. In these works, the artificial ants that travel across the network are only used to configure the routing tables in the nodes, helping to solve only a part of the RWA problem.

I. Tomkos et al. (Eds.): ONDM 2007, LNCS 4534, pp. 299–308, 2007.

In this paper, we present a novel algorithm to solve the dynamic RWA problem in an optical network with distributed control, where each network node cooperates with the others to solve the problem. The new algorithm, which will be described in Section 2, uses ACO not only to configure the routing tables in the nodes, but also to collect information about wavelength availability in the links of the network, thereby providing complete information about the RWA problem, and leading to lower blocking probability and lower mean setup time than other distributed algorithms, as we will show in Section 3 of the paper.

2   The New Algorithm, DACOS

The new algorithm, named DACOS (Distributed Ant Colony Optimization Scheme for the RWA problem), is designed to operate in an optical network without wavelength converters where bidirectional connections are dynamically established and released. The algorithm comprises two parts: *the ACO system*, which is in charge of building local routing tables with information about wavelength availability; and *the reservation system*, which processes establishment and release requests. DACOS involves the transmission of control packets through the network, which could be transported using a specific wavelength reserved for that aim.

2.1   The ACO System: Building the Local Routing Tables

The first part of DACOS is an ACO system which consists in the network nodes periodically sending artificial ants (management packets) to other nodes. These artificial ants imitate the behavior of natural ants, and so choose the following node of their route according to the concentration of artificial pheromone, which is a value associated to each link for each destination node (Fig. 1). When an ant arrives at a node, the pheromone is increased in the link by which it arrived, but only for the nodes that it has previously visited (as a way of indicating that those nodes previously traversed by the ant can be reached going through that link). The pheromone is increased as a function of both the distance (in number of hops) traversed by the ant from each of those nodes, and the availability of common wavelengths in the links traversed by the ant. For instance, let us consider the network shown in Fig. 1, and assume that an ant departs from node 4, traverses node 5 and arrives at node 1. Since the ant arrives through link l_5, and has visited nodes 4 and 5 in its trip, the pheromone would be increased for that link and those destinations (represented in bold in the table in Fig. 1). The pheromone amounts in the table are normalized for each row, so their sum must be one for each destination node. In this way, the pheromone amounts are used as a probability distribution to choose the outgoing link that the artificial ants arriving at that node will traverse, and —as we will explain later—, as a way to estimate the most appropriate next node for a specific connection request. As an example of the first issue, if an ant is processed at node 1 and has node 2 as destination, the ant will be very probably forwarded through link l_1, as it has the largest amount of pheromone for the row corresponding to the destination node 2. However, with the aim of expanding the exploration of new routes across the network, a small percentage of artificial ants select the next node to visit according to a uniform distribution, instead of having more probability to follow the link with the largest amount of pheromone.

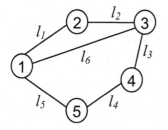

Destination	Link		
	l_1	l_5	l_6
Node 2	0.6	0.1	0.3
Node 3	0.2	0.1	0.7
Node 4	0.1	**0.5**	0.4
Node 5	0.1	**0.6**	0.3

Pheromone table at node 1

Fig. 1. Sample network and example of pheromone table at node 1.

The artificial ants are also used to configure the routing tables of the nodes by combining the pheromone tables with information about wavelength availability. In this way, the routing tables indicate the most appropriate outgoing link for establishing a connection with each possible destination (that having more pheromone), and contain updated data on the availability of each wavelength to reach each destination. To accomplish this latter issue, an ant not only collects information while traversing the network such as the availability of common wavelengths in that route, but it also retrieves information about wavelength availability from the nodes of its route that was deposited by previous ants following alternate routes, and propagates this information to the next nodes that it visits.

2.2   The Reservation System: Processing Establishment and Release Requests

The second part of DACOS is a reservation system, which is in charge of processing establishment and release requests. When a node desires to establish a lightpath with another node, the information of the local routing table is used to select both the optimal outgoing link and the wavelength that has more probability to be available until the destination node. First of all, the link with the highest pheromone value (for the destination node) is selected. Then, the routing table is again consulted to determine the set of wavelengths that are available until the destination node of the connection through that link (or more precisely, that were available according to the information carried by the last ant that arrived at this node from the destination node). If there are several wavelengths available, the one selected is that which is free in a lower number of links connected to the source node, as a heuristic to minimize the blocking probability of future connections traversing that node. If at this point there are several wavelengths with the same number, one of them is chosen randomly. If there are no available wavelengths, that link is ruled out and the same process is repeated on the next link with highest pheromone value. Once the outgoing link and the wavelength for the connection have been selected in the source node, a control packet is sent to the following node, and the selected wavelength is reserved. Since the network does not have wavelength converters, the next node must select an outgoing link where the selected wavelength is available. First of all, the link with the highest pheromone value for the destination node is selected. Then, using the information of the local routing table, it is checked whether the selected wavelength for the connection is expected to be available in each link across some route to the destination node. If found, that link is chosen to send the control packet. Otherwise, that link is ruled out and the next link with highest pheromone value is checked. The process that we have just mentioned is repeated in the following nodes of the route until the destination node is reached.

It is worthy to note that the wavelength is chosen by the source node and it is reserved in all the links that the control packet crosses. However, the sequence of nodes that will be traversed by the control packet (and hence used to establish the connection) is not predetermined in the source node but selected according to the availability of the wavelength chosen for the connection. Moreover, each control packet of the reservation system includes an ant, so that it also updates the routing tables of the nodes when moving across the network.

If the connection cannot be established due to lack of resources, a probe packet is sent from the point where the connection gets blocked towards the destination node, and a release packet is sent backwards to free the reserved wavelengths in the visited links. When the probe packet reaches the destination, the connection establishment process can be repeated from there (since connections are bidirectional, they can be established in any direction). In this way, if a connection is blocked due to wavelength unavailability, it can be reattempted for a number of times if required, at the expense of increasing the setup time of the connection.

On the other hand, when a connection is to be released, a control packet is sent along the route in order to free all the reserved resources.

3   Performance Evaluation

In order to evaluate the performance of DACOS in terms of both blocking probability and setup time, we have compared it by means of simulation with two distributed RWA algorithms. The first one uses Fixed Routing (FR), where the routes are precalculated using an ACO method that searches the shortest routes. For wavelength assignment a backward reservation method [5] is used together with the First Fit (FF) heuristic [6]. From now on, this algorithm will be named FR-FF. The second algorithm is a modified version of the ACO-based algorithm presented by Ngo et al. [3]. The original algorithm used centralized control, so we have adapted it to a distributed environment. Like the previous algorithm, a backward reservation method has been implemented for wavelength assignment together with the First Fit (FF) scheme. Henceforth we will name it Ngo algorithm. In DACOS, a maximum of four attempts were allowed for each connection request.

The simulator has been implemented in OPNET Modeler [7], and the algorithms have been evaluated in the NSFNet topology [8] in terms of blocking probability, and mean and maximum setup time. Connection requests were generated at each node following a Poisson process with arrival rate λ_{node}, and the destination nodes for the connections were randomly chosen according to a uniform distribution. Thus, the total arrival rate in the whole network was $\lambda_{total} = N \cdot \lambda_{node}$, where N is the number of network nodes. The connection holding time was exponentially distributed with mean T. Therefore, the normalized traffic load is given by $\lambda_{total} T / [N \cdot (N-1)]$.

3.1 Blocking Probability

Fig. 2 shows the mean blocking probability for DACOS, FR-FF and Ngo algorithms, for different numbers of wavelengths per link. It can be seen that DACOS outperforms the other algorithms, and as the network links become saturated, its blocking probability converge with the other ones. For a moderate load and an adequate number of wavelengths, between 16 and 32, the blocking probability of DACOS is around 10^{-3}, while the other algorithms have a blocking probability higher than 10^{-2}. Fig. 2 also shows that there is an interval of traffic loads for which the blocking probability of the FR-FF and Ngo algorithms decrease. Due to the method that we have used to increase the traffic load in the simulation, which has consisted in extending the duration of the connections, the network state changes less frequently, and FR-FF and Ngo algorithms improve their performance in that scenario. That is due to the backward reservation method, which leads to lower backward blocking probability since the information about wavelength availability carried by the control packet requesting the connection, and received by the final node, is less outdated in these scenarios [5]. Obviously, if the traffic load is further increased, the lack of resources becomes dominant and the blocking probability rises again.

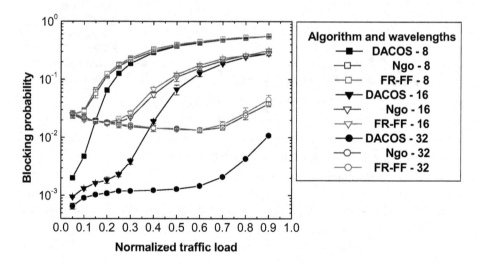

Fig. 2. Blocking probability for DACOS and other distributed algorithms. 95% confidence intervals are shown.

3.2 Mean Setup Time

Fig. 3 shows the mean setup time, i.e., the mean time elapsed between the generation of a connection request at a node and the moment in which both end nodes receive confirmation of the connection being completely established. DACOS again outperforms the other algorithms. The mean setup time for DACOS is around 28 ms, while the others are around 40 ms in the low traffic load region. DACOS gets a lower mean setup time because in FR-FF and Ngo three trips of control packets are always required to establish a connection (one to check wavelength availability, one to reserve a wavelength, and another one to confirm the reservation) whereas in DACOS only two trips are required if the connection is established in the first attempt (one to reserve a wavelength and another one to confirm the reservation). Fig. 3 also shows that as the traffic load increases, the mean setup time decreases. The reason is that as links become saturated, the longest connections are the most likely ones to be blocked, so the mean time to establish a connection decreases as the connections become shorter (and hence the control packets travel shorter distances). However, the mean setup time initially increases for DACOS, because when the network becomes slightly saturated, it selects longer routes that are less congested, in an attempt to not to increase the blocking probability. Due to this behavior a peak in the figure can be noticed, showing a displacement to higher traffic loads as the number of wavelengths increases.

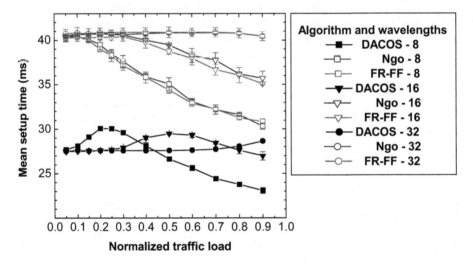

Fig. 3. Mean setup time for DACOS and other distributed algorithms. 95% confidence intervals are shown.

3.3 Maximum Setup Time

Fig. 4 shows the maximum setup time, i.e., the time elapsed between the generation of a connection request at a node and the moment in which both end nodes receive confirmation of the connection being completely established in the worst case. DACOS presents a higher and more unpredictable maximum than the other two algorithms. FR-FF and Ngo get a very stable worst case around 90-100 ms, because they always require three trips of control packets to establish a connection. Thus, the only affecting factor is the length of the path. For this reason, FR-FF yields a slightly lower maximum, since it always searches for the shortest paths, while Ngo lets longer less congested paths to be used to establish the connections. DACOS, besides being more flexible with the length of the paths, allows up to four attempts for each connection request, so the setup time can vary notably and be high, around 180 ms with 95% confidence intervals even wider than 60 ms. Like in the mean setup time case, a peak can be noticed in DACOS graph, showing a displacement to higher traffic loads as the number of wavelengths increases. On the other hand, it is worthy to note that, if required, the maximum number of connection attempts in DACOS can be reduced in order to decrease the maximum setup time and its variability.

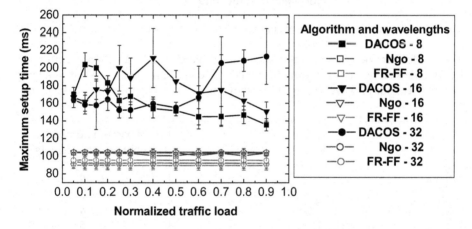

Fig. 4. Maximum setup time for DACOS and other distributed algorithms. 95% confidence intervals are shown.

3.4 Impact of the Number of Connection Attempts

The FR-FF and Ngo algorithms can only attempt to establish the connection once. However, DACOS can repeat the process several times if required (in the analyzed scenario, up to four times). Thus, Fig. 5 shows the mean number of connection attempts used by DACOS to establish a connection. As shown in the figure, the higher the number of wavelengths in the links, the lower the mean number of connection attempts that DACOS requires. Even when considering the worst case, with eight wavelengths, the mean is low (1.14), so that most of the connections are established in the first attempt. This fact proves that the algorithm is very efficient in using the information gathered by the artificial ants to select the wavelength and the route of the connections.

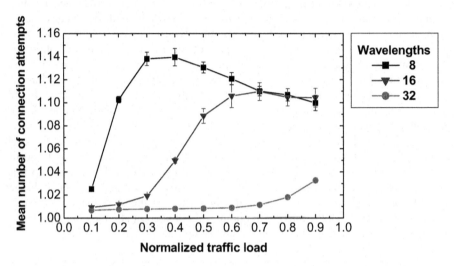

Fig. 5. Mean number of connection attempts for DACOS. 95% confidence intervals are shown.

As a further view about this issue, Fig. 6 shows the blocking probability for DACOS depending on the maximum number of connection attempts, when equipping the NSFNet with 16 wavelengths. Fig. 7 shows the mean setup time for that scenario.

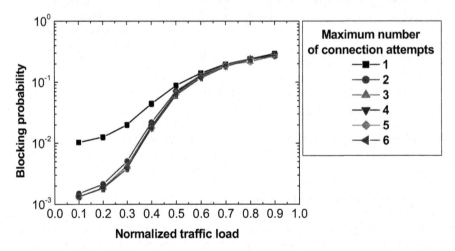

Fig. 6. Blocking probability for DACOS depending on the maximum number of connection attempts, with 16 wavelengths. 95% confidence intervals are shown.

As shown in Fig. 6, the blocking probability obtained with DACOS, when only one attempt is allowed to establish a connection, is similar to that obtained with FR-FF and Ngo algorithms (Fig. 2). However, when setting a maximum of two attempts, there is a significant decrease in blocking probability, getting very close to the results obtained with a maximum of four attempts (which was the value used in the simulation results previously shown). In fact, as shown in the figure, there is no significant improvement for further increases on the maximum number of attempts.

On the other hand, allowing several attempts has very little penalty on the mean setup time, as shown in Fig. 7. In fact, even with six attempts (Fig. 7), the mean setup time is lower than that obtained with FR-FF and Ngo algorithms (Fig. 3).

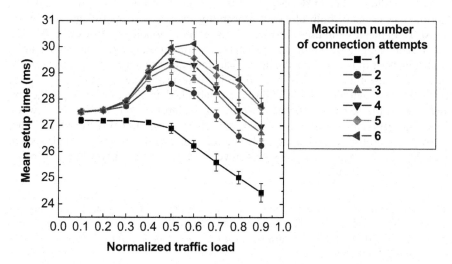

Fig. 7. Mean setup time for DACOS depending on the maximum number of connection attempts, with 16 wavelengths. 95% confidence intervals are shown.

In summary, if DACOS is configured with a single attempt, the performance in terms of blocking is similar to that of Ngo and FR-RR, but much faster, as FR-FF and Ngo require three trips of control packets to establish a connection, but DACOS only requires two. If DACOS is configured with more than one attempt, the blocking probability significantly decreases while still being notably faster than Ngo and FR-RR.

4 Conclusions

We have proposed a new algorithm, DACOS, which efficiently solves the dynamic and distributed RWA problem by means of ant colony optimization.

DACOS presents some completely original characteristics when compared with other methods based on ant colony optimization. The ants are not only limited to make the tasks of configuring routing tables in the network, but they also provide updated information about wavelength availability. In this way, the reservation system takes advantage of that information in the source node of each connection to select both the next node and the wavelength with highest probability of being available in all the links until the destination node. While the wavelength is selected by the source node, the route is not predetermined, but it is adaptively selected by the intermediate nodes, using more updated information than that available in the source node, and thus making possible to cope with changes in the state of the network, by taking alternative routes to establish the connection when required. Moreover, a hybrid reservation mechanism combining ideas from forward and backward reservation is used, which allows to make manifold attempts of connection if needed.

We have shown that DACOS outperforms other distributed and dynamic RWA algorithms in terms of both blocking probability (over an order of magnitude in the NSFNet) and mean setup time (about 30% faster in that topology). However, DACOS also presents a higher and more unpredictable maximum setup time, mainly due to its variable number of connection attempts. That technique leads to a significant decrease of the blocking probability with very little increase in the mean setup time. That is due to the fact that most connections are established in the first attempt (thereby having very little setup time as only two trips of control packets are involved, while other less efficient algorithms require three), and additional connections, that otherwise would be blocked, are established (of course at the expense of higher setup time). For those reasons, DACOS gets lower blocking probability while still being faster than other algorithms.

Acknowledgments. This work has been supported by the Spanish Ministry of Education and Science and Technology (*Ministerio de Educación y Ciencia*) under grant TEC2005-04923 and by *Consejería de Educación de la Junta de Castilla y León* under grant VA028B06.

References

1. Bonabeau, E., Dorigo, M., Theraulaz, G.: Inspiration for Optimization from Social Insect Behaviour. Nature, Vol. 406 (2000) 39-42
2. Garlick, R.M., Barr, R.S.: Dynamic Wavelength Routing in WDM Networks via Ant Colony Optimization. In: Proceedings ANTS'02 (2002) 250-255
3. Ngo, S.H., Jiang, X.H. and Horiguchi, S.: An Ant-Based Approach for Dynamic RWA in Optical WDM Networks. Photonic Network Communications, Vol. 11 (2006) 39-48
4. Sousa, G., Waldman, H.: Evaluation of an Ant-based Architecture for All-Optical Networks. In: Proceedings ONDM'06. Copenhagen, Denmark, (2006)
5. Lu, K., Xiao, G. and Chlamtac, I.: Analysis of Blocking Probability for Distributed Lightpath Establishment in WDM Optical Networks. IEEE/ACM Transactions on Networking, Vol. 13, No. 1 (2005) 187-197
6. Zang, H., Jue, J.P., Mukherjee, B.: A Review of Routing and Wavelength Assignment Approaches for Wavelength-Routed Optical WDM Networks. Optical Networks Magazine, Vol. 1, No. 1 (2000) 47-60
7. OPNET Modeler, http://www.opnet.com
8. Baroni, S., Bayvel, P.: Wavelength Requirements in Arbitrarily Connected Wavelength-Routed Optical Networks. Journal of Lightwave Technology, Vol. 15, No. 2 (1997) 242-251

Optical IP Switching for dynamic traffic engineering in next-generation optical networks

Marco Ruffini, Donal O'Mahony, Linda Doyle

Centre for Telecommunication Value-Chain Research
University of Dublin, Trinity College
Dublin 2, Ireland
{ruffinim@tcd.ie, Donal.OMahony@cs.tcd.ie,lodoyle@tcd.ie}

Abstract. WDM technology has increased network link capacity dramatically, moving the network bottleneck from the transport to the routing layer. Hybrid electro-optical architectures seem at the moment a reliable and cost-effective solution for near-future implementations of the routing/switching layer. In this paper we present a novel approach to dynamic optical circuit switching based on the Optical IP Switching (OIS) model. OIS nodes classify IP packets by destination prefix, aggregating them into dynamically created optical paths. We report simulation results based on real traffic traces collected from the pan-European GÈANT network.

Keywords: Dynamic Optical Circuit Switching, Optical Traffic Engineering.

1 Introduction

Wavelength division multiplexing (WDM) and erbium-doped fiber amplifiers (EDFA) technologies, 10 years ago, have started a deep revolution in networking. Combined together, they increased the overall bandwidth availability and drastically reduced the cost of data transfer. These technologies allow the transport of hundreds of gigabits of data on a single fiber for distances over 1000km without need of electro-optical conversion. Such dramatic progress has not yet occurred at the routing layer; as a result the network bottleneck has moved from the optical-transport to the routing layer, as conventional electronic routing does not seem capable of offering a cost-effective solution to the increasing bandwidth demand.

During the past 10 years a lot of work has been done on optical packet switching, with the aim of bringing transparent optical operation to the network layer. However, even though many solutions have appeared in different optical laboratories around the world, what is missing is a break-through technology capable of delivering a cost effective implementation suitable for large scale deployment. More and more people in the research community begin to believe that it is unlikely that all-optical switching will reach the market in the near future.

I. Tomkos et al. (Eds.): ONDM 2007, LNCS 4534, pp. 309–318, 2007.

Under these circumstances, the idea of a hybrid electro-optical solution becomes instead more feasible: many optical architectures have already been proposed that implement the concept of dynamic circuit switching.

The basic idea behind dynamic optical circuit switching is to group and switch all the packets sharing a common route into dedicated all-optical channels, bypassing some of the intermediate IP hops. This process can produce consistent cost saving, as expansive router cards can be replaced by much cheaper transparent optical ports. Such savings however are only possible if data is efficiently aggregated into the optical paths.

In this paper we first investigate how existing optical architectures implement dynamic optical circuit switching. Then we introduce the Optical IP Switching concept, presenting a novel method of forwarding IP traffic through dynamically established optical circuits. In section 5 we report the results of our simulations, based on real traces, analyzing the efficiency of OIS to switch data at the optical layer. Finally we conclude the paper.

2 State of the art

Optical bypass of the IP layer is a well-known technique that allows saving router resources and OEO conversions, bringing potential economic advantages to service providers. This technique has been used in the past few years on Sonet/SDH networks where the deployment of optical add-drop multiplexers (OADM) enabled the addition or extraction of selected wavelengths from an incoming WDM bundle.

Add drop multiplexers however could not be dynamically configured, and network operators could not adapt their topology to the actual traffic demand, since any modification or update required a large amount of time and manual interaction.

Recent advances in optical technology brought to market new devices like reconfigurable OADM (ROADM) and MEMs-based photonic switches. Their capability of switching wavelength and fibers on a sub-second scale is ideal for fast network reconfiguration and allows to implement novel bandwidth on-demand services. However reconfiguring a network topology is a critical task, which needs to be supported by a robust control plane. The Internet Engineering Task Force is currently working on this issue, carrying on the standardization of the Generalized Multiprotocol Label Switching, a protocol suite for the optical control plane.

GMPLS provides intra and inter-domain discovery and signaling allowing dynamical, on-demand creation and deletion of data circuits (either in the electrical or optical domain). We remind the reader however that GMPLS is not a traffic engineering tool: it provides the signaling features needed to modify the topology at different layers, but does not include a planning capability that automatically suggests how connections should be updated. Researchers and network operators have so proposed different approaches to dynamic optical circuit switching, using GMPLS as signaling protocol.

In [1] for example a hybrid electro-optical architecture is presented, where electrical routers use GMPLS to create new optical paths, when the existing ones become congested.

NTT has produced a prototype, the Hikari router [2],[3], based on similar concepts: the router uses a photonic switch to create new optical paths, when the existing ones cannot allocate any other MPLS circuit.

Dynamic optical circuit switching seems also to be the key towards the implementation of grid network architectures. The idea of grid networks was developed to support the interaction of high-end applications distributed around the globe that need to exchange information at ultra-high data rates: distributed computing, e-VLBI, High-energy physics, e-Health applications, only to mention a few. Reconfigurable optical networks seem to suit this concept very well, as dedicated optical lightpaths can be established on demand to satisfy the large bandwidth demand of such applications. DRAGON [4] and MUPBED [5] are an example of optical grid networks [6]. The OptIPuter [7], in particular, is capable of creating dedicated end-to-end lightpaths in real-time, either by analyzing IP flows or following direct requests from applications.

The fast reconfiguration offered by ROADM and MEMs switches has also triggered the development of more unconventional approaches: in [8] for example, the authors propose the idea of an ad-hoc optical network, where nodes can connect to use and offer network services, with a plug-and play approach. Auto-discovery, self-configuration and signal monitoring are in this case essential features for the correct operation of the network.

In [9],[10] we have proposed an architecture, called Optical IP Switching (OIS), that creates and deletes optical paths depending on the traffic encountered at the IP layer. One of the distinguishing features is that optical paths are created in a distributed fashion, based only on local decisions. We believe that this approach better satisfies the requirements of inter-domain networking, since different domains can implement their own policies to decide, for example, if an incoming signal should be transparently switched or locally terminated. This is in our opinion more realistic compared to the idea of creating end-to-end optical paths, where each domain is supposed to accept modifications to their logical and physical topology demanded by competing network operators.

3   Prefix-based Optical IP Switching

In the OIS network architecture we have introduced in [9], each IP router constantly analyzes the traffic searching for large, long-lived IP flows. When a suitable aggregate of flows is identified the router activates the photonic switch to create an optical cut-through path between its upstream and downstream neighbours. Only three nodes participate initially to the new optical path, which can then be extended by other nodes in a distributed fashion.

In this paper we focus on a novel approach to Optical IP Switching that aggregates packets based on routing destination prefixes instead of considering distinct IP flows.

The main idea behind prefix-based OIS is to classify the forwarded IP traffic using the network prefixes stored in the IP routing table.

Packets are classified as follows:

1. First we differentiate the packets depending on their arrival (source) and departure (destination) interfaces: this is necessary because photonics switches do not have the capability of grooming traffic optically. The classification is operated building a matrix with number of lines and rows equal respectively to the number of incoming and outgoing interfaces (we assume for simplicity that different interfaces are linked to different destinations[1]). The generic matrix cell (i,j) will identify traffic incoming from interface i, relayed through interface j.

2. Within each of the previous classes we operate a finer classification by destination prefix: in each cell of the matrix we build up a list of destination prefixes reachable through the corresponding interface. This classification is also necessary because the upstream node, source of the new optical path, needs to be informed on which network prefixes it should route into the new lightpath.

In the second classification it is possible to include a "prefix threshold" that will discard all the prefixes bringing data below a certain value. As we will show in section 4, this technique can be used to diminish the amount of information exchanged and save resources at the upstream router.

Each OIS node examines incoming data sampling packets at a rate up to 1/1000, a value that, according to [11], allows a good statistical traffic characterization. For each packet the router checks its destination address and determines the output interface, using the longest prefix matching algorithm. This information is sufficient to classify the packets: the size of the packet payload adds up to the total amount of data carried by its matching prefix. In this implementation we only consider the packet size, but other attributes, for example a timestamp, could be used to improve the traffic characterization.

Each router collects data during an "observation" period before taking a decision. At decision time, the router analyzes the statistics collected, summing up the amount of data brought by the different prefixes within each cell. Only cells whose cumulative data is over a pre-established "path threshold" (100 Mbps in our case) are considered for Optical IP Switching.

The path creation process (Figure 1) begins considering the generic cell (i,j) with the highest data rate: the router signals the upstream and downstream neighbors (using interfaces i and j) checking their capability to support a new optical cut-through path and proposing a suitable wavelength. After both neighbors have acknowledged the request, the router passes upstream the list of prefixes to be switched through the new optical path. Once the path is created the upstream node updates its routing table and starts injecting the suitable packets into the new optical path. The same operation repeats for the remaining matrix cells with traffic above the path threshold.

The advantage of using a prefix-based approach compared to the flow-based one is twofold: first it lowers the amount of information exchanged between the router and its upstream neighbor, as each prefix counts for a large number of flows. Second, it simplifies the routing for the upstream node. While the flow-based approach requires adding class D IP addresses to the routing table (which might create a problem in

---

[1] Interfaces linked to the same destination would be considered as a unique entry in the matrix.

terms of the size of the routing table), the prefix-based approach only requires a reordering of the IP table. The prefix based approach works as a highly dynamic traffic engineering tool, maximizes the amount of switched data and generally improves the QoS levels. On the other hand however it lacks the granularity of the flow-based method, and cannot be used to guarantee a deterministic QoS for individual flows.

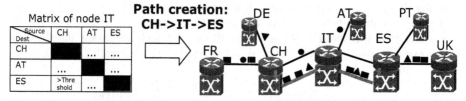

Matrix of node IT			
Source \ Dest	CH	AT	ES
CH		...	...
AT	...		...
ES	>Thre shold	...	

Path creation:
CH->IT->ES

Figure 1 Path creation process

3.1 Path extension mechanism

As previously stated, the decision of creating an optical cut-through path is local, as it only involves a node and its direct neighbors. The path can be set up only if all 3 nodes give a positive acknowledge (this decision could depend on hardware capabilities, inter-domain policies and SLAs between the operators).

A path is created in three different circumstances:

1. The path is newly established. In this case the operation proceeds exactly as described in the previous paragraph.
2. The outgoing interface of the selected cell is the source of an already existing path. In this case (Figure 2) the node creates an upstream extension to an already existing path.
3. The incoming interface of the selected cell is the destination of an already existing optical path. This mechanism is complementary to the one above and creates a downstream path extension.

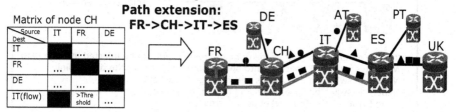

Matrix of node CH			
Source \ Dest	IT	FR	DE
IT		...	...
FR	...		...
DE	...	...	
IT(flow)		>Thre shold	...

Path extension:
FR->CH->IT->ES

Figure 2 Path extension mechanism

The path extension mechanism presents some differences from the path creation. The purpose of the extension algorithm is to select a subset of the prefixes switched by the original paths. Only this subset will be carried by the new extended path, while the remaining data will be routed through the default links.

The extension algorithm plays an important role in the trade-off between length of the optical path and amount of data carried by the path. An optical cut-through path can aggregate together only packets sharing a common path. When an existing path is extended, statistically, only a subset of the original packets will share the new longer path, diminishing the amount of data transported by the optical channel and consequently the channel efficiency. On the other hand however longer cut-through paths increase the number of transparent hops, enhancing the cost-saving potentials of optical switching. From this perspective, the extension algorithm has the task to optimize the cost-efficiency problem delineated by this trade-off.

We have identified two algorithms for deciding whether a path should be extended:

- **Relative threshold extension algorithm.** The path extension threshold is expressed as a percentage of the data in the existing path, following the formula:

$$\text{Threshold}(\%) = \frac{\text{PrevNode} - 2}{\text{PrevNode} - 1} \cdot 100 \qquad (1)$$

where PrevNode is the number of nodes in the existing path. Considering that packets excluded from the extended path need to be routed electronically over default routes, this approach makes sure that the extension process never increases the amount of data routed at the IP level.

Figure 3 reports an example of a cut-through path extended from 3 to 4 nodes, using a threshold selected according to (1).

- **Absolute threshold extension algorithm.** In this case the path can be extended if the amount of data switched after the extension is above the original path threshold. The original path can also be maintained if the data carried in it after the extension remains over the path threshold. This algorithm maximizes the total amount of switched data.

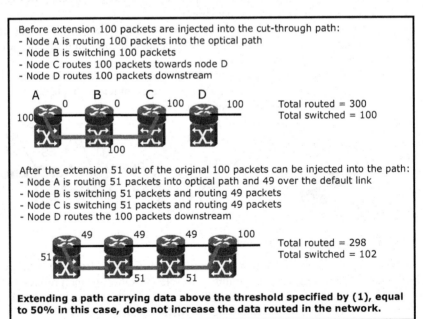

Figure 3 Example of relative threshold extension algorithm.

4 Elephant prefixes

The flow-based optical IP switching method introduced in [9] was developed considering the heavy tailed distribution of the Internet traffic, where a small number of "Elephant" flows carries most of the traffic [12],[13].

We have found a similar heavy tailed distribution in the routing prefixes: in a router's IP table a small number of prefixes routes most of the data. We can use these results to reduce the number of prefixes in the optical paths, with little impact on the amount of switched data.

We conducted our test on a dataset collected from the pan-European GÈANT network, using traces dating back to May 2005[2].

We have studied the network prefixes in the routing tables, classified them by the data rate at which they routed packets, and summed up the amount of data routed by all prefixes in each class. The results are shown in Figure 4[3]: the percentage of data carried by the prefixes routing traffic above a certain data rate (reported in the x axis), diminishes with much slower pace respect to number of prefixes considered. This implies that a large number of prefixes route a minor percentage of the total data. Excluding these prefixes from the optical path will save routing and network resources, without noticeably affecting the switched data.

In Figure 4, for example, we see that a threshold of 100 Kbps, would reduce the number of prefixes considered by 84%, causing only 8% of the data to be excluded from the optical path.

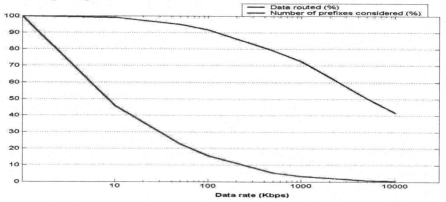

Figure 4 Heavy tail distribution of routing prefixes in GÈANT.

5 Simulation results

We have taken the pan-European GÈANT network as a reference model for our simulations, using empirical data collected from the access points. The GÈANT

---

[2] More details on the GÈANT dataset will be given in section 5.
[3] From a 15 minutes trace collected on the 4/5/2005 at 15.45.

dataset appeared to be especially suited for our case, as it complements the traffic traces with the BGP routes collected from the border routers. Data are made available by researchers from the Computing Science and Engineering dept. at the University of Louvain-la-Neuve, who also provide C-BGP[14], a network simulator capable of reconstructing the BGP network from the routes included in the dataset. The tool uses a clustering algorithm that reduces the number of BGP entries by more than two orders of magnitude, making it possible to simulate a real network scenario. The traffic traces, collected using Netflow with sampling rate of 1/1000, are summarized depending on their source/destination prefix and only the total number of bytes over a 15 minutes period is provided. This has the two-fold effect of saving storage space for the data files while keeping the traces anonymous. The disadvantage is that information about the precise timing of the flows is lost: a condition that however does not influence our study, since the mechanism that creates optical cut-though paths averages the observed traffic over a period of some minutes.

We have simulated the prefix-based OIS approach considering traffic traces and BGP tables on 4 different days, spaced approximately a week from each other. The dynamic optical links simulate 1 Gbps channels and the path threshold was set to 100 Mbps.

In Figure 5 we report the percentage of data that our OIS approach can switch optically using either of the extension algorithms introduced in paragraph 3. The results are averaged over 4 different traces and reported for different traffic levels, obtained by multiplying the original traces by progressively increasing factors.

The plot shows that the absolute threshold algorithm performs better compared to the relative threshold one. A higher traffic level moreover increases the number of channels above the path threshold, allowing creation of new optical paths. However, once all the possible paths have been created, the ratio between routed and switched traffic stabilizes.

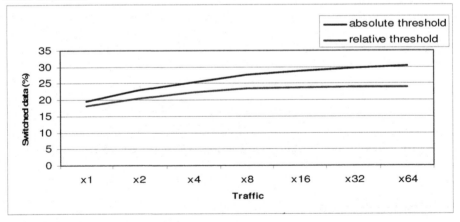

Figure 5 Data optically switched by the OIS architecture

Figure 6 reports the average optical channel usage. Under this perspective the relative threshold algorithm outperforms the absolute threshold one, allowing better exploitation of the optical bandwidth. According to these results we can state that the absolute threshold algorithm maximizes the total amount of switched data while the relative threshold makes better use of the channels it creates. Which of the two would represent a better approach in a practical implementation depends on the cost associated to electronic routing, optical switching and wavelength channels.

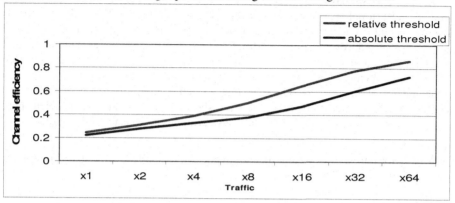

Figure 6 Average channel occupancy comparison.

In order to determine whether optical IP switching is more successful in networks where the node degree is lower, we altered the network topology deleting links to lower the average node degree. We have removed 5 main links from the original GÈANT topology, lowering the average node degree from 3.2 down to 2.8. Figure 7 shows the topology comparison, considering both the total switched traffic and the average channel usage (using the absolute threshold algorithm). The topology with fewer links switches over 20% more traffic, while the channel efficiency improves between 2 and 13%.

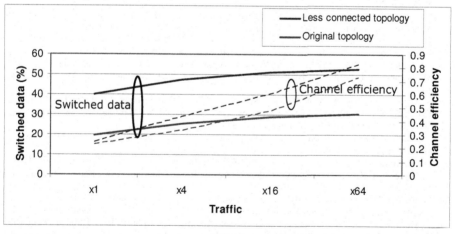

Figure 7 Comparison of different network topologies.

From this analysis we can deduce that the network topology can have a significant impact on the amount of traffic switched by the Optical IP Switching architecture. In particular a less connected topology increases the average number of IP hops, favoring the aggregation of packets at the optical level.

6 Conclusions

In this paper we have presented a novel optical architecture capable of adapting the optical layer topology to the real traffic demand at the IP layer. The traffic is aggregated into dynamically created optical paths using a destination prefix based approach that reduces signaling overhead and saves router resources. Our simulation results, modeled on the pan-European GÈANT network, show that the OIS approach can switch optically about 20% of the total data, with the current network topology unchanged. Taking into consideration that an actual implementation might require higher values to be cost-effective, we have shown that OIS can switch much more data (over 50%) as traffic increases and considering less connected topologies.

Acknowledgments. We would like to thank Steve Uhlig and Bruno Quoitin for providing the GÈANT dataset.

References

1. S. Kano, T. Soumiya, M. Miyabe, A. Chugo. A Study of GMPLS Control Architecture in Photonic IP Networks. Workshop on High Performance Switching and Routing: Merging Optical and IP Technologies, 26-29 May 2002.
2. K. Sato, N. Yamanaka, Y. Takigawa, M. Koga, S. Okamoto, K. Shiomoto, E. Oki, W. Imajuku. GMPLS-Based Photonic Multilayer Router (Hikari Router): Architecture An Overview of Traffic Engineering and Signaling Technology. IEEE Communications Magazine, Vol. 40 , No 3 , March 2002.
3. E. Oki, K. Shiomoto, D. Shimazaki, N. Yamanaka, W. Imajuku, Y. Takigawa. Dynamic Multilayer Routing Schemes in GMPLS-Based IP+Optical Networks. IEEE Communications Magazine, Vol. 43 , No 1 , Jan. 2005.
4. P. Szegedi, Z. Lakatos, J. Spath. Signaling Architectures and Recovery Time Scaling for Grid Applications in IST Project MUPBED. IEEE Communications Magazine, March 2006.
5. T. Lehman, J. Sobieski, B. Jabbari. DRAGON: A Framework for Service Provisioning in Heterogeneous Grid Networks. IEEE Communications Magazine, March 2006.
6. I. W. Habib, Q. Song, Z. Li, N. S. V. Rao. Deployment of the GMPLS Control Plane for Grid Applications in Experimental High-Performance Networks. IEEE Communications Magazine, March 2006.
7. N. Taesombut, F. Uyeda, A. A. Chien, L. Smarr, T. A. DeFanti, P. Ppadopoulos, J. Leigh, M. Ellisman, J. Orcutt. The OptIPuter: High-Performance, QoS-Guaranteed Network Service for Emerging E-Science Applications. IEEE Communications Magazine, May 2006.
8. I. Cerutti, A. Fumagalli, R. Hui, A. Paradisi, M. Tacca. Plug and Play Networking with Optical Nodes. Proceedings of ICTON 06, Nottingham, UK, 18-22 Feb, 2006.
9. M. Ruffini, D. O'Mahony, L. Doyle. A Testbed Demonstrating Optical IP Switching (OIS) in Disaggregated Network Architectures. Proceedings of IEEE Tridentcom 2006,Barcelona, Spain, 1-3 March, 2006.
10. M. Ruffini, D. O'Mahony, L. Doyle. A cost analysis of Optical IP Switching in new generation optical networks. Proceedings of Photonics in Switching 2006. 16-18 October 2006, Herakleion, Greece.
11. T. Mori, M. Uchida, R. Kawahara, J. Pan, S. Goto. Identifying Elephant Flows Through Periodically Sampled Packets. IMC 04, Oct 25-27, 2004. Taormina, Sicily, Italy.
12. N. Brownlee, KC Claffy. Understanding Internet Traffic Streams: Dragonflies and Tortoises. IEEE Communications Magazine, October 2002.
13. K. Papagiannaki, N. Taft, S. Bhattacharya, P. Thiran, K. Salamatian, C. Diot. On the Feasibility of Identifying Elephants in Internet Backbone Traffic. Sprint ATL technical report, Sprint Labs, November 2001.
14. B. Quoitin, S. Uhlig. "Modeling the Routing of an Autonomous System with C-BGP". IEEE Networks Magazine, Nov/Dec. 2005.

An Efficient Virtual Topology Design and Traffic Engineering Scheme for IP/WDM Networks

Namik Sengezer and Ezhan Karasan

Department of Electrical and Electronics Engineering
Bilkent University, Ankara 06800, TURKEY,
`namik,ezhan@ee.bilkent.edu.tr`

Abstract. We propose an online traffic engineering (TE) scheme for efficient routing of bandwidth guaranteed connections on a Multiprotocol label switching (MPLS)/wavelength division multiplexing (WDM) network with a traffic pattern varying with the time of day. We first consider the problem of designing the WDM virtual topology utilizing multi-hour statistical traffic pattern. After presenting an effective solution to this offline problem, we introduce a Dynamic tRaffic Engineering AlgorithM (DREAM) that makes use of the bandwidth update and rerouting of the label switched paths (LSPs). The performance of DREAM is compared with commonly used online TE schemes and it is shown to be superior in terms of blocked traffic ratio.

Keywords: Traffic Engineering, Virtual Topology Design, MPLS, LSP, Dynamic Routing.

1 Introduction

MPLS architecture provides powerful features for traffic engineering [1]. The encapsulated Internet Protocol (IP) packets can be routed explicitly along virtual connections called LSPs. The explicit routing capability in MPLS combined with enhanced link state interior gateway protocols (IGPs) and resource reservation protocol (RSVP), enables the routing of the traffic flows taking into account both the quality of service and bandwidth requirement of the traffic flows and current network state such as traffic load and available capacity on the links. The extensions to RSVP allows the dynamic updates of the LSP bandwidth and construction of the new explicit route before tearing down the old route, which is useful for LSP reroutings [2, 3].

The rerouting of virtual connections is frequently referred in the literature as a means of TE. When it is performed on the underlying WDM layer [4], the virtual topology seen by the upper layer changes with the TE actions. When it is carried out solely on the MPLS layer, the virtual topology is kept intact and the LSPs are rerouted on the existing lightpaths [5, 6].

In this work, we develop a new MPLS layer TE scheme, DREAM, for dynamic connections and show that with a properly designed virtual topology, it can enhance the network performance to a large extent compared to other TE

I. Tomkos et al. (Eds.): ONDM 2007, LNCS 4534, pp. 319–328, 2007.

schemes, without changing the virtual topology. We model the dynamic traffic demand using a Multi-Hour traffic pattern which will be discussed in Section 2.1. In Section 2.2, the Multi-Hour Virtual Topology Design (MVTD) problem is investigated. A heuristic algorithm utilizing tabu search (TS) is proposed to solve this problem and the solutions are compared with upper bounds obtained by using mixed integer linear programming (MILP). DREAM is explained in Section 3, and its performance is compared with other commonly used TE schemes. Section 4 includes an extension of DREAM where multiple parallel LSP tunnels can be provisioned between any node pair.

2   Multi-Hour Virtual Topology Design

Multi-hour virtual topology design problem can be identified as designing the optimum MPLS layer virtual topology for a daily traffic pattern that changes over time [7]. We assume that, some basic information on the hourly traffic statistics is available. That statistical information is the estimate of the maximum traffic rate in every hour, between each node pair. In this work, it is assumed that there exists a non-zero traffic demand between any two nodes in the network for all hourly periods.

2.1   Traffic Model

Since flow-level traffic models are better-suited for TE purposes than packet-level models, flow level traffic statistics is utilized in this work. To model the traffic, an approach that is proposed in [8] is used, which is suitable for generating the traffic when the nodes are spread over a large geographical area. In this approach, for each node i, a time zone offset τ_i and a traffic generation ratio $tgen_i$ are defined. $tgen_i$ represents the maximum instantaneous traffic generated by node i. An activity function $act(i,t)$ depending on these values is defined as

$$
act(i,t) = \begin{cases} 0.2 & \text{if } t_{\text{local}}(i,t) \in [0:00; 6:00) \\ 1 - 0.8 \left(\cos \left(\frac{(t_{local}(i,t)-6)\pi}{18} \right) \right)^{10} & \text{if } t_{\text{local}}(i,t) \in [6:00; 24:00) \end{cases}
$$

In this formula, t is the coordinated universal time in hours and the function $t_{\text{local}}(i,t)$ is defined as, $t_{\text{local}}(i,t) = (t + \tau_i + 24) \bmod 24$. The activity function is illustrated in Fig. 1 for node i which belongs to the universal time zone.

For a node pair (i,j), the expected instantaneous traffic between these nodes in time t is calculated as

$$
T_{\text{expected}}(i,j,t) = tgen_i \times act(i,t) \times tgen_j \times act(j,t) \tag{1}
$$

The actual instantaneous traffic between nodes i and j is obtained by adding a zero mean Gaussian random variable to this expected value. The standard deviation of the Gaussian random variable is given by k times the expected traffic, i.e.,

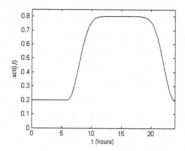

Fig. 1. Activity function

$$T_{\text{actual}}(i,j,t) = T_{\text{expected}}(i,j,t) + N(0, k \times T_{\text{expected}}(i,j,t)) \qquad (2)$$

The parameter k is a measure of the deviation of the actual traffic from its estimate and will be used as a parameter for evaluating the robustness of the TE schemes with respect to uncertainty in traffic estimations. The traffic generation ratio $tgen_i$ is assumed to be uniform for all the nodes in the network and will be referred as $tgen$ throughout the paper. In the proceeding sections, the lightpath capacities will be expressed as the ratio of their bandwidth to $tgen^2$.

2.2 Tabu-Search Based Multi-Hour Virtual Topology Design(TS-MVTD)

The objective in designing the topology is to maximize the total amount of routed traffic during the day while satisfying the optical layer interface constraints. We assume that the total number of lightpaths in the virtual topology is given, and constructing the virtual topology corresponds to deciding between which source destination pairs to place these lightpaths subject to the nodal interface constraints. The nodal interface constraints correspond to the maximum number of lightpaths emanating at a node.

We develop a heuristic algorithm based on the Tabu Search procedure to solve this problem. Tabu Search is an iterative search procedure utilizing adaptive memory [9]. Its distinguishing feature from other search procedures is that non-improving moves are also allowed to escape local optima. Entrapment in cycles is avoided by declaring the visited solutions *tabu* for a number of iterations and forbidding the moves leading to a tabu solution.

The statistical traffic information that is input to the algorithm consists of H $n \times n$ matrices where H is number of hours in the considered interval and n is the number of nodes. The traffic information for hour h is represented by matrix T^h and the entry $T^h_{i,j}$ in this matrix is the maximum value of the expected traffic between nodes i and j during the hth hour.

The objective function is the total amount of traffic that can be routed on the current topology in every hour. To calculate the objective function, an offline routing algorithm based on Dijkstra's shortest path approach is utilized. For each single hour h, the entries in T^h are sorted according to their bandwidth requirement in a descending order and they are served in that order. The dynamic link weights used for computing the shortest paths are inversely proportional to the residual capacities on the links. The offline routing algorithm tries to route the traffic flows starting from the largest flow using paths with the maximum residual capacity. If more than one move give the highest objective value, a tie breaker parameter is used to choose the best move. The tie breaker parameter is calculated as $\sum_{i,j \in V} \sum_{h=1}^{H} s_{ij} T_{i,j}^h$ where V is the set of nodes and s_{ij} denotes the number of hops on the shortest path between i and j in the resulting topology. Between moves giving the same objective value, the one with the smaller tie breaker value is chosen.

The flowchart of the TS-MVTD algorithm is given in Fig. 2. A move is defined as tearing down an existing lightpath and setting up a new lightpath between two nodes which do not share a common lightpath. A valid move is defined as a move that is not in the tabu list and results in a topology that satisfies the interface constraints. The algorithm stops if there is no improvement in the objective value for I iterations.

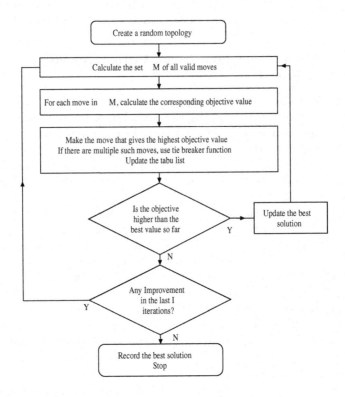

Fig. 2. Flowchart of the TS-VTDM algorithm

To evaluate the performance of the proposed algorithm, an upper bound on the percentage of routed traffic for a given number of lightpaths that can be achieved by any algorithm is obtained by formulating the problem as an MILP problem and relaxing the integrality constraints on the flow variables. The percentages of traffic carried by each algorithm are given in Table 2.2 for different number of lightpaths. These results are obtained using a lightpath capacity $C = 3tgen^2$. According to the lower bounds, at least 15 lightpaths are needed to route all the traffic demands. Proposed TS-MVTD algorithm achieves this using 17 lightpaths, however it can route close to 100% of the traffic demands for 15 and 16 lightpaths.

Table 1. The percentages of routed traffic by the TS-MVTD algorithm and the upper bound for different number of lightpaths

# of lightpaths	13	14	15	16	17
upper bound	93.403	97.886	100	100	100
TS-MVTD result	86.897	93.885	98.816	99.971	100
optimality gap(%)	6.966	4.087	1.184	0.029	0

3 Online Traffic Engineering with Single LSP Between Two Nodes

3.1 Modeling Traffic Flows

As stated in Section 2.1, a flow model is used to represent the traffic in our work. In this section, a single LSP is constructed between each source-destination pair and the changes in the traffic amount between these nodes is represented by changing the bandwidth requirement of the constructed LSP. The changes in the bandwidth requirements of the LSPs are modelled by Bandwidth update Requests (BRs) with a Poisson arrival model having a fixed rate of $\lambda = 30$ arrivals/hour. At an arrival of a BR at time t, the new bandwidth requirement of the LSP is calculated from $T_{actual}(i, j, t)$ according to (1).

3.2 LSP Rerouting

The online TE schemes investigated in this work are based on rerouting the LSPs to optimize the network performance. For each LSP to be constructed between a source-destination pair, a number of shortest paths are calculated beforehand using a K-shortest path algorithm. When a BR arrives for an LSP, among the paths belonging to that LSP, the best one is chosen according to the employed TE scheme and the LSP is (re)routed on that path. If there is not sufficient residual capacity along the path to accommodate the LSP, all of the available capacity is dedicated to the LSP and the amount of traffic that cannot be routed is assumed to be blocked.

DREAM. DREAM is an alternate routing scheme which chooses the best path according the number of hop lengths and instantaneous residual capacities of the candidate paths. Each of the candidate paths is assigned a dynamic cost that is calculated by a cost function that utilizes the instantaneous residual capacity information. Then, the path with the minimum cost is chosen for rerouting the LSP. The cost function is designed in a way to choose the shorter paths when the network is lightly loaded and the paths with higher residual capacity when the network is heavily loaded. The cost function for path p, is given by:

$$F_{\text{cost}}(p) = L_p + A^{u - \frac{C^p_{\text{residual}}}{C}} \qquad (3)$$

where, L_p is the number of hops on path p and C^p_{residual} is the residual capacity of the path , i.e. the minimum residual capacity of the links (lightpaths) along the path, after the LSP is routed along that path and C is the lightpath capacity. The performance of DREAM is tested on several networks with 10 nodes for various values of the parameters A between 1 and 20 and u between 0 and 1, and the best results were obtained for the values of $A = 10$, $u = 0.5$.

The performance of DREAM is compared with various other schemes in the literature: Shortest Path routing (SP), Available Shortest Path routing (ASP) and Widest Shortest Path routing (WSP) [10, 11]. In SP, no traffic engineering is applied and the LSPs are routed along the fixed shortest paths. ASP only consider the paths with sufficient residual capacity to accommodate the LSP. The shortest among these paths is chosen. If there are multiple such paths, one is chosen randomly. In WSP, similar to ASP, the shortest one is chosen among the paths with sufficient residual capacity. If there are multiple such paths, the one with the highest residual capacity is chosen. In both ASP and WSP, if there is a path with sufficient residual capacity to accommodate the LSP, it is rerouted along the path with the highest residual capacity and the amount of traffic that cannot be routed is assumed to be blocked.

Simulations are run for two different networks each having 10 nodes. The logical topologies are designed using the TS-MVTD algorithm with lightpath capacity $C = 2tgen^2$ with 14 lightpaths and $C = 4tgen^2$ with 23 lightpaths, respectively. These lightpath numbers are the minimum values needed by the TS-MVTD algorithm to fulfill all the connection requests with the given capacity. The link capacities are then overprovisioned by $OP > 0$, i.e. the capacities C used in TS-VTDM are multiplied by $(1+OP)$, so that the uncertainty in the bandwidth requests can be accommodated.

The performances of the TE schemes are compared in terms of blocking ratio which is the ratio of the amount of blocked traffic to the total amount of offered traffic. The comparisons are made for $OP=10\%$ and 25%. The results are presented in Figs. 3 and 4 with networks of 14 and 23 lightpaths, respectively. In the figures, the x axis is the value of the parameter k, which is the ratio of the standard deviation of the offered traffic to its expected value as explained in Section 2.1. The simulations are run for different values of k.

The significant difference in blocking ratios of the dynamic rerouting schemes and shortest path routing emphasizes the benefits of traffic engineering. Among

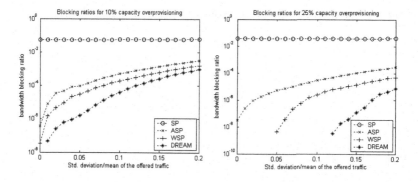

Fig. 3. Blocking ratios of the TE schemes for a network with 14 lightpaths

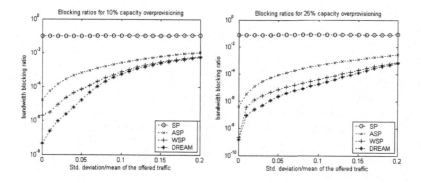

Fig. 4. Blocking ratios of the TE schemes for a network with 23 lightpaths

the dynamic rerouting schemes, ASP uses only the availability information of the paths (having or not having the sufficient residual capacity), while WSP and DREAM uses the full capacity information of the lightpaths. As a result, these two schemes outperform ASP. Among all the rerouting schemes, DREAM performs best.

3.3   LSP Rerouting with Time Limit

Although DREAM achieves the best blocking performance, it may result in a high frequency of LSP reroutings. Each time a BR arrives, the best route is calculated and the LSP is rerouted if the calculated route is different than the current one. To prevent excessive number of LSP reroutings, we introduce a minimum time limit between two consecutive reroutings of the same LSP. When a BR comes for an LSP, if sufficient time has not passed after the last rerouting of that LSP, only the bandwidth dedicated to that LSP is updated, however the route is not changed. We investigate the performance of DREAM with different time limits of 0.1 to 0.5 hour using the 23 lightpath network. Both the blocking ratios and the maximum number of reroutings per hour per LSP are depicted in Fig.5.

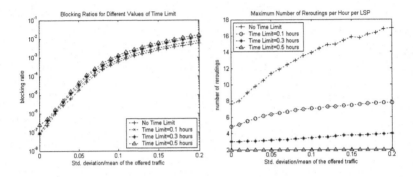

Fig. 5. Blocking ratios and number of reroutings for different values of time limit

As it can be seen from the results, DREAM can generate more than 16 reroutings per hour for an LSP in the worst case. Decreasing the number of reroutings by implementing a time limit causes an increase in the blocking ratio. However, it is possible to optimize the blocking ratio and the frequency of reroutings by choosing a time limit that keeps the frequency of reroutings in an acceptable range while not decreasing the throughput below a desired level.

4   Online Traffic Engineering with Multiple Parallel LSPs

If the total traffic between two nodes can be treated as the sum of multiple uninterruptible traffic flows, multiple parallel LSPs can be set up between the source and the sink nodes, and the total traffic can be distributed among these LSPs without splitting the individual flows over multiple LSPs. To model the traffic for the case of L LSPs per node pair, the instantaneous traffic rate of each LSP is calculated independently by using the traffic generation function described in Section 2.1, with an expected value of $1/L$ times the total expected value and a standard deviation of $1/\sqrt{L}$ times the total standard deviation. Hence, the mean and the standard deviation of the total traffic for each node pair is same as the single LSP case. The arrivals of the BRs for the LSPs are modeled as independent Poisson processes with a rate of λ/L, where λ is the arrival rate that is used in the case of single LSP per source-destination pair. The blocking ratios of the DREAM are presented for different values of L in Fig. 6, for a network with 17 lightpaths and $OP = 10\%$.

The results show that increasing the number of the parallel LSPs brings a significant improvement on the network performance. Increasing the number of LSPs decreases flow granularity and DREAM can make better use of the residual capacities on the lightpaths at the expense of the additional control plane complexity.

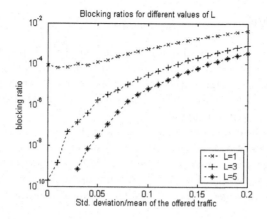

Fig. 6. Blocking ratios for different values of L

4.1   Selecting the Overprovisioning Ratio

In this part, we investigate how much overprovisioning is needed to guarantee a target blocking ratio for different number of parallel LSPs per node pair. Fig. 7 gives the required OP as a function of the desired blocking probability for various values of L. These results are for a network having 10 nodes and 16 lightpaths. The lightpath capacities without overprovisioning are 3 $tgen^2$ and the traffic parameter k is 0.15. It is seen from the results that to keep the blocking ratio below 10^{-6}, $OP > 40\%$ is required if a single LSP is set up between every node pair. If 5 parallel LSPs are set up, the required OP decreases below 20%. Using this figure, the network provider can choose the overprovisioning ratio according to the target blocking ratio and the number of parallel LSPs.

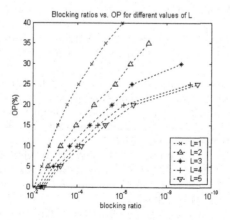

Fig. 7. Blocking ratio vs. OP

5   Conclusions

In this paper, we have presented a new and efficient TE scheme, DREAM, based on rerouting of the LSP connections. DREAM makes use of the available bandwidth information of the lightpaths. It favors the shorter paths when the network is lightly loaded and the paths with more residual capacity are preferred when the load on the lightpaths is higher.

The problem of designing a virtual topology for hourly changing traffic demands is also investigated and a heuristic is developed producing quite satisfactory solutions. On a topology produced by the proposed heuristic, DREAM provides a good blocking performance without topology updates and is superior to similar TE schemes also using the available bandwidth information.

Acknowledgements. This work is supported in part by the Scientific and Technological Research Council of Turkey (TUBITAK) under project EEEAG-104E047 and by the IST-FP6 NoE project e-Photon/ONe+. Namik Sengezer is supported in part by the graduate scholarship of TUBITAK.

References

1. Xiao, X., Hannan, A., Bailey, B., Ni, L.: Traffic Engineering with MPLS in the Internet. Network, IEEE **14**(2) (2000) 28–33
2. Awduche, D., Malcolm, J., Agogbua, J., O'Dell, M., McManus, J.: RFC2702: Requirements for Traffic Engineering Over MPLS. Internet RFCs (1999)
3. Awduche, D., Berger, L., Gan, D., Li, T., Srinivasan, V., Swallow, G.: RFC3209: RSVP-TE: Extensions to RSVP for LSP Tunnels. Internet RFCs (2001)
4. Bouillet, E., Labourdette, J., Ramamurthy, R., Chaudhuri, S.: Lightpath Reoptimization in Mesh Optical Networks. IEEE/ACM Transactions on Networking (TON) **13**(2) (2005) 437–447
5. Bhatia, R., Kodialam, M.S., Lakshman, T.V.: Fast Network Re-optimization Schemes for MPLS and Optical networks. Computer Networks **50**(3) (2006) 317–331
6. Iovanna, P., Sabella, R., Settembre, M.: A Traffic Engineering System for Multilayer Networks Based on the GMPLS Paradigm. Network, IEEE **17**(2) (2003) 28–37
7. Pióro, M., Medhi, D.: Routing, Flow, and Capacity Design in Communication and Computer Networks. Elsevier/Morgan Kaufmann Amsterdam (2004)
8. Milbrandt, J., Menth, M., Kopf, S.: Adaptive Bandwidth Allocation: Impact of Traffic Demand Models for Wide Area Networks. In: 19th International Teletraffic Congress (ITC19), Beijing, China (2005)
9. Glover, F., Laguna, M.: Tabu Search. KLUWER ACADEMIC PUBL (1997)
10. Suri, S., Waldvogel, M., Warkhede, P.: Profile-Based Routing: A New Framework for MPLS Traffic Engineering. Quality of Future Internet Services, Lecture Notes in Computer Science **2156** (2001)
11. Wang, B., Su, X., Chen, C.: A New Bandwidth Guaranteed Routing Algorithm for MPLS Traffic Engineering. Proceedings of ICC **2** (2002) 1001–1005

Optical Packet Buffers with Active Queue Management

Assaf Shacham and Keren Bergman

Columbia University, Department of Electrical Engineering, New York, NY 10027
bergman@ee.columbia.edu,
http://lightwave.ee.columbia.edu

Abstract. Active queue management (AQM) is an important function in today's core routers that will be required in the future optical internet core. A recently reported novel architecture for optical packet buffers is extended by implementing necessary AQM functions. The suggested AQM scheme is validated and explore via simulations.

1 Introduction

One of the key challenges to the implementation of all-optical routers is the difficulty of realizing optical packet buffering. Contemporary internet routers use very large packet buffers, which store millions of packets, to efficiently utilize expensive long haul links. These large buffers are clearly impractical for implementation using photonic technology. Recent studies have indicated that by sacrificing some of the link utilization, buffer sizes can be reduced dramatically to the capacity of approximately 20 packets [3].

Photonic packet buffers of this capacity have the potential of realization. Numerous optical buffer architectures have been suggested (see, for example, Refs. [1, 2, 5], among others). However, many of the suggested architectures suffer from fundamental drawbacks that prohibit scaled implementations or make them unusable in optical routers. For example, some are designed to store a single packet and others require a complex and nonscalable control schemes. In previous work [8, 9] we have presented a new buffer architecture that is modular, scalable, extensible, and transparent and therefore provides significantly improved performance.

In this paper we extend this work by considering an active queue management (AQM) scheme which can be straightforwardly implemented on the optical buffer architecture. AQM is a technique used for congestion control in packet-switched routers. In a typical AQM technique, known as random early detect (RED) [4], packets are **deliberately dropped** even when the router's buffers are not completely full, to provide an early congestion notification (ECN) signal to the TCP terminals and the network endpoints.

In the scheme suggested here, AQM is employed to solve another problem: packet-loss may occur when a low-capacity buffer is operated under high load. While this may be a price that a designer is willing to pay in an optical router

I. Tomkos et al. (Eds.): ONDM 2007, LNCS 4534, pp. 329–337, 2007.

with small buffers [3], the buffer is operated in a nearly full state most of the time, thus incurring an unnecessarily high queueing latency on all packets. AQM can be used to reduce the queueing latency while increasing the packet loss rate (PLR) by a small amount.

This paper is organized as follows: In Section 2 the buffer architecture is reviewed and explained. The suggested AQM scheme and its mapping on the existing architecture are presented in Section 3. A simulation based exploration of the AQM parameters then follows in Section 4 and a concluding discussion is provided in Section 5.

2   Architecture Overview

The optical packet buffer is comprised of identical buffer building-block modules that are cascaded to form a complete buffer (see Fig. 1). Each building-block module has two input ports and two output ports and is capable of storing a single packet on a fiber delay line (FDL). A pair of ports (*Up-in* and *Up-out*) connects the module to the next module in the cascade, and the *Down-in* and *Down-out* ports are connected to the previous module. In the root module the *Down-in* and *Down-out* ports are used as the system input and output ports, respectively. Each module is also connected to the next module in the cascade by an electronic cable, for the transmission of *Read* signals.

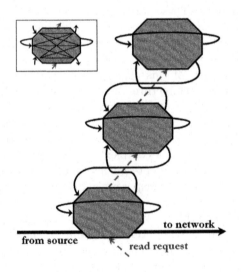

Fig. 1. The buffer's cascade structure (large image). A 3×3 non-blocking switch, of which a reduced version is used to implement the modules (inset) [9]

Writing packets to the buffer is performed implicitly: optical packets arrive into the *Down-in* port of the root module, aligned with system timeslots. The packets are either stored locally if the internal FDL buffer is empty, or routed to the *Up-out* port to be stored in the next module in the chain. Each packet is forwarded in this manner up the cascade, and is stored in the first empty module it encounters, which is necessarily the last position in the queue.

The read process is completely independent from the write process: when a *Read* signal is received, the locally stored packet is transmitted from the *Down-out* port and a *Read* signal is sent to the next module in the cascade to retrieve the next packet. With each *Read* signal all the packets in the chain move a step closer to the output port, while maintaining the packet sequence.

This distributed modular structure has several advantages: (1) no central management is required – all modules follow an identical simple set of rules; (2) the buffer capacity can be increased simply by connecting additional modules at the end of the cascade; (3) packet dropping, in the case of overflow, is cleanly executed by routing packets to the *Up-out* port of the last module, which is not connected.

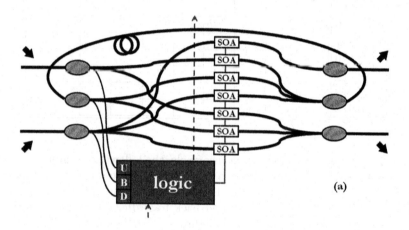

Fig. 2. Schematic diagram of the building-block module with ovals for couplers; *U*, *B*, and *D* for their respective low-speed receivers; dashed lines for read request signals [9]

2.1   Building-block module structure

Each building-block module in the buffer is implemented as an SOA-based 3×3 non-blocking optical switch (Fig. 2). At each of the three inputs (*Up-in*, *Down-in*, and *Buffer-in*), the packet's power is split by a coupler. A portion of the packet power is directed to a low-speed *p-i-n*-TIA optical receiver acting as an envelope detector. The envelopes of the packets from all input ports along with the input electronic *Read* signal are used in an electronic decision circuit to set the state of the SOA-based switch. The decision rule used by the electronic circuitry is represented by the truth table in Table 1. The table should be read as follows: the four left columns R,U,B,D are the *Read* signal, and the three inputs representing the existence of packet on the input ports (*Up-in, Buffer-in, Down-in*), respectively. The next nine columns represent the switching states of the SOAs (e.g. U2B means *Up-in* to *Buffer-out*, etc.) Finally, the last column (ER) represents when a *Read* signal is transmitted to the next module in the chain.

Table 1. Truth table for building-block module ($\emptyset$ represents a don't-care value)[8]

R	U	B	D	U2U	U2B	U2D	B2U	B2B	B2D	D2U	D2B	D2D	ER
$\emptyset$	0	0	0										
0	0	0	1							1			
0	0	1	0					1					
0	0	1	1					1		1			
0	1	0	0		1								
0	1	0	1		1					1			
$\emptyset$	1	1	$\emptyset$	\multicolumn{10}{c}{*illegal states*}									
1	0	0	1								1		
1	0	1	0						1				1
1	0	1	1						1		1		1
1	1	0	0		1								1
1	1	0	1		1					1			1

To assist in understanding the decision rule, the following examples are given: line {0011} represents a case where a new packet is received from the *Down-in* port ($D=1$), while another packet is locally stored in the buffer ($B=1$). In that case the locally stored packet is routed back to *Buffer-out* (B2B=1) while the new packet is routed to the next module through the *Up-out* port (D2U=1). Line {1010} represents the case where a *Read* signal is received ($R=1$) and a packet is locally stored ($B=1$). In this case, the local packet is sent to the *Down-out* port (B2D=1) and a *Read* signal is sent to the next module (ER=1) to retrieve the next packet in the queue. Careful inspection of Table 1 reveals that some states can never occur if all the modules follow the rules. More importantly, two paths (*Up-in* to *Up-out* and *Buffer-in* to *Up-out*) are never used so the module can be implemented using only seven SOAs, as appearing in Fig. 2.

3   Introducing AQM to the Buffer

As explained in the introduction, the suggested buffer architecture, as any other FDL-based optical buffer architecture, is practical for the implementation of small buffers, capable of storing tens of packets, at most. In an optical router, typically, a buffer would be designed to store as many as 20 packets [3]. In these systems, when routers face periods of high load and congestion, it is reasonable to assume that buffers will fill up very quickly and packet dropping due to buffer overflow will be frequent. While higher layer mechanisms, such as TCP, are used to recover lost packets and regulate the transmission rates of network terminals in these cases, these mechanisms operate with fairly slow time-constants. Thus, when overflow-generated packet-loss is used as the sole mean of congestion notification, buffers remain full and overflowing for long periods of time, and the penalty in terms of packet loss rate (PLR) and end-to-end latency can be severe. Further, in the case of buffer overflow many packets are dropped at the same time and thus many sources are requested to throttle their transmission at the same time. This effect leads to the well-known and undesirable effects of terminal-synchronization, TCP oscillations, and failure to reach equilibrium [6].

AQM can be used, in optical routers, to improve the overall network performance by dropping packets even before the buffers reach overflow. By dropping these packets, an early congestion notification (ECN) signal is sent to the end terminals, thus causing rate throttling before the buffers reach full capacity. Further, when probability-based methods are employed to decide when to drop packets, ECN messages are sent to different TCP terminals in different times, oscillations are prevented and the network is more likely to reach equilibrium [6, 4].

The distributed structure of the buffer facilitates a simple AQM implementation. The digital decision circuit in every building block module is modified by the addition of a circuit emulating a Bernoulli random number generator (BRNG), or a biased coin-flip. In every slot, the result of the BRNG (a true/false value) is checked. If the BRNG result is **true** while a packet is about to be buffered in the module, the packet is discarded. This policy is implemented by performing a logical *AND* operations of the individual control signals of three SOA gates (U2B, B2B, and D2B) with the negation of the BRNG value:

$$U2B_{NEW} = U2B \land \neg BRNG$$

$$B2B_{NEW} = B2B \land \neg BRNG$$

$$D2B_{NEW} = D2B \land \neg BRNG$$

When a packet is dropped, a read signal is emitted to the next module up the cascade, so that the module is replenished and the packets in the queue advance by one spot:

$$ER_{NEW} = (U2B \lor B2B \lor D2B) \land BRNG$$

The BRNG in each module is programmed with a different Bernoulli parameter which is the probability that a packet buffered in that module will be dropped. The parameter in the kth module (The 0th module is the root module) is $p_d(k)$, or the **packet-dropping probability** function. Hence, in a given slot time a packet which is buffered in the kth spot in the queue is discarded with a probability $p_d(k)$.

The packet-dropping probability function is a monotonously non-decreasing function in k and $p_d(0) = 0$. Thus, packets are never dropped in the root module and the probability that a packet is dropped rises in the modules that are further down the queue. Constructing specific $p_d(k)$ functions is the topic of Section 4.

The feasibility and correctness of the suggested distributed AQM scheme is validated using simulations on a specifically developed simulator, built using the OMNeT++ event-driven simulation environment [7]. OMNeT++ provides support for modular structures and message exchange and the simulator is highly parameterized and offers complete configuration flexibility to simulate buffers of varying sizes and varying internal parameters. The simulator is also used to explore the design parameters of the AQM probability function in the next section.

4   Exploring AQM Parameters

The buffer performance, i.e. PLR and latency, is highly dependent on the chosen dropping-probability function. As described in Section 3, the function $p_d(k)$ is a monotonously non-decreasing function in k and $p_d(0) = 0$. To perform an initial study of the optical buffer with AQM, and to explore different functions we define the function $p_{AQM}(k)$ as follows:

$$p_{AQM}(k) = \left\{ \begin{array}{cc} 0 & k \leq x_1 \\ \frac{y_2 \cdot (k - x_1)}{C - x_1} & k > x_1 \end{array} \right\}$$

where C is the total number of modules in the buffer (i.e., the buffer capacity) and x_1 and y_2 are parameters defining a given function. Fig. 3 clarifies the $p_{AQM}(k)$ function and its parameters: x_1 is the *threshold capacity* above which packets start to be actively dropped and y_2 is the maximum drop probability under AQM.

In this section we explore the effect of the two parameters: the threshold capacity (x_1) and the maximum drop probability (y_2), examine their effect on the PLR and latency, and compare them to the case where AQM is not used. In the simulations conducted, the arrival and service processes are both Bernoulli. The goal of the simulations is perform an initial evaluation of the suggested AQM scheme. The simulated case is, therefore, constructed such that the buffer is heavily loaded and the PLR due to overflow, in the absence of AQM, is fairly high (nearly 3%). To achieve these conditions, the arrival parameter is chosen to be $p = 0.50$ and the service parameter is $q = 0.52$. The buffer capacity is chosen to be 10 packets.

Obviously, to fully evaluate the performance of an AQM scheme, the systems should be simulated with traffic comprised of a large number of multiplexed TCP flows, such that the interaction of the AQM mechanism and the TCP sources can be studied. This work is planned but is beyond the scope of this paper. The work here studies the effect of the AQM function on the latency and PLR under simpler traffic models.

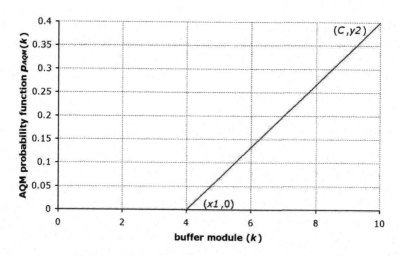

Fig. 3. The AQM dropping-probability function, defined by x_1 and y_2. In this case $x_1 = 4$, $y_2 = 0.4$ and the buffer capacity is $C = 10$.

4.1 Controlling AQM Maximum Dropping Probability

To evaluate the effect of the dropping-probability function, we first simulate different values of y_2, the maximum drop probability. This value can be seen as the *intensity* of the AQM scheme. When y_2 is increased the dropping probability due to AQM is increased and the dropping due to overflow is reduced. Simulation should be used, however, to verify that the AQM is not too strong such that it unnecessarily increases the total PLR. The results of the y_2 study are shown in Fig. 4. In these simulations, the threshold capacity value is $x_1 = 4$.

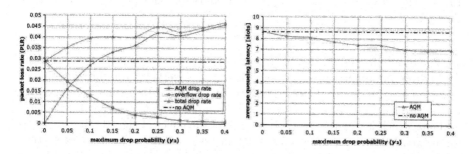

Fig. 4. The PLR and average latency of a 10-packet buffer with varying levels of maximum drop probability, y_2 ($x_1 = 4$).

The results in Fig. 4 show that with any degree of AQM applied to the buffer, the total PLR increases. As expected, as the AQM becomes more dominant, the PLR due to overflow is diminished. The negative effect of the higher PLR is accompanied by a favorable effect of lower queueing latency, resulting from the lower occupancy in the buffer due to the AQM. Another effect that should be considered is that we can view the AQM packet drops as *better* drops, because of the more gradual effect they have on the TCP sources at the network edges [6, 4].

4.2   Controlling the Threshold Capacity

The threshold capacity can be used as a mean of controlling the spread of the AQM function across the buffer. In this case, for a given maximum dropping-probability level, the AQM threshold and the slope of the AQM function are varied. When x_1 is low, the function becomes smoother: packet dropping starts at a low capacity and the dropping-probability function rises slowly. When, conversely, x_1 is high, the AQM does not have an effect until the buffer occupancy is high, and the dropping-probability function, then, rises sharply. The results of the x_1 value study are shown in Fig. 5. In these simulations, the maximum drop probability value is $y_2 = 0.20$.

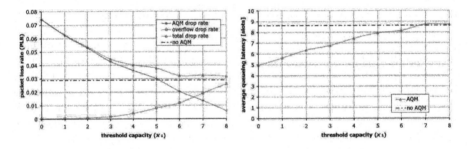

Fig. 5. The PLR and average latency of a 10-packet buffer with varying levels of threshold capacity, x_1 ($y_2 = 0.20$).

The results in Fig. 5 resemble the ones in Fig. 4. As x_1 moves down, the total AQM drop probability increases, suppressing the overflow drop rate, but making a contribution to the total PLR. The latency is reduced by a value corresponding to the rising PLR. When we attempt to compare the two scenarios, trying to evaluate which is a better method to control the effect of the AQM, we see that for a given PLR – the corresponding latency is equal in both scenarios, so there is no conclusively better method.

5   Conclusions and Future Work

In this paper we extend previous work on the design of a scalable optical packet buffer architecture by adding AQM capabilities. Several versions of AQM-RED are used in contemporary electronic internet routers to complement TCP and overcome some of its inherent problems. It is expected that the implementation of AQM will be required in the optical routers in the future internet core.

The presented buffer architecture, which is scalable, extensible and transparent, also lends itself to a simple implementation of AQM. This implementation was presented in this paper and was validated through extensive simulations. The design of the AQM dropping-probability function determines the performance of the AQM system. In this paper we present a simple function and explore its parameters using simulations. The simulations reveal that while the suggested AQM scheme does reduce latency, it does it at the expense of increasing the PLR.

In future work, the optical packet buffer with the suggested AQM scheme and other schemes will be simulated under a large number of multiplexed TCP streams. This complex modeling of the actual traffic and environment will facilitate the development of AQM schemes which are appropriate for the future optical routers in the internet core.

References

1. C.-S. Chang, Y.-T. Chen, and D.-S. Lee. Constructions of optical FIFO queues. *IEEE Trans. Inform. Theory*, 52(6):2838–2843, June 2006.

2. I. Chlamtac et al. CORD: Contention resolution by delay lines. *IEEE J. Select. Areas Commun.*, 14(5):1014–1029, June 1996.

3. M. Enachescu, Y. Ganjali, A. Goel, N. McKeown, and T. Roughgarden. Part iii: Routers with very small buffers. *SIGCOMM Comput. Commun. Rev.*, 35(3):83–90, 2005.

4. C. Hollot, V. Misra, D. Towsley, and W.-B. Gong. On designing improved controllers for AQM routers supporting TCP flows. In *20th Annual Joint Conference of the IEEE Computer and Communications Societies (INFOCOM 2001)*, volume 3, pages 1726–1734, Apr. 2001.

5. D. K. Hunter and I. Andonovic. Optical contention resolution and buffering module for ATM networks. *Electronic Letters*, 29(3):280–281, Feb. 1993.

6. C. Jin, D. X. Wei, and S. H. Low. Fast TCP: Motivation, architecture, algorithms, performance. In *20th Annual Joint Conference of the IEEE Computer and Communications Societies (INFOCOM 2001)*, volume 4, pages 2490–2501, Mar. 2004.

7. OMNeT++ discrete event simulation system. available online at http://www.omnetpp.org/.

8. A. Shacham, B. A. Small, and K. Bergman. A novel optical buffer architecture for optical packet switching routers. In *European Conference on Optical Communications (ECOC '06)*, Sept. 2006.

9. B. A. Small, A. Shacham, and K. Bergman. A modular, scalable, extensible, and transparent optical packet buffer. *J. Lightwave Technol.*, 25(4), Apr. 2007.

Segmentation-based Path Switching Mechanism for Reduced Data Losses in OBS Networks

Anna V. Manolova[1], Jakob Buron[1], Sarah Ruepp[1], Lars Dittmann[1], and Lars Ellegard[2]

[1] COM•DTU, Oersteds Plads, Building 345V, Technical University of Denmark, 2800 Kgs. Lyngby, Denmark, tel:+45 45256609
[2] Tellabs Denmark A/S, Lautrupvang 3C, 2750 Ballerup, Denmark
{avm, jbu, sr, ld}@com.dtu.dk
lars.ellegard@tellabs.com

Abstract. The Optical Burst Switching (OBS) technology emerged as an alternative switching paradigm for the optical transport layer. Its biggest disadvantage (high data losses) has been the focus of numerous research papers. This paper proposes a data loss reduction technique, which relies on the combination of global network coordination between network nodes and local contention resolution. Via full-scale network simulation it is demonstrated that the proposed scheme has improved performance in terms of data losses and resource utilization, compared to its constituent mechanisms alone. Furthermore, the effectiveness of the mechanism has proven to be less sensitive to load variations in the medium and high load ranges. Additionally, its complexity and deployment cost are low, due to the absence of immature optical components in the network.
Keywords: OBS, segmentation, load balancing, contention resolution

1 Introduction

Optical Burst Switching (OBS) is an emerging technology introduced approximately six years ago. In one of the first publications [1], the technology is described as a promising solution for the direct mapping of the bursty IP traffic over the optical transmission media (IP over WDM). There are numerous advantages of OBS over the traditional packet switching (OPS) and channel switching (OCS) paradigms. OBS uses medium granularity of the switching entities (bursts) and resides between OCS and OPC. Its specific way of operation results in bypassing the existing bottleneck in the electronic routers and poses more moderate requirements for optical buffering. OBS is highly dynamic, and achieves a great degree of statistical multiplexing. Additionally, the technology provides complete transparency of the switching node regarding the code, the format and the speed of the client traffic and the used upper-layer protocols by providing all-optical transport of the client data.

The basic idea of the OBS network is fairly simple [2]. At the edge of the OBS domain the client data is assembled into bursts depending on its destination.

I. Tomkos et al. (Eds.): ONDM 2007, LNCS 4534, pp. 338–347, 2007.

Each burst sent in the network is preceded by a special Burst Header Packet (BHP), which carries all needed routing and signaling information and uses a dedicated Control Channel. The burst is released in the network with or without waiting for an acknowledgment from the egress node. In the former case a one-way resource reservation is used, i.e. this OBS network is a "best-effort" type of network. The latter uses Tell-and-Wait (TAW) resource reservation [2], making the OBS network effectively connection-oriented. The majority of research work, as well as this paper, focuses on connectionless OBS networks. Despite its obvious advantages, the OBS technology is still unapplicable in real optical networks because it suffers from one major disadvantage: increased data losses. This paper discusses this issue and proposes an effective loss reduction technique called Segmentation-based Path Switching (SPS).

The rest of this paper is organized as follows. Section 2 offers an overview of the stated problem and summarizes the existing solutions found in the literature. Section 3 analyzes the performance of two contention resolution techniques and introduces the SPS mechanism. Section 4 provides numerical results obtained from simulations and their analysis. Section 5 concludes the paper.

2   Data Losses in OBS Networks

Contention refers to the situation, when two or more bursts request the same resource at the same time and is the main source of data losses in an OBS network. In connectionless OBS networks the contention issue is severe due to the the lack of optical memory and the applied one-way resource reservation technique, which does not guarantee burst delivery. In fact, the contention resolution problem involves many different aspects such as routing, signaling, channel scheduling, burst assembly and available hardware components.

There are basically two classes of contention resolution techniques: reactive and proactive. The reactive techniques aim at minimizing the losses after a contention has occurred. The main strategy is deflection, which can occur in time (buffering), space (deflection routing) or/and frequency (wavelength conversion) domains. Another strategy is the so called *soft contention resolution* policy. Such policies are Segmentation [3], look-ahead contention resolution and short-est/latest drop policies [4]. The proactive mechanisms try to avoid congestion in the network by applying smart traffic routing and balancing [5], admission control [6], and optimal burst assembly (e.g. composite burst assembly [3]).

In both proactive and reactive mechanisms the trade-off between complexity and efficiency is obvious. The contention resolution strategies are considered to be highly dependent on the traffic load [7], in particular the deflection routing strategy, which tends to decrease severely in performance when the network load increases. The proactive solutions involve feedback information distribution and/or sophisticated algorithms for smart routing and traffic shaping, which increase the complexity in the network.

Most of the research work done in this area focuses separately on prevention and resolution schemes. In this paper a combination of two mechanisms is offered. The operation of the mechanism is described in the next section and its performance is tested via the OPNET Modeler software [8].

3   Loss Reduction Techniques in OBS Networks

Even though the contention resolution mechanisms suffer from decreased performance at high loads they are an integral part of the overall effort to reduce the packet loss ratio in an OBS network. In order to overcome this drawback we propose to combine a reactive contention resolution scheme with a proactive scheme, which will address this problem. This scheme is called Segmentation-based Path Switching (SPS) and combines the Segmentation contention resolution (S) technique, proposed in [3], and the Congestion-Based Static-Route Calculation contention prevention (CP) scheme, proposed in [5]. S and CP are referred to as the constituent mechanisms of SPS. They are chosen because of their simplicity and because they do not require wavelength convertors or fiber delay lines. This makes the SPS mechanism low-cost and relatively simple.

3.1   Evaluation of the Constituent Loss Reduction Techniques

The Segmentation technique was first proposed in [9]. Its basic operation is as follows: when two bursts contend for resources, depending on the applied policy, only part of one of the bursts is dropped instead of dropping a whole burst. The authors of [9] offer to drop the head of the later-arriving burst to be dropped, whereas the authors of [3] suggest to drop the tail of the earlier-arriving burst. The technique has been further investigated in several research papers and has proven to be effective, but all results confirm the fact that at high loads the mechanism's performance degrades [3], [10], [11].

The CP mechanism, proposed in [5], needs a congestion situation (i.e. higher loads) in the network in order to be efficient. Each ingress node statically calculates two link disjoint paths towards each destination. Each core node periodically floods link-status advertisements, indicating whether the load on its outgoing links has passed a certain threshold. When there is a burst ready to be sent, the ingress node calculates the congestion level of the overall path and chooses the least congested path to the destination. There are two important parameters influencing the performance of this scheme: the threshold value, which indicates when a link is considered to be overloaded, and the observation period, over which the load status of a link is calculated. It must be noted that the performance of this CP technique is dependent on the network topology. In general, schemes which use adaptive routing techniques are dependent on the connectivity in the network and the overall topology. Nevertheless, the overall performance in terms of data losses is improved due to the load balancing achieved in the network. The biggest disadvantage of this scheme is that it does not support different classes of service.

Both mechanisms involve additional signaling in the network. In case Segmentation is applied, the downstream nodes are informed about the change in the burst length via update packets. The more segmentations there are in the network, the higher the signaling overhead will become. The CP mechanism is a feedback-based mechanism, which means that feedback information is constantly exchanged. When the network load increases it is obvious that feedback information will be exchanged more often.

3.2   Combined Segmentation-based Path Switching Technique

Our proposed loss reduction scheme combines the best of both described mechanisms: the contention prevention is responsible for better load balancing in the network, while the contention resolution reactively reduces the packet losses. Unlike the scheme proposed in [12], which also consists of global and local loss reduction strategies, the SPS scheme does not require fiber delay lines, wavelength convertors or complicated channel scheduling algorithms. This makes our strategy applicable in the near future and reduces the capital expenditure of deployment.

The complexity of SPS is distributed among two main modules: a local contention resolution module (LCRM) and a global contention prevention module (GCPM). LCRM executes the Segmentation and only has knowledge of the local node condition, whereas GCPM has a global overview of the network status which is stored in every node and is updated periodically. These two modules are completely independent, which increases the reliability of the mechanism. They have a different scope of work and a failure in one of them will not disturb the operation of the other. Each module is also related to a different plane of the OBS network. The GCPM operates on the control plane of the network by affecting the routing and the offset-time management, whereas the focus of the LCRM is on the data plane. This clear separation of both modules is compliant with the separation of the data and control planes of the OBS network and makes the mechanism fit properly in the overall OBS architecture (see Fig. 1).

The independence of both constituent mechanisms allows the SPS to fully exploit one significant advantage - support for different class differentiation techniques. Class differentiation can be done by using the built-in class support of the Segmentation technique [3]. It offers class differentiation based on prioritization of the traffic and by applying different segmentation policies. Another option is introducing differentiated class handling based on the CP scheme. For example only high priority class traffic can be switched to the secondary path in case of a congested primary one, or different link and node disjoint paths can be calculated separately for delay and for loss sensitive traffic. Since nothing prevents different combinations of class differentiating schemes, SPS turns into a highly flexible class-supporting mechanism. At this stage, the proposed SPS technique accommodates only the built-in Segmentation capabilities. The suggested CP-based methods are subject to further study.

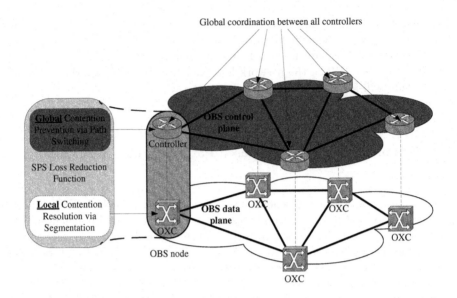

Fig. 1. Separation of the different modules in the SPS mechanism.

4   Numerical Results

In this section, the results obtained through OPNET simulations are presented. A model of an OBS node (see Fig. 2 (a)) was applied in the NSFNET network, shown in Fig. 2 (b). Each OBS node can serve as both Edge and Core node. There are four wavelengths for data transfer and one for the signaling channel on each optical fiber. No wavelength conversion, nor fiber delay lines are employed. Based on [3], the BHP processing time is chosen to be $2.5\mu s$, the cross-connect configuration time is $10\mu s$ and upon contention, the tail of a burst is segmented and dropped. Each data burst consists of 100 segments and has a fixed length of $100\mu s$. The burst arrival process has exponential distribution and the generated traffic is uniformly distributed among all sender-receiver pairs. Minimum Delay Shortest Path routing is used for path computation. Whenever Segmentation is applied, the minimum allowed burst length is chosen to be 10 segments. The threshold value for the CP part of the mechanism is 0.5 and the observation period (i.e. the period between updates for the status of a link) is 5 ms. Each data point is obtained after 30 seconds of simulation.

The performance of the SPS mechanism and its constituent mechanisms (Segmentation and CP) will be compared to a Basic Case (BC), in which no loss saving techniques are used. The main performance measures are the Packet Loss Ratio (PLR), the End-to-End Delay (E2ED), the Resource Utilization (RU) and the Improvement Factor (IF) of the mechanisms. In the following, PLR will be used instead of BLR, because of the ambiguity of BLR when some packets are segmented out of a burst during transmission. The presented results will not in-

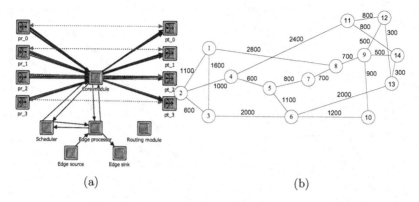

(a)   (b)

Fig. 2. (a) OBS node model and (b) NSFNET with 14 nodes and 21 bidirectional links (numbers indicate distance in km).

clude class differentiation because the scope of this paper is limited to the general SPS performance compared to the performance of the constituent mechanisms. Due to the same reason, none of the earlier-specified mechanisms' parameters are be varied during simulations. It is clear that the performance of all of the mechanisms is dependent on these values, but using constant values will give a common ground for an objective and clear evaluation and comparison.

4.1   Packet Loss Ratio

The PLR values for all tested mechanisms and for the BC can be observed in Fig. 3. It is clearly visible that our proposed mechanism achieves the lowest PLR for medium and high loads. It is interesting to observe that the PLR characteristic of the SPS mechanism resembles the one for the CP scheme. At low loads the CP mechanism causes insignificant change in the PLR value, compared to the BC and for some loads its performance is even worse. This is due to the fact that the alternative link-disjoint paths are almost always significantly longer than the primary ones, which increases the probability for contention along the paths. In the middle load range, the operation of the mechanism stabilizes and the benefit of the load balancing becomes visible. This effect can also be observed for the SPS mechanism, which means that at low loads the combined SPS approach is not as good as the Segmentation one. It must be mentioned that the PLR for the SPS mechanism is strongly influenced by the values of the threshold and the observation periods of the GCPM part of the mechanism and the applied Segmentation policy for the LCRM part.

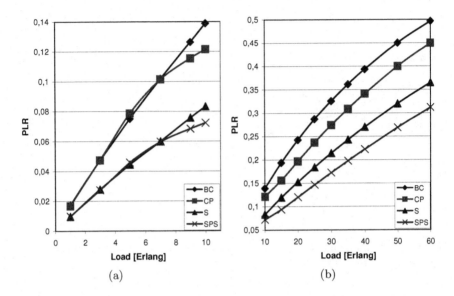

Fig. 3. PLR for low (a) and high (b) loads for all tested loss reduction mechanisms.

4.2   End-to-End Delay and Resource Utilization

Applying path switching will inevitably result in increased E2ED value and
increased RU. A high RU value does not necessarily indicate improved network
performance, since a dropped data burst close to the receiver will increase the
RU. Figure 4 shows the E2ED and the RU values for all tested cases throughout
the entire tested load range. The first noticeable result is the increased E2ED for
all applied loss reduction mechanisms. There are two contributing components:
the longer alternative paths, used in the CP and the SPS mechanisms, and
the fact that bursts, which need to pass more nodes along their way to the
destination, have a higher chance of surviving in the cases where Segmentation
is applied. This results in the highest E2E Delay for the SPS mechanism. The
higher value of the RU for the Segmentation and the SPS mechanisms indicate
more saved bursts in the network, compared to the CP and the BC (see Fig. 3).
From the figure it can be concluded that the improved PLR parameter for the
SPS mechanism does not come at the expense of increased resource utilization.
In fact, for the middle load range the RU parameter is slightly lower (about
1%). This, combined with the lower PLR value, indicates more proper resource
utilization.

4.3   Improvement Factors of the Investigated Mechanisms

The IF parameter, calculated according to (1), is a relative performance measure,
and indicates a given mechanism's improvement or reduction in PLR compared

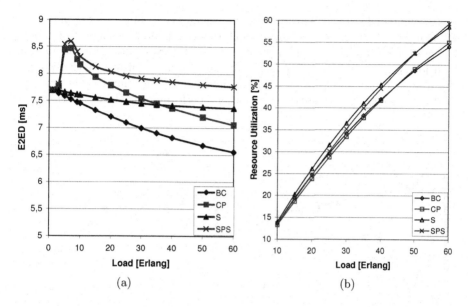

Fig. 4. End-to-End Delay (a) and Resource Utilization (b) vs. Load for all simulated mechanisms.

to the Basic Case.

$$IF_i = \frac{PLR_{BC} - PLR_i}{PLR_{BC}} * 100 \ [\%], \tag{1}$$

IF_i – the improvement factor of a given mechanism i; $i \in \{$Segmentation, CP, SPS$\}$.

The IF for all three loss saving mechanisms is presented in Fig. 5 (a). From the figure it can be seen that for medium loads the CP mechanism has increasing performance factor and for the low and very high loads the performance of the mechanism degrades. This also applies to the SPS technique. Assuming the improvement, achieved by the SPS mechanism, is the maximum gain that can be achieved, it can be seen that the Segmentation technique can attain more than 70% of this maximum gain. If the increased complexity in the network does not justify the remaining 30% gain of the improvement factor, then the Segmentation technique is the best choice. However, according to Fig. 4 (b), the SPS mechanism has more effective resource utilization and if this is the main performance measure, the SPS mechanism should be preferred. The last observation is that the performance of all mechanisms degrade at very high loads. From the results presented in Fig. 5 (a), it can be concluded that the SPS mechanism is the most appropriate for medium and high loads, and for the low load range, one should consider Segmentation only.

Since the SPS mechanism is a combination of two mechanisms, one can expect that adding up the improvement factors achieved by the component mechanisms separately, should result in a higher improvement factor than the one achieved by the combined scheme. This comes from the fact that some bursts can be saved with either of the two mechanisms, which reduces the improvement from adding the second mechanism. Figure 5 (b) illustrates this expectation for medium and high loads. At very high loads the IF of the SPS mechanism is actually higher than the sum of the IFs for the component schemes, meaning that at very high loads both mechanisms reinforce each other's performance. The last performance

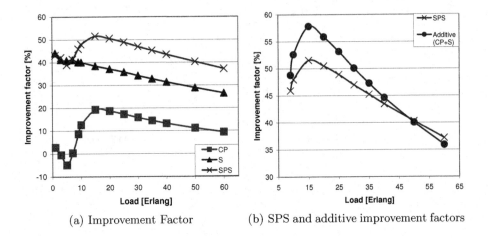

(a) Improvement Factor   (b) SPS and additive improvement factors

Fig. 5. Improvement factors vs. Load.

evaluation technique is the comparison of the speed of degradation of the mechanisms. Applying a simple linear approximation to the PLR graphs (Fig. 3 (b)) and calculating the inclination angle of the approximation gives the speed of the PLR increase, which is an indication of the speed of the degradation of the mechanisms' performance. The obtained results for the inclination angles are:

- 33.17° for the CP technique,
- 29.41° for the Segmentation technique,
- 25.63° for the SPS technique.

It can be seen, that the SPS mechanism has the smallest value of the inclination angle and thus the slowest speed of degradation, whereas the CP and the Segmentation mechanisms degrade faster. This is a very strong advantage of the SPS mechanism, as it indicates that the mechanism will be more stable under dynamic load conditions, compared to its component mechanisms alone.

5   Conclusion

In this paper a combined loss reduction strategy for OBS networks, called SPS, is proposed. The mechanism combines a local contention resolution mechanism

(Segmentation) and a global contention prevention mechanism (Congestion-Based Static-Route Calculation). Its performance is evaluated in realistic IP-over-WDM full-scale network scenario. The technique overcomes the disadvantages of its constituent mechanisms and leverages their strengths at the expense of slightly increased signalling in the network. Our approach does not require wavelength conversion, fiber delay lines for buffering or complex channel scheduling algorithms and is hence more eligible for deployment in the near future. SPS shows improved PLR, better utilization of resources for medium and high loads, and offers great flexibility for differentiated traffic handling. At low loads the achieved improvement does not justify the increased complexity in the network and the Segmentation strategy alone proves to be more efficient. In addition, the mechanism proved to be less susceptible to traffic load variations for medium and high loads, which makes it suitable for dynamic traffic patterns. The conducted simulations reveal that the SPS mechanism is most appropriate to be used at the middle load ranges, where the effectiveness of the mechanism is the highest and the resource utilization is optimal.

References

1. Qiao, C., Yoo, M.: Optical Burst Switching (OBS) - A New Paradigm for an Optical Internet. J. High Speed Networks, Vol. 8, No. 1, pp. 69 − 84, 1999.
2. Jue, J.P., Vokkarane, M.V.: Optical Burst Switched Networks. Springer Science + Business Media Inc., 2005.
3. Vokkarane, M.V., Jue, J.P.: Prioritized Burst Segmentation and Composite Burst-Assembly Techniques for QoS Support in Optcal Burst-Switched Networks. IEEE Journal on Selected Areas in Communications, Vol. 21, No. 7, September 2003.
4. Farahmand, F., Jue, J.P.: Look-ahead Window Contention Resolution in Optical Burst Switched Networks. IEEE Workshop on HPSR, Italy, pp. 147 − 151, 2003.
5. Thodime, G., Vokkarane, M.V., Jue, J.P.: Dynamic Congestion-Based Load Balanced Routing in Optical Burst-Switched Networks. In Proc. of Globecom 2003, pp. 2628 − 2632.
6. Maach, A., Bochmann, G.V., Mouftah, H.: Congestion Control and Contention Elimination in Optical Burst Switching. Telecommunication Systems Journal, Vol. 27, No. 2, pp. 115 − 131, October 2004.
7. Gauger, C.M., Kohn, M., Schafr, J.: Comparison of Contention Resolution Strategies in OBS Network Scenarios. ICTON 2004, Wroclaw, Poland, July 2004.
8. OPNET Modeler Educational Version 11.5, http://www.opnet.com
9. Detti, A., Vramo, E., Listanti, M.: Performance Evaluation of a New Technique for IP Support in a WDM Optical network: Optical composite Burst Switching (OCBS). J. Lightwave Technol., Vol. 20, No. 2, February 2002.
10. Anpeng, H., Linzhen, X.: A Novel Segmentation and Feedback Model for Resolving Contention in Optical Burst Switching. Photonic Network Communications, 6 : 1, 61 − 67, 2003.
11. Zalesky, A. et al.: OBS Contention Resolution Performance. Performance Evaluation (2006), doi:10.1016/j.peva.2006.06.002
12. Wen, H., Song, H., Li, L., Wang, S.: Load Balancing Contention Resolution in OBS Networks Based on GMPLS. J. High Performance Computing and Networking, Vol. 3, No. 1, 2005.

Towards Efficient Optical Burst-Switched Networks without All-Optical Wavelength Converters

João Pedro[1,2], Paulo Monteiro[1,3], and João Pires[2]

[1] D1 Research, Siemens Networks S.A., R. Irmãos Siemens 1, 2720-093 Amadora, Portugal
{joao.pedro, paulo.monteiro}@siemens.com
[2] Instituto de Telecomunicações, Instituto Superior Técnico, Av. Rovisco Pais 1,
1049-001 Lisboa, Portugal
jpires@lx.it.pt
[3] Instituto de Telecomunicações, Universidade de Aveiro, Campus Universitário de
Santiago, 3810-193 Aveiro, Portugal

Abstract. Optical Burst Switching (OBS) is a promising switching paradigm to efficiently support Internet Protocol (IP) packets over optical networks, under current and foreseeable limitations of optical technology. The prospects of OBS networks would greatly benefit, in terms of cost and ease of implementation, from limiting the wavelength conversion capabilities at the network nodes. This paper presents a framework for contention minimization/resolution combining traffic engineering in the wavelength domain to minimize contention in advance and optical buffering at the core nodes to resolve contention. Simulation results show that with the proposed contention minimization/resolution framework the large number of expensive all-optical wavelength converters used at the core nodes of an OBS network can be replaced by a moderate number of shared optical delay lines without compromising network performance.

Keywords: optical burst switching, traffic engineering, wavelength contention minimization, wavelength continuity constraint.

1 Introduction

The Optical Burst Switching paradigm [1] has attracted considerable interest as an optical networking architecture for efficiently supporting IP packet traffic, while exploiting the huge transmission capacity provided by optical fibres and Wavelength Division Multiplexing (WDM) technology [2]. OBS bandwidth utilization efficiency and technological requirements are in between those of coarse-grained Optical Circuit Switching (OCS) and fine-grained Optical Packet Switching (OCS). At the OBS edge nodes, multiple IP packets are assembled into bursts, which are the traffic units routed and switched inside the OBS core network. Thus, OBS provides sub-wavelength granularity, rendering higher bandwidth utilization efficiency than OCS in supporting IP traffic. Moreover, bandwidth for data burst transmission is reserved in advance and using out-of-band signalling, avoiding complex optical processing capabilities at the core nodes, which are mandatory in OPS networks.

I. Tomkos et al. (Eds.): ONDM 2007, LNCS 4534, pp. 348–357, 2007.

OBS networks use one-way resource reservation mechanisms for setting up the resources for each burst transmission [1]. Therefore, the burst's ingress node starts transmitting soon after the burst has been assembled, instead of waiting for an acknowledgment of successful resource reservation in the entire burst path. Hence, two or more bursts may contend for the same resources at a core node. Given that unsolved contention leads to burst loss, degrading the network performance, it is clear that efficient contention resolution strategies are of paramount importance in these networks. Wavelength conversion was shown to be the most effective contention resolution strategy for OBS/OPS networks [3] and, as a result, most OBS proposals and studies assume the use of full-range wavelength converters at the network core nodes to resolve contention. However, because all-optical wavelength conversion devices are still undergoing research and development [4], they remain complex and expensive. Consequently, avoiding wavelength conversion would greatly reduce the complexity/cost of implementing OBS networks in the near future.

Recently, strategies for minimizing wavelength contention in advance have been investigated [5], [6], and even tested in an OBS network demonstrator [7], with the aim of reducing burst loss in the absence of wavelength conversion and optical buffering. Still, despite the reported performance improvements, an OBS network in these conditions only achieves low burst blocking probabilities at the expense of very small offered traffic loads [6], thus becoming a less attractive networking solution.

In this paper, we propose a novel framework for contention minimization and resolution in OBS networks without all-optical wavelength converters at the network nodes. Simulation results show that the proposed framework enables the design of efficient OBS networks employing a moderate number of shared Fiber Delay Line (FDL) buffers instead of a large number of expensive wavelength converters.

The remainder of the paper is organized as follows. Section 2 outlines the OBS contention resolution strategies and corresponding node architectures. A contention minimization/resolution framework tailored for OBS networks without wavelength converters is described in section 3, whereas section 4 assesses the performance of this framework and compares its network resource requirements with those of a network with wavelength conversion, for the same objective performance. Finally, section 5 presents some concluding remarks.

2 OBS Contention Resolution and Node Architecture

The performance of OBS networks is mainly hampered by contention at the network nodes, which arises whenever two or more data bursts, overlapping in time, are directed to the same wavelength on the same output fibre. Contention can be resolved using strategies acting in one or several of three domains: wavelength, time, and space. Thus, contending bursts can be converted to other wavelengths available at the output fibre, delayed using optical buffers, or deflected to other output fibres/links of the node. Wavelength conversion requires all-optical wavelength converters, whereas optical buffering demands FDLs, which unlike Random-Access Memory (RAM) only provide fixed delays. Deflection routing does not require additional hardware, but its effectiveness is heavily dependent on network topology and traffic pattern.

The complexity of the OBS network nodes depends on the strategies employed for contention resolution in the wavelength and time domains. Fig. 1 (a) shows the key blocks of the simplest node architecture, comprising a fast space switch matrix with a number of transit ports given by two times MW, where M is the number of input and output fibres and W is the number of wavelengths per fibre. Fig. 1 (b) depicts the most common OBS node architecture, which includes a set of MW dedicated wavelength converters, one per output transit port. It has been shown in [8] that significant savings in the number of converters can be achieved, without noticeable performance degradation, by sharing the converters among all output transit ports, as illustrated in Fig. 1 (c). However, this is realized at the expense of increasing the size of the space switch matrix by adding $2C$ transit ports, where C is the number of shared converters. Fig. 1 (d) depicts a node using F shared FDL buffers for resolving contention in the time domain, which also requires a larger space switch matrix. Node architectures combining wavelength converters with FDL buffers can also be devised [3], [8].

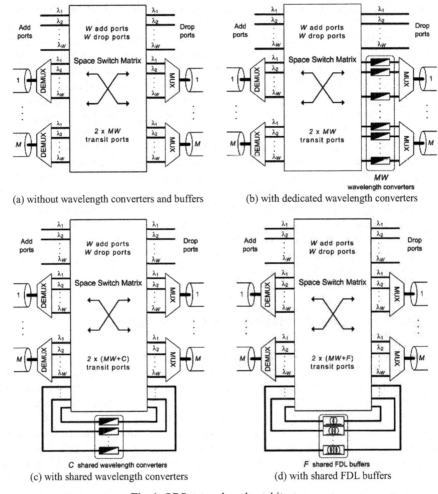

(a) without wavelength converters and buffers   (b) with dedicated wavelength converters

(c) with shared wavelength converters   (d) with shared FDL buffers

Fig. 1. OBS network node architectures.

In view of the above simplified node architectures, the complexity and cost of the OBS network depends of: (i) the switch matrices size, (ii) the number of wavelength converters, and (iii) the number of FDL buffers. Moreover, since a FDL buffer is basically an optical fibre with a length designed to provide a specific delay, it is simpler than an all-optical wavelength converter. This fact motivated us to investigate the feasibility of an OBS network using only shared FDL buffers to resolve contention. In the following, we describe a framework for contention minimization and resolution tailored for networks based on this architecture and show evidence that, using a moderate number of FDLs, it enables the same bandwidth utilization efficiency of an OBS network using dedicated full-range wavelength converters.

3 Contention Minimization and Resolution Framework for OBS Networks under the Wavelength Continuity Constraint

In all-optical networks, the absence of wavelength conversion at the network nodes imposes the wavelength continuity constraint on the data path, that is, the same wavelength must be used in all links of the path, from the ingress node to the egress node. The resulting wavelength assignment problem has been extensively studied in the context of OCS networks [9], where wavelength availability in the entire path is known and exploited by the wavelength assignment strategy. However, due to the one-way nature of resource reservation in OBS networks, the ingress node knows the wavelengths availability on its output links, but it is not aware of their availability on the downstream links of the burst path. Thus, bursts going through overlapping paths can be assigned the same wavelength by their ingress nodes, resulting in contention for that wavelength at some common link. The probability of wavelength contention can be minimized in advance by exploiting the following principle [5]: if two or more burst paths share one or more network links, contention on those links will be reduced if each burst path preferably uses wavelengths different from those preferred by the other overlapping burst paths. In practice, this is achieved by maintaining at the burst path's ingress node an optimized priority-based ordering of the wavelengths and using it to search for an available wavelength (on its output fibre link) for transmitting data bursts towards the egress node.

The work in [5] introduced a strategy for optimizing the wavelength orderings using information of network and traffic conditions that usually remain unchanged over relatively long time scales, such as network topology, routing paths, and average offered traffic load between nodes. Recently, we proposed a new strategy [6] that uses the same input information and was shown to be significantly more effective in minimizing wavelength contention than that of [5]. In the following, we describe a strategy simpler than that of [6], but that was found to attain similar performance.

Consider an OBS network with N nodes, L unidirectional links, and W wavelengths per link. Let Π denote the set of paths used to transmit bursts, and let E_i denote the set of links traversed by path $\pi_i \in \Pi$. Let also γ_i denote the average traffic load offered to burst path π_i. The extent of wavelength contention from bursts of a path π_i on bursts of a path π_j is expected to increase with both the average traffic load offered to the paths and the number of common links. Define the interference level of π_i on π_j as

$$I(\pi_i, \pi_j) = \gamma_i \mid E_i \cap E_j \mid, \; i \neq j, \tag{1}$$

where $|E_i \cap E_j|$ denotes the number of links shared by both paths.

Let $1 \leq P(\pi_i, \lambda_j) \leq W$ denote the priority of wavelength λ_j on path π_i, that is, the ingress node of π_i only assigns λ_j to a burst on this path if the first $W - P(\pi_i, \lambda_j)$ wavelengths, ordered by decreasing priority, are not available. Based on the problem inputs, the strategy determines the interference level of each path on every other path and reorders the paths of Π such that if $i < j$ the following condition holds

$$\sum_{\pi_k \in \Pi} I(\pi_i, \pi_k) > \sum_{\pi_k \in \Pi} I(\pi_j, \pi_k) \; \text{or} \; \sum_{\pi_k \in \Pi} I(\pi_i, \pi_k) = \sum_{\pi_k \in \Pi} I(\pi_j, \pi_k) \, \text{and} \mid E_i \mid > \mid E_j \mid . \tag{2}$$

Initially, no priorities are assigned to the wavelengths, that is, $P(\pi_i, \lambda_j)=0$ for all π_i and λ_j. The following steps are executed for every priority $1 \leq p \leq W$ in decreasing order, and for every ordered path $\pi_i \in \Pi$:

(S1)   Let $\Lambda = \left\{ \lambda_j : P(\pi_i, \lambda_j) = 0, \; 1 \leq j \leq W \right\}$ denote the initial set of candidate wavelengths containing all wavelengths that were not assigned a priority on π_i. If $|\Lambda|=1$ go to **(S7)**.

(S2)   Let $P_\Lambda = \left\{ p_k : \exists \pi_l, l \neq i, P(\pi_l, \lambda_j) = p_k, |E_l \cap E_i| > 0, \lambda_j \in \Lambda \right\}$ denote the set of priorities already assigned to candidate wavelengths on paths that overlap with π_i.

(S3)   Let $\rho = \min_{\lambda_j \in \Lambda} \left\{ \max \{ P(\pi_l, \lambda_j) : l \neq i, |E_l \cap E_i| > 0 \} \right\}$ be the lowest priority from the set of the highest priorities assigned to candidate wavelengths on the paths that use links of π_i. Update the set of candidate wavelengths as follows

$$\Lambda \leftarrow \Lambda \setminus \left\{ \lambda_j : \exists \pi_l, l \neq i, P(\pi_l, \lambda_j) > \rho, |E_l \cap E_i| > 0, \lambda_j \in \Lambda \right\}. \tag{3}$$

If $|\Lambda|=1$ go to **(S7)**.

(S4)   Let $C(e_m, \lambda_j) = \sum \left\{ \gamma_l : E_l \supset e_m, |E_l \cap E_i| > 0, P(\pi_l, \lambda_j) = \rho \right\}$ be the cost associated with wavelength λ_j on link $e_m \in E_i$. Thus, the minimum cost among the highest costs associated with the candidate wavelengths on the links of path π_i is $\alpha_e = \min_{\lambda_j \in \Lambda} \left\{ \max \{ C(e_m, \lambda_j) : e_m \in E_i \} \right\}$. Update the set of candidate wavelengths as follows

$$\Lambda \leftarrow \Lambda \setminus \left\{ \lambda_j : \exists e_m, C(e_m, \lambda_j) > \alpha_e, e_m \in E_i, \lambda_j \in \Lambda \right\}. \tag{4}$$

If $|\Lambda|=1$ go to **(S7)**.

(S5)   Let $C(\pi_i, \lambda_j) = \sum_{e_m \in E_i} C(e_m, \lambda_j)$ be the cost associated with wavelength λ_j on π_i. Thus, $\alpha_\pi = \min_{\lambda_j \in \Lambda} C(\pi_i, \lambda_j)$ is the minimum cost among the costs associated with the candidate wavelengths on π_i. Update the set of candidate wavelengths as follows

$$\Lambda \leftarrow \Lambda \setminus \left\{ \lambda_j : C(\pi_i, \lambda_j) > \alpha_\pi, \lambda_j \in \Lambda \right\}. \tag{5}$$

If $|\Lambda|=1$ go to **(S7)**.

(S6)   Update the set of priorities assigned to the candidate wavelengths as follows

$$P_\Lambda \leftarrow P_\Lambda \setminus \{p_k : p_k \geq \rho, p_k \in P_\Lambda\}. \tag{6}$$

If $|P_\Lambda| > 0$ go to (S3). Else, randomly select a candidate wavelength $\lambda_j \in \Lambda$.

(S7)   Assign priority p to the candidate wavelength $\lambda_j \in \Lambda$ on path π_i, that is

$$P(\pi_i, \lambda_j) = p. \tag{7}$$

Each execution of the above steps, selects the wavelength that will be used with priority p on path π_i. Therefore, these steps are executed $W \cdot |\Pi|$ times to determine all wavelength orderings. The highest priority wavelengths are the first to be selected and preference in wavelength selection is given to the burst paths with larger interference level over the other paths. The wavelength selection is made by first defining a set of candidate wavelengths, and then successively reducing their number until there is only one. The criterions used to reduce this set size are such that the candidate wavelength selected minimizes the *priority interference* on the links of the path. In other words, the wavelength selected is one that has preferably been assigned low priorities on paths that share links with the burst path π_i. In the following, this greedy strategy is named Heuristic Minimum Priority Interference (HMPI).

Upon assembling a data burst, directed to a given egress node, the ingress node will search for an available wavelength using the wavelength ordering optimized for the corresponding path. Since a burst transmitted over a high priority wavelength will have greater chances of avoiding wavelength contention, we use a proper scheduling algorithm at the ingress node to increase the amount of traffic carried on these wavelengths. The ingress burst scheduling algorithm is allowed to delay assembled bursts at the electronic buffer of the ingress node by any amount of time bounded to $D_{ingress}$. Hence, the algorithm will first try to assign the highest priority wavelength to the burst if this wavelength is available during the required time for burst transmission starting at some time in the future not exceeding the additional $D_{ingress}$ delay. If this wavelength is still not available, the algorithm will proceed in the same manner for the second highest priority wavelength and so on, until either an available wavelength is found or all wavelengths have been searched.

In view of the wavelength continuity constraint, contention at a transit node can only be resolved by delaying contending bursts at the shared FDL buffers. Several FDL buffer configurations have been presented [3]. Here, we adopt a pragmatic approach to optical buffering. Thus, only single-wavelength FDL buffers are used, which cannot accommodate bursts overlapping in time using different wavelengths, but avoid a larger switch fabric and extra multiplexing and demultiplexing equipment required by multi-wavelength buffers. Moreover, since a larger delay demands either a longer fibre or buffer recirculation, increasing optical signal degradation and the amount of extra hardware for compensating it, we impose a lower bound D_{min} and an upper bound D_{max} on the FDL delay, irrespective of the number of buffers per node, and inhibit buffer recirculation. Hence, in a node with F shared FDL buffers, the FDLs delay is uniformly distributed between D_{min} and D_{max}. Contention is resolved by delaying the contending burst using the smallest available FDL buffer that enables the burst to be transmitted through the next link of its path using the same wavelength.

4   Results and Discussion

The performance of OBS networks using dedicated wavelength converters, shared
wavelength converters, and shared FDL buffers, is evaluated in this section using
network simulation [10]. A 10-node ring topology with a uniform traffic pattern is
used in the performance studies. Moreover, the network employs Just Enough Time
(JET) resource reservation [1] and bursts are routed through the shortest paths. In all
cases, each of the L=20 unidirectional links has W=32 wavelengths, a wavelength
capacity of B=10 Gb/s, a switch fabric configuration time of 10 μs, and an average
burst size of 100 kB, which gives an average burst duration of T=80 μs. A negative
exponential distribution is used for both burst size and burst interarrival time. The
average offered traffic load normalized to the network capacity is given by

$$\Gamma = \frac{\sum_{\pi_i \in \Pi} \gamma_i \cdot |E_i|}{L \cdot W \cdot B}. \tag{8}$$

The OBS ring network using shared wavelength converters (SWC) or shared FDL
buffers (SFB) is designed to match the bandwidth utilization efficiency of using
dedicated wavelength converters (DWC). This is achieved by determining the number
of shared converters or FDL buffers required to attain the same average burst
blocking probability at the expense of the same average offered traffic load than that
of using dedicated converters. In this case, we set the objective average burst blocking
probability to 10^{-3} and 10^{-4}, which is obtained with dedicated wavelength converters
for an average offered traffic load of Γ=0.40 and Γ=0.47, respectively.

Fig. 2 and Fig. 3 plot the average burst blocking probability of the 10-node ring
network as a function of the number of shared wavelength converters and the number
of shared FDL buffers, respectively. In the latter case, the minimum and maximum
FDL delays are set to $D_{min}=T$ and $D_{max}=10T$. The average burst blocking probabilities
were obtained by simulating 20 independent burst traces for Γ=0.40 and Γ=0.47.

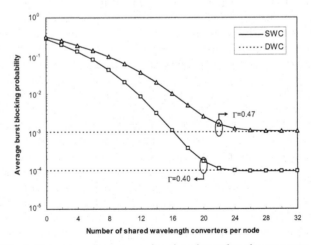

Fig. 2. Network performance using shared wavelength converters.

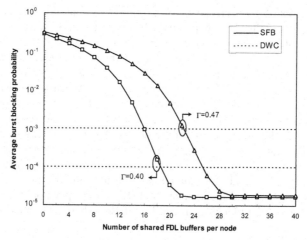

Fig. 3. Network performance using shared FDL buffers.

The curves in these plots show that as the number of shared wavelength converters or shared FDL buffers is increased, the average burst blocking probability is reduced and eventually the bandwidth utilization efficiency of using dedicated wavelength converters is achieved. The results also suggest that larger numbers of either shared converters or FDL buffers are needed when the objective burst blocking probability is higher. Moreover, in both cases the curves tend to a lower boundary of burst loss performance. In the SWC node architecture, this boundary corresponds exactly to the performance of DWC, whereas in the SFB node architecture the lower boundary is a result of imposing a maximum FDL delay irrespective of the number of FDL buffers.

Fig. 4 plots the performance of the 10-node ring network using shared FDL buffers and the HMPI strategy to optimize the wavelength orderings of each burst path. The minimum and maximum FDL delays are also set to $D_{min}=T$ and $D_{max}=10T$, whereas the maximum ingress delay is given by $D_{ingress}=T$.

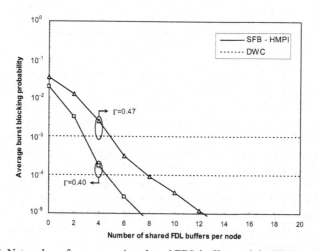

Fig. 4. Network performance using shared FDL buffers and the HMPI strategy.

These curves exhibit a decrease of several orders of magnitude in terms of average burst blocking probability, as compared to the ones of Fig. 3. More to the point, they show a remarkable decrease on the number of shared FDL buffers required to match the bandwidth utilization efficiency of using dedicated wavelength converters. For instance, for an objective average burst blocking probability of 10^{-4} the SFB node architecture needs around 19 shared FDL buffers, whereas with the use of wavelength orderings optimized with the HMPI strategy, this number is reduced to only 5. Thus, buffer requirements are reduced by almost a factor of 4. The reported FDL buffer savings are due to the effectiveness of the priority-based wavelength assignment and burst scheduling strategy in minimizing contention in the wavelength domain.

In order to gain further insight on the complexity of each of the node architectures under study, Fig. 5 presents the number of transit switching ports, the number of wavelength converters, and the number of FDL buffers per node that is needed to achieve an average burst blocking probability of 10^{-4} when the average offered traffic load is 0.40.

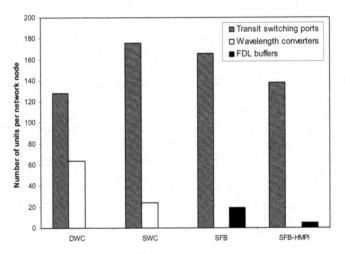

Fig. 5. Node resource requirements for an average burst blocking probability of 10^{-4} and an average offered traffic load of 0.40.

The node resource requirements plotted in Fig. 5 show that the use of dedicated wavelength converters requires a very large number of converters, although with the benefit of avoiding an increase on the number of transit switching ports. On the other hand, sharing the wavelength converters allows reducing their number to less than half, but demands a significant increase on the number of transit switching ports. The node architecture with shared FDL buffers needs slightly less transit switching ports and successfully replaces the expensive wavelength converters by simpler FDL buffers. However, the best compromise between the expansion of the space switch matrix and the number of either wavelength converters or FDL buffers is clearly given by using the priority-based wavelength assignment and burst scheduling strategy for minimizing wavelength contention combined with shared FDL buffers to resolve contention in the time domain.

5 Conclusions

In view of the current limitations of all-optical wavelength conversion technology, OBS networks would greatly benefit, in terms of cost and ease of implementation, from avoiding the use of wavelength converters at the network nodes. This paper has presented a framework from contention minimization/resolution combining traffic engineering in the wavelength domain to minimize contention in advance and optical buffering at the core nodes to resolve contention. Moreover, a performance study with node architectures that use either dedicated wavelength converters, shared wavelength converters, or shared FDL buffers to resolve contention was presented. Simulation results show that the proposed contention minimization/resolution framework only requires a moderate number of shared FDL buffers and a small increase in the space switch matrix to achieve the same bandwidth utilization efficiency of a network using dedicated full-range wavelength converters. Therefore, it can contribute to lower the complexity/cost of deploying OBS networks in the near future.

Acknowledgments. The authors acknowledge the financial support from Siemens Networks S.A. and Fundação para a Ciência e a Tecnologia (FCT), Portugal, through research grant SFRH/BDE/15584/2006.

References

1. C. Qiao and M. Yoo, "Optical Burst Switching (OBS) – A new paradigm for an optical Internet", *Journal of High Speed Networks*, vol. 8, no. 1, pp. 69-84, January 1999.
2. R. Ramaswami and K. Sivarajan, *Optical Networks: A Practical Perspective*, 2nd edition. San Francisco, CA: Morgan Kaufmann, 2002.
3. S. Yao, B. Mukherjee, S. Yoo and S. Dixit, "A unified study of contention-resolution schemes in optical packet-switched networks", *IEEE Journal of Lightwave Technology*, vol. 21, no. 3, pp. 672-683, March 2003.
4. A. Poustie, "Semiconductor devices for all-optical signal processing", in *Proc. of ECOC 2005*, paper We3.5.1, Glasgow, Scotland.
5. J. Teng and G. Rouskas, "Wavelength selection in OBS networks using traffic engineering and priority-based concepts", *IEEE Journal of Selected Areas in Communications*, vol. 23, no. 8, pp. 1658-1669, August 2005.
6. J. Pedro, P. Monteiro, and J. Pires, "Wavelength contention minimization strategies for optical burst-switched networks", in *Proc. of IEEE GLOBECOM 2006*, paper OPNp1-5, San Francisco, USA.
7. Y. Sun, T. Hashiguchi, V. Minh, X. Wang, H. Morikawa, and T. Aoyama, "Design and implementation of an optical burst-switched network testbed", *IEEE Communications Magazine*, vol. 43, no. 11, pp. s48-s55, November 2005.
8. C. Gauger, "Performance of converter pools for contention resolution in optical burst switching", in *Proc. of SPIE OptiComm 2002*, pp. 109-117, Boston, USA.
9. H. Zang, J. Jue, and B. Mukherjee, "A review of routing and wavelength assignment approaches for wavelength-routed optical WDM networks", *Optical Networks Magazine*, vol 1, no. 1, pp. 47-60, January 2000.
10. J. Pedro, J. Castro, P. Monteiro, and J. Pires, "On the modelling and performance evaluation of optical burst-switched networks", in *Proc. of IEEE CAMAD 2006*, pp. 30-37, Trento, Italy.

New Assembly Techniques for Optical Burst Switched Networks Based on Traffic Prediction

Angeliki Sideri and Emmanouel A. Varvarigos

Computer Engineering and Informatics Dept, University of Patras, Greece *

Abstract. We propose new burst assembly techniques that aim at reducing the average delay experienced by the packets during the burstification process in optical burst switched (OBS) networks, for a given average size of the bursts produced. These techniques use a linear prediction filter to estimate the number of packet arrivals at the ingress node in the following interval, and launch a new burst into the network when a certain criterion, which is different for each proposed scheme, is met. Reducing the packet burstification delay, for a given average burst size, is essential for real-time applications; correspondingly, increasing the average burst size for a given packet burstification delay is important for reducing the number of bursts injected into the network and the associated overhead imposed on the core nodes. We evaluate the performance of the proposed schemes and show that two of them outperform the previously proposed timer based, length based and average delay-based burst aggregation schemes in terms of the average packet burstification delay for a given average burst size.

1 Introduction

Optical Burst Switching (OBS) [1] aims at combining the strengths of packet and circuit switching, and is considered a promising technology for implementing the next generation optical Internet, required to cope with the rapid growth of Internet traffic and the increased deployment of new services (e.g., VoIP telephony, video on demand, grid computing, digital repositories). In OBS, bursts consisting of an integer number of variable size packets are switched through the optical network. The OBS network consists of a set of optical backbone nodes, responsible for the forwarding of the bursts, and a set of edge nodes, known as ingress and egress nodes, responsible for burst assembly and disassembly. When a burst is formed at an ingress node, a control packet is sent out through the backbone network to reserve the required resources, followed after a short offset time interval by the burst.

An important issue for the design of OBS networks is the burst assembly strategy used at the edge nodes. The burst assembly process starts with the arrival of the first packet and continues until a predefined threshold is met. The

---

* We would like to achknowledge the support of the Phosphorus and the e- Photon/ONE+ IST projects.

I. Tomkos et al. (Eds.): ONDM 2007, LNCS 4534, pp. 358–367, 2007.

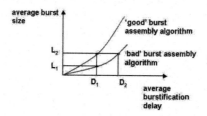

Fig. 1. Performance of a "good" and a "bad" burst assembly algorithm

burst aggregation policy influences the traffic characteristics in the network, but also the end-to-end performance. Reducing the packet burstification delay (so as to reduce total delay), and increasing the burst size (so as to reduce the number of bursts and the associated processing overhead at the core nodes) are two main performance objectives of the burst aggregation strategy. These objectives, however, contradict each other, since increasing the burst size also increases burstification delay (see Fig.1). A burst assembly algorithm, therefore, should be judged based on how well it performs with respect to one of these two performance metrics of interest, for a given value of the other performance metric. Given a burst assembly algorithm, choosing the desired balance between the burstification delay and the burst size then depends on the QoS requirements of the users, and the processing and buffering capabilities of the backbone nodes.

A number of burst assembly schemes have appeared in the literature, including the time-based algorithm (abbreviated T_{MAX} algorithm) and the length-based algorithm (abbreviated L_{MAX} algorithm) [3],[4], [5]. In the time-based algorithm, a time counter is started any time a packet arrives at an empty ingress queue, and the burst is completed when the timer reaches the threshold T_{MAX}; the timer is then reset to 0 and it remains so until the next packet arrival at the queue. Even though the time-based algorithm succeeds in limiting the average burstification delay, by limiting the maximum delay a packet can remain in the queue, it may generate very small bursts. In the length-based method, the burst is released into the network when its size reaches a threshold L_{MAX} (in fixed size packets, bytes, etc). This method produces bursts of a desired length, but may result in large burstification pdelays, especially when the traffic generation rate is low.

Hybrid schemes [2] have also been proposed, where a burst is completed when either the time limit T_{MAX} or the length limit L_{MAX} is reached, which ever happens first. An average delay-based algorithm (abbreviated T_{AVE} algorithm) was also introduced in [6], which aims at controlling the average burstification delay by letting out the bursts, the moment the average delay of the packets that comprise it reaches a threshold T_{AVE}. This method guarantees a desired average burstification delay, and also tends to minimize packet delay jitter. None of the burst assembly algorithms proposed so far, however, achieve an optimal tradeoff between the average burstification delay and the average burst size in the way described above.

In the present work we propose and evaluate several novel burst aggregation schemes that use traffic prediction to maximize the average length of the bursts produced for a given average burstification delay. Prediction of traffic characteristics has previously been examined in [7], [8] and [9]. In [7], specifically, it is demonstrated that despite the long-range dependence of Internet traffic, which would lead us to expect that we must look deep into the "past" for a precise estimation, a prediction filter of small order is sufficient for good performance. We use traffic prediction in order to estimate the number of packets that will arrive in the assembly queue in the near future and determine whether it would be beneficial for the burst assembly process to wait for these packets, or the burst should be sent immediately.

The performance measure we use to compare the algorithms proposed is the Average Packet Burstification Delay to Burst Size ratio (DBR) defined as

$$DBR = \frac{Average\ Packet\ Burstification\ Delay}{Average\ Burst\ Size}. \tag{1}$$

We find that two of the proposed schemes improve burstification efficiency, by reducing the average burstification delay by up to 33% for a given size of the bursts produced, compared to previously proposed schemes.

The remainder of the paper is organized as follows. In Section 2 we discuss the algorithms proposed. In Section 3, we examine the performance of the proposed schemes and compare it to that of the timer-based, length-based and average-delay-based algorithms. We also investigate the impact the various parameters involved have on performance.

2   The proposed burst assembly schemes

We assume that the time axis is divided into time frames of equal duration τ (see Fig. 2). During a frame, an edge OBS node assembles the IP packets arriving with the same destination address and the same QoS requirements (such packets are said to belong to the same Forwarding Equivalence Class or FEC) into a burst. At the end of each frame, a decision is taken about whether the burst should be sent out immediately and the assembly of a new burst should start, or the edge node should wait for another frame in order to include more packets in the current burst. This decision is taken by using a linear prediction filter to estimate the expected number $\hat{N}(n+1)$ of packet arrivals in the following frame $n+1$, and checking if a specific criterion (different for each algorithm proposed) is fulfilled. This criterion tries to quantify if the increase in the burst length expected by waiting for an extra frame is significant enough to warrant the extra delay that will be incurred.

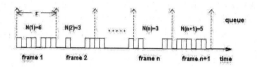

Fig. 2. Time frame structure. At the end of each frame n the algorithm decides if it should send out the burst immediately, or it should wait for another frame.

2.1 Fixed Additional Packets Threshold Algorithm (N_{MIN} algorithm)

In this proposed scheme, we define a lower bound N_{MIN} on the number of future arrivals above which we decide to wait for an extra frame before assembling the burst. At the end of frame n, the estimate $\hat{N}(n+1)$ produced by the linear predictor is compared to the threshold N_{MIN}, and if it is smaller than that, the burst leaves the queue immediately, otherwise it waits for another frame to be completed, at the end of which the same procedure is repeated. Therefore, the burst is sent out at the end of the n-th frame if and only if

$$\hat{N}(n+1) < N_{MIN}.$$

2.2 Proportional Additional Packets Threshold Algorithm (αL algorithm)

In this scheme, instead of using a fixed threshold value N_{MIN}, a fraction of the current length of the burst is used as the threshold. If α is the multiplicative parameter, the burst is completed at the end of the n-th time frame, if and only if

$$\hat{N}(n+1) < \alpha \cdot L(n),$$

where $L(n)$ is the burst length at the end of the n-th frame, and $\hat{N}(n+1)$ is the predictor's estimate for the number of packet arrivals expected during the following frame $(n+1)$.

2.3 Average Delay Threshold Algorithm (T_A algorithm)

This method tries to improve on the average-delay-based algorithm proposed in [6], which computes a running average of the packet burstification delay and lets out the burst, the moment the average delay of the packets that comprise it reaches a threshold T_{AVE}. The algorithm in [6] has two drawbacks: a) computing the running average introduces considerable processing overhead, and b) bursts may not be sent out at the optimal time, since the running average is non-monotonic in time and could drop in the future due to new packet arrivals. The T_A algorithm addresses these drawbacks using traffic prediction. At the end of each frame, it estimates the average burstification delay we expect to have at the end of the following frame, and launches the burst if this estimate exceeds some threshold value T_A.

The Average Packet Delay $D(n)$ of the packets in the burst assembly queue at the end of frame n is defined as

$$D(n) = \frac{\sum_{i=1}^{L(n)} T_i(n)}{L(n)}, \tag{2}$$

where $L(n)$ is the burst size at the end of frame n, $T_i(n) = n \cdot \tau - t_i$ is the delay of the i-th packet from the moment it enters the queue until the end of n-th frame, τ is the duration of the frame, and t_i is the arrival time of i-th packet.

Alternatively and more easily, we can compute $D(n)$ using the recursion

$$D(n) = \frac{L(n-1) \cdot D(n-1) + L(n-1) \cdot \tau + \sum_{i=1}^{N(n)} T_i(n)}{L(n-1) + N(n)}, \tag{3}$$

where $N(n)$ is the number of packet arrivals during frame n. If a burst was sent out at the end of the $(n-1)$-th frame then we take $L(n-1) = 0$ in Eq. 3.

To obtain an estimate $\hat{D}(n+1)$ of the Average Packet Delay at the end of frame $n+1$ we assume that the $\hat{N}(n+1)$ packets estimated by the predictor to arrive by the end of frame $n+1$ will have an average delay of $\tau/2$. Using Eq. 3, the estimated Average Packet Delay $\hat{D}(n+1)$ at the end of frame $n+1$ is

$$\hat{D}(n+1) = \frac{L(n) \cdot D(n) + \tau \cdot L(n) + \hat{N}(n+1) \cdot \frac{\tau}{2}}{L(n) + \hat{N}(n+1)}. \tag{4}$$

A burst is completed at the end of the n-th frame if and only if

$$\hat{D}(n+1) > T_A,$$

where T_A is the predefined threshold value.

2.4   Average Delay to Burst Size Ratio Improvement Prediction Algorithm (L_{MIN} algorithm)

The proposed Average Delay to Burst Size Ratio Improvement algorithm (abbreviated L_{MIN} algorithm) uses traffic prediction to compute an estimate $\hat{DBR}(n+1)$ of DBR at the end of frame $n+1$, and decides that the burst is completed, if this estimate is worse than the current $DBR(n)$. The average burstification delay to burst size ratio $DBR(n)$ at the end of frame n is defined as

$$DBR(n) = \frac{D(n)}{L(n)} = \frac{\sum_{i=1}^{L(n)} T_i(n)}{L^2(n)}. \tag{5}$$

Alternatively, and more easily, $DBR(n)$ can be found recursively as

$$DBR(n) = \frac{L(n-1) \cdot D(n-1) + L(n-1) \cdot \tau + \sum_{i=1}^{N(n)} T_i(n)}{(L(n-1) + N(n))^2} .$$
(6)

The Estimated Average Packet Burstification Delay to Burst Size ratio $D\hat{B}R(n+1)$ at the end of frame $(n+1)$ can be found as

$$D\hat{B}R(n+1) = \frac{L(n) \cdot D(n) + L(n) \cdot \tau + \hat{N}(n+1) \cdot \frac{\tau}{2}}{(L(n) + \hat{N}(n+1))^2} .$$
(7)

The algorithm decides that a burst is completed and should be sent out at the end of frame n if and only if

$$D\hat{B}R(n+1) < DBR(n) \ AND \ L(n) > L_{MIN}.$$

During the first frames following a burst assembly completion, there is a great likelihood that the right term of the preceding inequality will be quite large, making it difficult to fulfill. The threshold L_{MIN} is used as a lower bound on the length of the bursts, and also makes the algorithm parametric (as with all the previous algorithms examined) so that the desired tradeoff between the average burst size and the average packet burstification delay can be obtained.

3   Performance Analysis and Simulation Results

We used the Matlab environment to simulate the burst aggregation process at an ingress queue, in order to evaluate the performance of the proposed schemes and compare it to that of previously proposed schemes. We also quantified the impact the parameters N_{MIN}, α, T_A and L_{MIN} have on performance, and the effect of the frame size τ and the order h of the linear predictor.

It is worth noting that each of the proposed schemes corresponds to a different Burst Size versus Packet Burstification Delay curve (the reader is referred to Fig. 1), while the choice of the parameters N_{MIN}, α, T_A and L_{MIN} (or of the parameters T_{AVE}, T_{MAX}, L_{MAX} in previously proposed schemes) determines the exact points on each curve the burst assembly process is operating at.

3.1   Traffic Generating Source

In our experiments, the arrivals at the ingress queue were obtained from an Exponential-Pareto traffic generating source of rate r *bits/sec*. The traffic source generates superpackets (they can also be viewed as busy periods) with exponentially distributed interarrival times of mean $1/\lambda$ seconds. The size of each superpacket follows the Pareto distribution with shape parameter β. If a superpacket has size greater than l bytes, which is taken to be the size of the packets used in the network, it is split and sent as a sequence of packets of size l. The

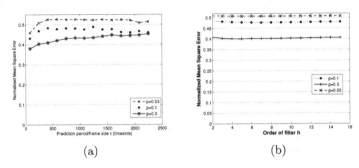

(a)   (b)

Fig. 3. LMS performance for various values of: (a) the prediction period τ, (b) the length of the predictor h.

time units used for displaying our results are measured in packet slots, where 1 slot $= l/r$ (the transmission time of a packet).

The values of the parameters we used in our experiments are the following: β=1.2, r=1Gbps, l=1500 bytes, $1/\mu$=60KB, and $1/\lambda$=1.6 msec or 4.8msec, yielding corresponding load utilization factors p=0.1 and p=0.3. The parameter β determines the Hurst parameter $H = (3 - \beta)/2$, which takes values in the interval $[0.5, 1)$ and defines the burstiness of the traffic. The closer the value of H is to 1, the more bursty the traffic can be characterized.

3.2 Linear Predictor LMS

The *Least Mean Square Error (LMS)* predictor, described in [10] and also used for traffic prediction in [8] and [9], has been chosen as the linear predictor in our burst assembly schemes. It is simple, fast and effective without great computational cost. The estimate $\hat{N}(n + 1)$ of the number of packet arrivals during the $(n + 1)$-th frame is generated according to the relationship

$$\hat{N}(n + 1) = \sum_{i=1}^{h} w_i \cdot N(n - i + 1) \ ,$$

where $w_i, i \in 1, \ldots, N$, are the filter coefficients and h is the length of the filter. The error $e(n)$ of the n-th frame is calculated as $e(n) = N(n) - \hat{N}(n)$, and the coefficients of the filter are updated at each iteration according to

$$w_i(n + 1) = w_i(n) + \delta \cdot e(n) \cdot N(n - i + 1) \ ,$$

where δ is the step size.

3.3 Predictor Performance

The accuracy of the estimations produced by the LMS predictor can be assessed by the *Normalized Mean Square Error (NMSE)* parameter, defined as:

$$NMSE = \frac{MSE}{P_{INPUT}} = \frac{\sum_{n=1} e^2(n)}{\sum_{n=1} N^2(n)} \ .$$

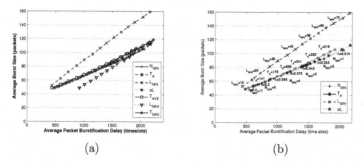

(a)   (b)

Fig. 4. Performance of the proposed algorithms for traffic load $p = 0.03$: (a)comparison of the proposed schemes with previously proposed algorithms, (b) parameters applied on the proposed algorithms

The first set of results presented examines the dependence of the performance of the LMS predictor on the frame duration τ, the order of the prediction filter h, and the traffic load p. Fig. 3(a) shows the way $NMSE$ varies with the frame duration τ for bursty traffic ($H=0.9$). Note that short frame durations result in smaller values of $NMSE$, which can be justified by the fact that for bursty traffic, the characteristics of its behaviour remain static for only shorts periods of time. For light traffic, the predictor's performance is worse than it is for heavy traffic. This can also be seen in Fig.3(b), which illustrates the impact the order of the filter has on $NMSE$. This figure also indicates that there is very little improvement when the order of the filter is increased beyond a certain value. This is in agreement with the results in [7], where it was argued that the performance of linear predictors for internet traffic is dominated by short-term correlations, and we don't have to "look deep" into the history of traffic arrivals to obtain a valid estimation. A small order of filter is, therefore, preferable, since it also results in smaller computation overhead. As the frame size τ increases, the $NMSE$ remains steady after a certain value ($\tau > 0.005sec$) when the traffic is light ($p = 0.1$, $p = 0.03$), while it worsens slightly for heavier traffic ($p = 0.3$).

3.4   Comparison Between the Burst Assembly Schemes

In this section we compare the performance of the proposed schemes to that of the previously proposed $T_{AVE}, T_{MAX}, L_{MAX}$ burst assembly schemes. The results reported here were obtained for bursty traffic ($H = 0.9$) and varying load utilization p. The length h of the LMS predictor was set to 4, while the frame size τ varied depending on the traffic load. The parameters of all the schemes have been chosen so as to produce average burstification delays that lie in the same range so that the resulting burst sizes can be compared. Time delays are measured in slots. Figures 4(a), 5(a) and 6(a) illustrate the average burst size produced versus the average packet burstification delay when the traffic load utilization factor is $p = 0.03$, 0.1 and 0.3, respectively. The labels

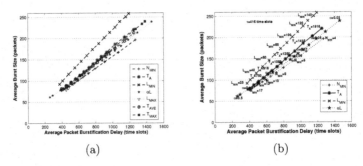

Fig. 5. Performance of the proposed algorithms for traffic load $p = 0.1$: (a)comparison of the proposed schemes with previously proposed algorithms, (b) parameters applied on the proposed algorithms.

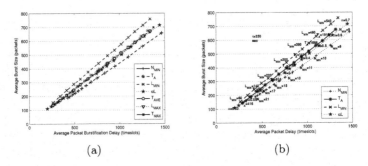

Fig. 6. Performance of the proposed algorithms for traffic load $p = 0.3$: (a) comparison of the proposed schemes with previously proposed algorithms, (b) parameters applied on the proposed algorithms.

in Figs. 4(b), 5(b) and 6(b) display the details on the values of the parameters N_{MIN}, T_A, L_{MIN} and α that give the corresponding results.

From Figs. 4(a), 5(a) and 6(a), we can see that the L_{MAX} algorithm exhibits (as expected) the worst performance for light load ($p = 0.003$), while its performance becomes relatively better for heavier load ($p = 0.1$ and 0.3). The opposite is true for the N_{MIN} algorithm, which behaves relatively worse for heavy traffic ($p = 0.1$ and 0.3), while its performance improves for light traffic ($p = 0.003$). For a given traffic load, the N_{MIN} algorithm shows worse relative performance when the parameter N_{MIN} is set at low values so as to produce large bursts (this is because for small values of N_{MIN}, the algorithm is rather intolerable to estimation errors). The αL algorithm always gives better results than the N_{MIN} algorithm, but does not succeed in outperforming some of the other algorithms considered. For a given traffic load, its relative standing compared to the other algorithms does not change with the choice of the parameter α (small values of α produce longer bursts as it can be seen in Figs. 4(b), 5(b) and 6 (b)). Among the burst assembly algorithms already proposed in the literature (that is, the T_{MAX}, L_{MAX}, T_{AVE} algorithms), the T_{AVE} algorithm gives the best performance. The proposed T_A algorithm outperforms the T_{AVE} algorithm, even though the improvement is rather small, as shown in Fig. 6(a). The improvement is more pronounced when the T_A algorithm generates longer bursts and when the traffic load is heavier.

The best performance is consistently demonstrated by the L_{MIN} algorithm, which achieves a 33% improvement over the T_A algorithm (the second best) for light traffic load ($p = 0.003$ and 0.1) and a 8% improvement for heavier traffic load ($p = 0.3$). The L_{MIN} algorithm can be considered a variation of the L_{MAX} algorithm, enhanced with the ability to predict the time periods where the value of DBR is expected to improve because of a large number of future packet arrivals. Note that in most of the figures, the curve that corresponds to the L_{MIN} algorithm is parallel to and above that of the L_{MAX} algorithm.

To conclude, the L_{MIN} algorithm seems to be the algorithm of choice when the average burstification delay (for a given burst size) or the average burst size (for a given burstification delay) is the criterion of interest. One should note, however, that the T_{AVE} and the T_A algorithms may be preferable when the delay jitter [6] is the main consideration (both of these algorithms also give a satisfactory average burstification delay to average burst size ratio).

References

1. T. Battestilli and H. Perros, "An Introduction to Optical Burst switching", *IEEE Optical Communications*, August 2003.
2. X. Yu, Y. Chen, and C. Qiao, "Study of traffic statistics of assembled burst traffic in optical burst switched networks," Proc. Opticomm, 2002, pp. 149-159.
3. V. Vokkarane, K. Haridoss, J. Jue, "Threshold-Based Burst Assembly Policies for QoS Support in Optical Burst-Switched Networks", Proc. Opticomm, pp. 125-136, 2003.
4. A. Ge, F. Callegati and L. Tamil, "On Optical Burst Switching and Self-Similar Traffic", *IEEE Communications Letters*, Vol. 4, No. 3. March 2000.
5. X. Cao, J. Li, Y. Chen and C. Qiao, "Assembling TCP/IP Packets in Optical Burst", Proc. IEEE Globecom, 2002.
6. K. Christodoulopoulos, E. Varvarigos and K. Vlachos, "A new Burst Assembly Scheme based on the Average Packet Delay and its Performance for TCP Traffic", under 2nd revision, Optical Switching and Networking, Elsevier, 2006.
7. Sven A. M. Östring, H. Sirisena, "The Influence of Long-range Dependence on Traffic Prediction", Proc. ICC, 2001.
8. D. Morató, J. Aracil, L. A. Díez, M. Izal, E. Magaña, "On linear prediction of Internet traffic for packet and burst switching networks", Proc. 10th Int. Conf. Computer Communications Networks, 2001, pp. 138-143.
9. J. Liu, N. Ansari, T. Ott, "FRR for Latency reduction and QoS Provisioning in OBS Networks", *IEEE JSAC*, Vol. 21, No. 7, Sept. 2003.
10. J.R. Treicher, C.R. Johnson Jr. and M. G. Larimore, *Theory and Design of Adaptive Filters*, New York: Wiley 1987.

A Novel Burst Assembly Algorithm for Optical Burst Switched Networks Based on Learning Automata

T. Venkatesh, T. L. Sujatha and C. Siva Ram Murthy*

Department of Computer Science and Engineering
Indian Institute of Technology Madras, Chennai - 600036, India
{tvenku,lsujatha}@cse.iitm.ernet.in, murthy@iitm.ac.in

Abstract. Optical Burst Switching (OBS) is widely believed to be the technology for the future core network in the Internet. Burst assembly time at the ingress node is known to affect the traffic characteristics and loss distribution in the core network. We propose an algorithm for adapting the burst assembly time based on the observed loss pattern in the network. The proposed Learning-based Burst Assembly (LBA) algorithm uses learning automata which probe the loss in the network periodically and change the assembly time at the ingress node to a favorable one. We use a discrete set of values for the burst assembly time that can be selected and assign a probability to each of them. The probability of selecting an assembly time is updated depending on the loss measured over the path using a Linear Reward-Penalty (L_{R-P}) scheme. The convergence of these probabilities eventually leads to the selection of an optimal burst assembly time that minimizes the burst loss probability (BLP) for any given traffic pattern. We present simulation results for different types of traffic and two network topologies to demonstrate that LBA achieves lower BLP compared to the fixed and adaptive burst assembly mechanisms existing in the literature.

1 Introduction

Wavelength Division Multiplexing (WDM) technology has become widely popular for deployment in the core network to meet the ever-increasing demand for bandwidth in the Internet. In WDM networks, a single fiber can support concurrent transmission of multiple wavelengths resulting in complexity of switching compared to the traditional networks. There are three important switching paradigms in WDM networks, namely, optical circuit switching, optical packet switching (OPS) and optical burst switching (OBS). Among these, OBS has attracted attention of the researchers due to its advantages among the other paradigms [1].

In an OBS network, IP packets are assembled into bursts at the ingress node and a control burst is sent out before each data burst. The control burst

---

* Author for correspondence. This work was supported by Microsoft Corporation and Microsoft Research India under the Microsoft Research India PhD Fellowship Award.

I. Tomkos et al. (Eds.): ONDM 2007, LNCS 4534, pp. 368–377, 2007.

carries necessary information about the arrival time and the duration of the burst. Wavelengths are reserved only for the duration of the bursts. A time gap, called *offset time*, is maintained between the control burst and the data burst to enable reservation. At the egress node, the bursts are disassembled into IP packets again. Due to the dynamic reservation mechanism bursts are dropped without any information to the ingress node, whenever contention occurs at the core nodes. Thus, contention becomes the main source of the losses in an OBS network.

There are several factors responsible for contention losses in OBS networks such as, burst assembly, scheduling algorithm, offset time and routing and wavelength assignment algorithms. Though several solutions have been proposed for contention resolution in OBS networks (see [2] for a survey on this work), it is an active area for research due to inherent complexity of the problem. The main objective of research in OBS networks is often reduction of the burst loss probability (BLP) which is done by careful choice of mechanisms for the aforementioned factors. Burst assembly time at the ingress node is an important parameter in OBS networks that can affect the dynamics of the network. The assembly mechanism changes the burst size, inter-arrival time, distribution of traffic at the core node and hence the loss distribution [3].

In this work we try to reduce the BLP by a learning-based burst assembly mechanism at the ingress node instead of the fixed assembly schemes used earlier. The motivation for this work arises from the observation made in [4] that an adaptive burst assembly algorithm that considers the variation in the traffic, outperforms the schemes that set assembly time independent of the traffic. We use the loss rate in the path as a feedback to learn appropriate assembly time over the time and hence, achieve reduction in the BLP. The rest of the paper is organized as follows. Section 2 gives some links to the work in the literature on the burst assembly mechanisms and provides the motivation for our work. We give a brief introduction to the theory of Learning Automata and the Linear Reward Penalty (L_{R-P}) scheme in Section 3. In Section 4, we propose the Learning-based Burst Assembly (LBA) algorithm and demonstrate the improvement in BLP through simulations in Section 5. Finally, we conclude the paper in Section 6.

2   Related Work and Motivation

The impact of burst aggregation on the BLP, primarily for TCP traffic, has been studied in the literature earlier [4], [5], [6] and it was observed that assembly time affects the BLP as well as throughput in the network. This is primarily because, the size of the bursts as well as their inter-arrival times at the core node depend on the assembly mechanism used at the ingress node. The impact of burst assembly time on the TCP throughput was studied in [7]. The existence of an optimal burst assembly time for which the TCP throughput is maximum was proved theoretically as well as through simulations.

Intuitively there are two ways to aggregate the bursts: either define a fixed burst size or define a burst aggregation time. In [4], the authors studied the effect of a time-based Fixed Assembly Period (FAP) algorithm and a combination of size-based and time-based algorithm, called Min-Burst Length-Max-Assembly-Period (MBMAP) algorithm on TCP traffic. Since the FAP is independent of the traffic rate, it was found to adversely impact the network performance. Therefore, the authors of [4] proposed an Adaptive Assembly-Period (AAP) algorithm that can dynamically change the value of assembly period (AP) at every ingress node according to the length of burst recently sent. It was shown that AAP is best among the three assembly algorithms, because it matches with the TCP rate control mechanism.

However, the AAP algorithm can be defined only for TCP traffic. It assumes that the rate of the incoming traffic varies linearly and uses an equation to adjust the assembly period proportional to the average burst length. The recently arrived burst is given a higher weight while computing the assembly period, to enable synchronization of the assembly algorithm with TCP. Such a scheme works well mainly with a linearly increasing traffic like TCP but not with other traffic sources like, ON/OFF, CBR or a combination of these. For non-TCP traffic, adapting burst assembly according to the arrival rate is infeasible because, the traffic does not follow a regular pattern. Instead of using rate of the traffic to change the assembly time, we observe the variation in the BLP along the path to learn the optimal assembly time. Since it was proved in [8] that any algorithm that arrives at an optimal policy over time through exploration outperforms the schemes based on fixed policy, we propose a learning automata-based solution for the burst assembly.

3   Learning Automata

In a learning automata system, a finite number of actions can be performed in a random environment. When a specific action is performed, the environment provides either a favorable or an unfavorable random feedback. The objective in the design of the learning automata system is to determine how the previous actions and responses should affect the choice of the current action to be taken, and to improve or optimize some predefined objective function. A learning automaton comprises of a learning module which learns from the feedback provided by the environment, and a decision module that makes decisions based on the assimilated knowledge of the environment. At any stage, the choice of action could be either deterministic or stochastic. In the latter case, probabilities are maintained for each possible action to be taken which are updated with each response from the environment.

A learning automaton can be formally described as a triple $\{\alpha, \beta, A\}$, where $\alpha(n)$ is the output or action of the automaton at time instant n, $\beta(n)$ is the input to the automaton at time instant n, and A is called the *updating algorithm* or the *reinforcement scheme*. The updating algorithm determines the $\alpha(n + 1)$ in terms of the network state and $\beta(n)$, and could be either deterministic or stochastic. Determining the updating algorithm for a stochastic automaton makes for a very important design choice. The updating function can be either linear or non-linear. Well known linear updating functions include reinforcement schemes such as the Linear Reward-Penalty (L_{R-P}) scheme, the Linear Reward-Inaction (L_{R-I}) scheme, and the Linear Reward-ϵ-Penalty ($L_{R-\epsilon P}$) scheme [9]. The objective of the updating function is to enable the automaton to *learn* the state of the environment based on the feedback obtained and choose the best possible action at any point of time. It should be able to efficiently guide the automaton to quickly adapt to the changes in the environment. The updating function needs to be simple, and yet efficient, especially when the environment is known to change rapidly. For a detailed description of learning automata, readers are referred to [9] and references thereof.

The L_{R-P} Scheme: For the burstification problem, we use the L_{R-P} scheme [9] to update the automaton action probabilities. When a positive response is obtained for an action, its probability is increased and the probabilities of all the other actions are decreased. If a negative feedback is received for an action, the probability of that action is decreased and that of others is increased. For a multi-action system with S states, the updating algorithm can be written as follows, where $P_i(n)$ represents the probability of choosing action i at time instant n, and a and b are the reward and penalty parameters, respectively:

- When a positive feedback is obtained for action i,

$$P_i(n+1) = P_i(n) + a(1 - P_i(n))$$
$$P_j(n+1) = (1-a)P_j(n), j \neq i$$

- When a negative feedback is obtained for action i,

$$P_i(n+1) = (1-b)P_i(n)$$
$$P_j(n+1) = \frac{b}{S-1} + (1-b)P_j(n)), j \neq i$$

and similarly for all $i = 1, 2, \ldots, S$, where $0 < a < 1, 0 \leq b < 1$. In our case, an increased or decreased BLP are the two kinds of feedback returned to the automaton residing on an ingress node for the action of selecting a burst assembly time. Convergence of the S-model L_{R-P} scheme has been discussed in detail in [9].

4   Learning-based Burst Assembly Mechanism

Since the action selection is only made at the ingress node, the learning automata are placed only at the edge nodes. The burst manager at the ingress node and the learning agent comprise the decision module which decides the assembly time to be used. For every time period T, the learning agent decides the action to be chosen which here corresponds to selecting a burst assembly time from a discrete set of values. The bursts for the next period are transmitted with the selected assembly time. The BLP value for each ingress-egress pair obtained from the acknowledgment packet is used as the feedback from the environment.

Algorithm 1 Learning-based Burst Assembly Algorithm

Let $\mathcal{T} = \{t_1, t_2, \ldots, t_n\}$ be the vector of assembly times and i be the index of the burst assembly time chosen.

Let P_i be the probability of selecting the ith burst assembly time, T_i.

Let $P_i = 1/n$ be the initial probabilities $\forall, 1 \le i \le n$.

    if (loss = zero) **then**

        $P_i = P_i + a * (1 - P_i)$

           for all $T_j, \ni j \ne i$

        $P_j = 1 - a * P_j$

    endif

    elseif (loss > zero) **then**

        $P_i = (1 - b) * P_i$

        for all $T_j, \ni j \ne i$

           $P_j = \frac{b}{n-1} + (1 - b) * P_j$

        endfor

    endif

where a is the reward and b is the penalty.

Depending on whether the response is favorable or not (decreased/increased BLP), the next assembly time (action) is selected. The period T for which the assembly time is changed, is selected such that the overhead incurred in collecting the feedback is minimized and any transient network conditions are ignored by the agent. A low value of T increases the probe overhead and also leads to transient conditions due to frequent changes in burst size. However, if T is set to a large value the response to congestion is slow but the overhead is low. Hence, T can be selected judiciously depending on the burstiness of the traffic. We assume that the time taken to collect feedback is much smaller than the time of adaptation. This is valid since the probe packets take negligibly small time for transmission and processing compared to the period in which decision is made.

We assume that a set of values for burst assembly time is available at the burst manager. Initially all the values have equal probability. The main idea in this mechanism, is to estimate the BLP for a selected burst assembly time and use the one with lowest BLP. Due to the L_{R-P} updation scheme, the assembly time corresponding to the lowest BLP has the highest probability. The estimated BLP is initially set to a predefined value. Each node in the network maintains the loss information for every connection (traffic request between an ingress node and an egress node). When a burst loss occurs at the node, the local BLP value for that flow is updated. The agent at every ingress node periodically sends a probe packet to the egress node. The probe packet collects information on the BLP along the path and the egress sends a negative feedback through acknowledgment packet if the BLP is higher than a threshold.

When a burst loss occurs at the node, the local BLP value for that flow is updated and this information is sent to the destination along with the probe packet. When the BLP is higher than a threshold, the source gets a negative feedback. On receiving the BLP value, the source node updates the probability for the corresponding burst assembly time using the L_{R-P} scheme and accordingly selects the assembly time for the next burst. We reserve a control wavelength for the probe and acknowledgment packets and assume no losses in control plane. This also avoids contention losses of control bursts. To avoid the implementation complexity at the core node to generate a control packet, we assume that the feedback is only sent from the egress node and the core nodes along the path update the cumulative BLP in the probe packet.

The LBA algorithm that uses L_{R-P} equations given in Section 3 to update the probability associated with a value of assembly time, depending on the resulting BLP is given in Algorithm 1. This problem wherein an agent selects an action from a finite set at each time instant to optimize an objective function over the long run is similar to the multi-armed bandit problem in the literature of reinforcement learning [10] for which learning automata provides an optimal solution. This is the motivation to design a learning automata-based solution for burst assembly. Here the reward for an action (selecting a burst assembly time) is characterized by a decreasing BLP value. So the learning automaton tries to select the optimal burst assembly time that minimizes the BLP over a period of time. Every time it selects an assembly time, depending on whether the BLP increases or decreases over the next time period, the agent selects an assembly time that minimizes BLP in the subsequent period.

5   Simulation Results

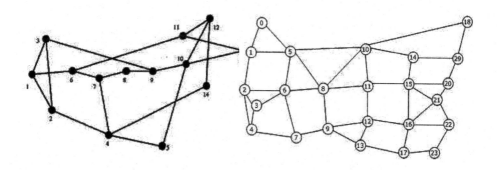

Fig. 1. NSFNET Topology.   **Fig. 2.** USIP Backbone Topology.

In this section, we study the performance of LBA algorithm and compare it with AAP and MBMAP algorithms for Just-Enough-Time (JET) reservation scheme. Architectures for core node and edge node as proposed in [11] with Latest Available Unused Channel with Void Filling (LAUC-VF) scheduler were used. We use variation in the BLP with the network load (percentage of the maximum load), as a measure for the performance of the algorithms. A *connection* wherever used means a demand generated between a pair of edge nodes at

higher layer that results in a flow of bursts for a time period between them. A set
of 20 values equally spaced between 0.1 *ms* and 1 *ms* constitute the action set.
The time period of updating assembly time was selected keeping the trade-off
between the learning rate and control overhead. We assign +10 for reward and
−10 for penalty. We use two topologies, NSFNET topology (Fig. 1) and USIP
network (Fig. 2) with 32 *wavelengths per link* (8 control wavelengths) in all our
simulations. All the results were obtained at 95% confidence level.

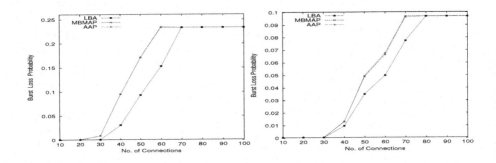

Fig. 3. Variation in the BLP for LBA, **Fig. 4.** Variation in the BLP for LBA,
AAP and MBMAP as the load increases AAP and MBMAP as the load increases
for CBR traffic on NSFNET.   for CBR traffic on USIPNET.

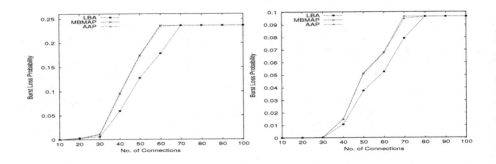

Fig. 5. Variation in the BLP for LBA, **Fig. 6.** Variation in the BLP for LBA,
AAP and MBMAP as the load increases AAP and MBMAP as the load increases
for ON/OFF traffic on NSFNET.   for ON/OFF traffic on USIPNET.

5.1   CBR Traffic

In this section we use several CBR connections generated randomly between
source-destination pairs each with a rate selected randomly between 1 MB and

8 MB. Fig. 3 and Fig. 4 show the variation of BLP as the load increases for NSFNET and USIPNET, respectively. Initially when load is less there is no loss with all the algorithms. But as the load increases, contention losses increase in the network. We can observe that for upto 70% load LBA has a lower BLP than AAP. However, as the load grows beyond, both LBA and AAP have the same BLP. This is because contention losses cannot be controlled by varying the burst assembly time beyond a certain point. At this point contention losses occur only because of lack of wavelengths (severe congestion) and the burst size has lower effect on the contention losses after this.

5.2   ON/OFF Traffic

We use ON/OFF traffic with Pareto distribution with shape factor 1.5. Fig. 5 and Fig. 6 compare the BLP for the three algorithms in case of NSFNET and USIPNET, respectively as the load increases. Unlike in CBR traffic, where the packets are sent with constant rate, packets are sent in bursty fashion at certain times. So, even for a low load, there is contention loss in the network. However, we can observe that LBA has lower BLP than AAP. The reason is because the AAP algorithm is designed for TCP traffic and hence does not adapt with losses for any other traffic pattern. Even here we can observe that AAP and MBMAP have the same loss probability for ON/OFF traffic whereas LBA has the lowest BLP among all the schemes.

5.3   TCP Traffic

With increasing percentage load, the variation of BLP for all the three algorithms is compared in Fig. 7 and Fig. 8 for NSFNET and USIPNET topologies, respectively. In CBR and ON/OFF traffic, loss increases as the load increases. But, in TCP it is not the case because TCP controls the rate of flow of packets based on the loss events. However, we can observe that LBA and AAP have almost equal loss probabilities on an average. This shows that LBA performs atleast as good as AAP, which was specially designed for TCP traffic. We can observe that LBA and AAP algorithms are far better than MBMAP algorithm for TCP traffic. Hence, we can conclude that an adaptive assembly algorithm is essential to control the losses in an OBS network.

5.4   Mixed Traffic

In this section, the traffic is composed of all the types of traffic mentioned above (hence called mixed traffic). Fig. 9 and Fig. 10 compare the BLP for LBA, AAP and MBMAP with NSFNET and USIPNET, respectively. We see that LBA algorithm outperforms the AAP algorithm significantly at higher network load. We can observe that upto 30% load, AAP and LBA have almost equal loss probabilities but lower than MBMAP. When the load is between 30% and 60%, the BLP for AAP increases faster than that with LBA. Beyond that AAP cannot

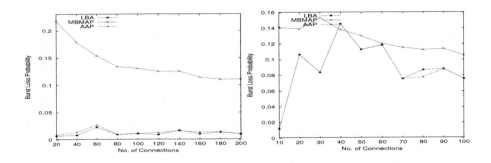

Fig. 7. Variation in the BLP for LBA, AAP and MBMAP as the load increases for TCP traffic on NSFNET.

Fig. 8. Variation in the BLP for LBA, AAP and MBMAP as the load increases for TCP traffic on USIPNET.

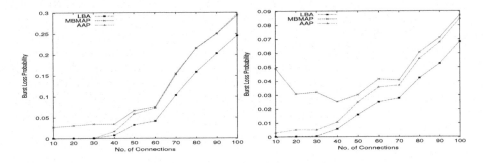

Fig. 9. Variation in the BLP for LBA, AAP and MBMAP as the load increases for mixed traffic on NSFNET.

Fig. 10. Variation in the BLP for LBA, AAP and MBMAP as the load increases for mixed traffic on USIPNET.

control the loss and it is as good as MBMAP. But LBA is better than AAP in spite of the higher loss rate for the mixed traffic type. In [4], the authors show that AAP performs better than MBMAP for TCP traffic. But we can observe from the results here that AAP and MBMAP have same loss rate for CBR and ON/OFF traffic and that AAP is advantageous only for TCP traffic and not for other types of traffic. Even for TCP traffic, LBA performs atleast as good as AAP algorithm.

6   Conclusion

In this paper, we proposed a new algorithm for adaptive burst assembly for OBS networks that use learning automata namely, Learning-based Burst Assembly (LBA). Since it was shown in the literature that the choice of burst assembly time affects both the throughput and the loss rates in OBS networks, we came up with a simple scheme to learn the optimal choice of burst assembly time. Upon convergence, the LBA algorithm learns the optimal burst assembly time for any traffic pattern. We demonstrated through simulations on two networks topologies that BLP is indeed reduced compared to the other adaptive assembly mechanism (AAP algorithm) in the literature. Further, the AAP algorithm available in the literature has advantage only for TCP traffic whereas, the LBA algorithm works well irrespective of the type of traffic (comparable to AAP for TCP). Since AAP algorithm uses the rate of incoming traffic to adapt the assembly time, it is not suitable to be used for non-TCP traffic and has almost same BLP as that of MBMAP algorithm for other traffic types. LBA algorithm performs well independent of the traffic type and captures the explicit dependence of the BLP on assembly time. We justified the motivation for designing a loss-aware assembly mechanism through this work and showed the advantage of an autonomic burst assembly algorithm.

References

[1] Qiao, C., Yoo, M.: Optical Burst Switching (OBS)-a new switching paradigm for optical Internet. Journal of High Speed Networks **8**(1) (1999) 69–84
[2] Farahmand, F., Jue, J.P.: Look-ahead window contention resolution in optical burst switched networks. In: Proceedings of IEEE Workshop on High Performance Switching and Routing (HPSR). (2003) 147–151
[3] Detti, A., Listanti, M.: Impact of segments aggregation of TCP Reno flows in optical burst switching networks. In: Proceedings of IEEE INFOCOM. (2002) 1803–1812
[4] Cao, X., Li, J., Chen, Y., Qiao, C.: Assembling TCP/IP packets in optical burst switched networks. In: Proceedings of IEEE GLOBECOM. (2002) 84–90
[5] Malik, S., Killat, U.: Impact of burst aggregation time on performance of optical burst switching networks. In: Proceedings of Optical Network Design and Modeling (ONDM). (2005) 19–25
[6] Hong, D., Poppe, F., Reynier, J., Bacelli, F., Petit, G.: Impact of burstification on TCP throughput in optical burst switching networks. In: Proceedings of 18th International Teletraffic Congress (ITC). (2003) 89–96
[7] Yu, X., Qiao, C., Liu, Y., Towsley, D.: Performance evaluations of TCP traffic transmitted over OBS networks (2003) Tech. Report 2003-13, CSE Dept., SUNY Buffalo.
[8] Littman, M., Boyan, J.: A distributed reinforcement learning scheme for network routing. In: Proceedings of the International Workshop on Applications of Neural Networks to Telecommunications. (1993) 45–51
[9] Narendra, K.S., Thathachar, M.: Learning Automata: An Introduction. Prentice Hall, New Jersey (1989)
[10] Sutton, R.S., G.Barto, A.: Reinforcement Learning : An Introduction. The MIT Press, Cambridge, Massachusetts (1998)
[11] Xoing, Y., Vandenhoute, M., Cankaya, C.: Control architecture in optical burst-switched WDM networks. IEEE Journal of Selected Areas in Communications **18**(10) (2000) 1838–1851

Fast and effective dimensioning algorithm for end-to-end optical burst switching networks with ON-OFF traffic model

Reinaldo Vallejos[1], Alejandra Zapata[1], Marco Aravena [2,1]

[1] Telematics Group, Electronic Engineering Department, Universidad Técnica Federico
Santa María, Chile
reinaldo.vallejos@usm.cl, alejandra.zapata@usm.cl
[2] Computer Science Department, Universidad de Valparaíso, Chile
marco.aravena@uv.cl

Abstract. A novel algorithm for fast dimensioning of end-to-end optical burst switching networks is proposed. The proposed method determines the number of wavelengths for each network link according to the traffic load, the routing algorithm and the required blocking probability per connection. The burst input traffic is modeled by an ON-OFF alternating renewal process, which is more realistic for OBS networks than the typically used Poisson model. Compared to the two most typically used dimensioning approaches, the proposed method results in significant lower wavelength requirements whilst achieving the same target blocking probability. Additionally, the proposed method takes less than one second to dimension the network links which makes it several orders of magnitude faster than the conventional simulation approach.

Keywords: WDM networks, OBS networks, ON-OFF traffic, network dimensioning, blocking probability.

1 Introduction

The high bandwidth requirement imposed by the ever increasing amount of Internet traffic can only be met by using Wavelength Division Multiplexing (WDM) networks, currently allowing bandwidths of the order of 1-10 Tb/s per fibre and already used in transport networks as high-speed transmission channels throughout the world.

At the moment, however, the potential of WDM channels could not be fully utilised by already successful networking approaches as packet-switching because optical buffering and processing is not mature enough [1,2]. Alternatively, these tasks could be carried out electronically, but electronic operation cannot match the high speed of optical transmission.

As a result, a new switching paradigm –feasible in the short term- has been proposed for WDM networks: optical burst switching (OBS) [3-6], which integrates the fast transmission speed of optical channels and the electronic capability of electronic nodes. OBS networks aim to decrease the electronic speed requirements - imposed by the optical transmission rate and the packet-by-packet operation- by

I. Tomkos et al. (Eds.): ONDM 2007, LNCS 4534, pp. 378–387, 2007.

aggregating packets at the edge of the network. The electronic aggregation of packets is into a container, so called burst, and intermediate optical network resources are allocated burst-by-burst. Therefore, the demanding switching and processing speed requirements (which are served electronically) are relaxed and data transmission is carried out at maximum speed. The overhead incurred by operating in this way is the time spent in the reservation of optical resources whenever a burst must be transmitted (and the time spent in the corresponding resource release).

OBS networks can operate in two ways: in a hop-by-hop or in an end-to-end basis. In the first case, the burst is optically transmitted through the network just after a control packet has configured switches on a hop-by-hop basis [3,4]. However, bursts can be dropped at any point along the path to destination due to wavelength contention. In the second case, the burst is transmitted only after all the resources from source to destination have been reserved [5,6]. In [5] this type of network was termed WR-OBS (wavelength-routed optical burst switching) network. This paper focuses on end-to-end OBS networks as it has been shown to significantly reduce the burst loss rate with respect to a hop-by-hop OBS network of equivalent complexity [7].

Given that the number of wavelengths impacts significantly the network cost, the efficient dimensioning of every link in an end-to-end OBS network is of paramount importance. Usually, the dimensioning of OBS networks has been done assuming that all the network links have the same number of wavelengths, see for example [8-10]. However, this approach leads to over-dimensioned networks with under-utilised links. An alternative approach for networks using fixed routing has been to allocate as many wavelengths to a link as connections using such link (this is equivalent to the capacity required in an equivalent static network), see for example [11,12]. However, this method does not take advantage of the statistical multiplexing of connections, especially at low traffic loads. Finally, in the context of hop-by-hop OBS networks, the application of the Erlang-B (under the assumption of a Poisson process arrival of bursts) formula to quantify the wavelength requirements of the network links as a function of the link blocking has been used [12, 13]. However, a Poisson process (for which a well developed theory exists) is not representative of the real burst traffic offered to the optical core because one main property of the Poisson process is that its arrival rate does not change on time. This makes the Poisson model unsuitable for applications where this property does not hold. In an OBS network the total number of bursts that might be simultaneously sharing a given link is small (no more than a few tens in networks of practical interest) and the arrival rate at any instant to any link depends on the number of connections (from the total using such link) that are inactive. Thus, the arrival rate changes significantly when a single connection switches its state, which makes the Poisson model unsuitable for the burst traffic of OBS networks. Instead, the burst arrival process naturally follows an ON-OFF process: a source is in OFF mode during burst aggregation and in ON mode during burst transmission [14-16].

In this paper, the dimensioning of end-to-end OBS networks is carried out by means of a new algorithm which considers the ON-OFF nature of burst traffic, the required blocking probability of connections and the routing algorithm. Results show that the proposed method achieves a much lower wavelength requirement than conventional approaches whilst guaranteeing that the blocking probability per connection does not exceed a specified threshold. Additionally, the proposed method is much faster than any conventional approach exploiting the statistical multiplexing gain (typically based on the sequential execution of simulation experiments). These features make the proposed method a natural choice for implementation in future dynamic OBS networks.

The remainder of this paper is as follows: section 2 presents the network and traffic model and introduces the notation used throughout the paper; section 3 presents the proposed method whilst numerical results are presented in section 4. Section 5 summarizes the paper.

2 Network and traffic model

2.1 Network architecture

The architecture studied corresponds to an optical OBS network with end-to-end resource reservation and full wavelength conversion capability, as it has been shown that wavelength conversion improves the blocking performance of dynamic WDM networks and it is key for dynamic networks to achieve wavelength savings with respect to the static approach, see for example [17, 18].

In such an optical network architecture each node consists of an optical switch locally connected to an electronic edge router where incoming packets are classified according to destination (see upper section of Fig. 1). Every edge router is equipped with one buffer per destination, where burst aggregation is carried out. During the burst aggregation process a resource request is generated once a pre-defined condition is met (e.g. latency or burst size), and such request is sent through the network to reserve resources for the burst (see lower section of Fig. 1). In this paper requests are assumed to be processed according to a fixed routing algorithm [19]. If such algorithm finds an available route, an acknowledgement is sent to the corresponding

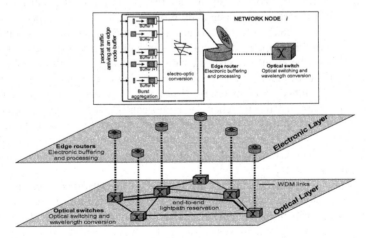

Fig. 1. Schematic of an end-to-end OBS network

source node, the network is configured to establish the end-to-end route and the burst is released into the optical core. Otherwise, the request is dropped with a NACK (Negative Acknowledgement) message sent to the source node which blocks the burst from entering the network.

2.2 Network Model

The network is represented by a directed graph $\mathcal{G}=(\mathcal{N},\mathcal{L})$ where $\mathcal{N}$ is the set of network nodes and $\mathcal{L}$ the set of uni-directional links (adjacent nodes are assumed connected by two fibers, one per direction). N and L represent the cardinality of the set $\mathcal{N}$ and $\mathcal{L}$, respectively. The capacity of the link $l \in \mathcal{L}$, in number of wavelengths, is denoted by W_l. The aim of this paper is to determine the minimum value for $\sum_{\forall l \in \mathcal{L}} W_l$ such that the value for the blocking probability of each connection is guaranteed not to exceed a given threshold.

2.3 Dynamic traffic model

The burst aggregation process transforms the original packet traffic into a burst traffic which can be described by a source which switches its level of activity between two states: ON (burst transmission) and OFF (time between the end of the transmission of a burst and the beginning of the following one). During the ON period, the source transmits at the maximum bit rate (corresponding to the transmission rate of one wavelength). During the OFF period, the source refrains from transmitting data. This behaviour has been modelled by an ON-OFF alternating renewal process, for example in [14-16], and it is also used in this paper to characterise the burst traffic generated at the edge nodes.

In the network there are $N(N-1)$ different connections (source-destination pairs). In this paper, each connection is mapped to a number –denoted by c- between 1 and $N(N-1)$. Each connection c is associated to a route, r_c, determined by a fixed routing algorithm and established only during the ON period of connection c.

According to the usual characterization of an ON-OFF traffic source [20], the demand of connection c is defined by two parameters: **mean OFF period duration** (denoted by t_{OFF}) and **mean ON period duration** (denoted by t_{ON}), equal for all connections (homogeneous traffic). The mean ON period includes burst transmission as well as the time required to reserve/release resources for the burst. The latter depends on the reservation mechanism used, which is out of the scope of this paper. According to the alternating renewal process theory [21], the traffic load offered by one connection, ρ, is given by $t_{ON}/(t_{ON}+t_{OFF})$. The traffic load can be also though as the stationary state probability that a given connection is in ON state.

3 Dimensioning algorithm

The dimensioning algorithm proposed in this paper aims to guarantee that the blocking B_c for each connection c does not exceed a given threshold B (typically agreed in the Service Level Agreement between carriers and customers).

By assuming link blocking independence, the blocking probability of connection c is given by the following expression:

$$B_c = 1 - \prod_{\forall l \in r_c} (1 - B_l); \quad 1 \leq c \leq N(N\text{-}1) \tag{1}$$

where l denotes a link in the route r_c and B_l is the blocking probability of link l.

Let H_c be the length (in number of hops) of the route r_c. There are many factors (of the form $(1\text{-}B_l)$, $\forall l \in L$) satisfying Eq. (1), including the one where all the H_c factors have the same value. To facilitate analytical treatment of Eq. (1) without modifying the target of the dimensioning process (to guarantee that the blocking B_c , $1 \leq c \leq N(N\text{-}1)$, does not exceed a given value B) the same value of B_l for all links in the route is used. This leads to the following expression:

$$1 - (1 - B_l)^{H_c} \leq B; \quad 1 \leq c \leq N(N\text{-}1) \tag{2}$$

where H_c is the length (in number of hops) of the route r_c.

Because the different connections using a given link l might have different route lengths, to guarantee that all connections (even the longest) achieve a blocking probability lower than B, the value of H_c corresponding to the longest route using link l is used in Eq. (2). Thus, the link l must be dimensioned so the following condition is met:

$$B_l = 1 - \sqrt[\hat{H}_l]{(1 - B)}; \quad \forall\, l \in L \tag{3}$$

where $\hat{H}_l$ corresponds to the length (in number of hops) of the longest route using link l.

3.1. Link blocking evaluation

Let W_l be the number of wavelengths of link l. Then, the blocking probability of link l is given by the following expression [22, 23]:

$$B_l = \lambda_l(W_l) \cdot P_l(W_l) / \lambda_l; \quad \forall\, l \in L \tag{4}$$

where $\lambda_l(w)$, $0 \leq w \leq W_l$, is the mean arrival rate of connection requests to link l when w wavelengths are in use, λ_l the mean arrival rate of connection requests to link l and $P_l(w)$, $0 \leq w \leq W_l$, the probability of w wavelengths being used in link l. That is, the link blocking corresponds to the ratio between the mean number of requests arrived (per time unit) when the link is fully used ($\lambda_l(W_l) \cdot P_l(W_l)$) and the mean number of requests arrived to the link (λ_l). Next, expressions for: a) $\lambda_l(w)$, b) λ_l and c) $P_l(w)$ in Eq. (4) are given.

a) Let T_l be the total number of connections whose routes use link l (in general, $T_l > W_l$, due to the statistical multiplexing gain) then:

$$\lambda_l(w) = (T_l - w) \cdot \lambda; \quad 0 \le w \le W_l, \forall l \in L \tag{5}$$

where $\lambda = 1/t_{OFF}$ corresponds to the mean burst arrival rate of one connection (when it switches to the OFF state).

b) From the mean value definition:

$$\lambda_l = \sum_{w=0}^{W_l} \lambda_l(w) \cdot P_l(w); \quad \forall l \in L \tag{6}$$

c) Finally, $P_l(w)$, $0 \le w \le W_l$, can be approximated from:

$$P_l(w) = P^*(w/T_l) \left[\sum_{w=0}^{W_l} P^*(w/T_l) \right]^{-1}; \quad 0 \le w \le W_l, \forall l \in L \tag{7}$$

where $P^*(w/T_l)$ is the probability that w wavelengths, $0 \le w \le W_l \le T_l$, are in use in the link l, given that the link has unlimited capacity (i.e. T_l wavelengths). The error introduced by this approximation is negligible, as shown in [23].

Because all connections have the same traffic load, $P^*(w/T_l)$ follows a Binomial distribution with parameters T_l and ρ. That is:

$$P^*(w/T_l) = \binom{T_l}{w} \rho^w (1-\rho)^{T_l - w}; 0 \le w \le W_l, \forall l \in L \tag{8}$$

3.2. Fast link dimensioning

Let B_{l_TARGET} be the value of B_l obtained from Eq. (3). Thus, each network link is dimensioned according to the following iterative procedure:

```
For each link l in the network:
        Allocate w₁=T₁ wavelengths to link l
        Evaluate B₁   (using Eq. (4))
        while (B₁<B₁_TARGET) {w₁=w₁-1; Evaluate B₁ (using Eq. (4)) }
        w₁=w₁+1;
End of procedure
```

We denote this method as ***Traffic-dependent Link-Based*** (**TLB**) dimensioning algorithm.

4   Numerical Results

The proposed dimensioning algorithm was applied to different network topologies for values of the target blocking B equal to 10^{-3} and 10^{-6}. For comparison purposes, the following two additional dimensioning algorithms were also included:

Static Link-Based (SLB) dimensioning. Each link l is allocated as many wavelengths as connections using link l (the number of connections using link l is determined by the fixed routing algorithm) [11-12]. This algorithm requires the minimum number of wavelengths to achieve zero blocking probability (in a wavelength-convertible network) and thus, it is used as a reference in this paper.

Homogeneous dimensioning (HD). All the network links have the same number of wavelengths. Initially, this number of wavelengths corresponds to the maximum number of wavelengths required by any link in the network when applying the static dimensioning. Then, the blocking probability is evaluated (typically, by means of simulation) and the original number of wavelengths is decreased by one until at least one of the connections achieves a blocking probability higher than the target. This type of dimensioning algorithm is commonly found in the literature; see for example [8-10].

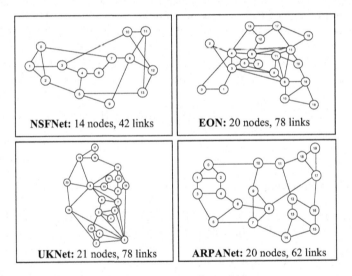

NSFNet: 14 nodes, 42 links	EON: 20 nodes, 78 links
UKNet: 21 nodes, 78 links	ARPANet: 20 nodes, 62 links

Fig. 2. Network topologies

As way of illustration, figures 3a)-d) show the results for the total network cost (that is, $\sum_{\forall l} W_l$, the sum of all the wavelengths required in the network links) achieved by the different dimensioning algorithms applied to the NSFNet and EON topologies with a target blocking B=10^{-3} and to the UKNet and ARPANet topologies for a target blocking B=10^{-6}.

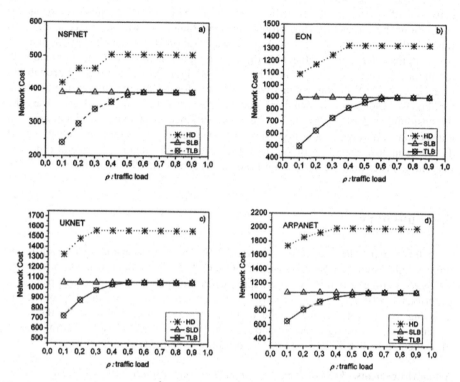

Fig. 3. Cost of a) NSFNet $B=10^{-3}$, b) EON $B=10^{-3}$, c) UKNet $B=10^{-6}$ and d) ARPANet $B=10^{-6}$ networks for the different dimensioning algorithms as a function of the traffic load

It can be seen that the TLB (Traffic-dependent Link-Based) dimensioning algorithm achieved lower or equal network cost than the SLB dimensioning algorithm. For example, for $B=10^{-3}$ the cost of TLB for traffic loads under 0.6 was lower than the reference SLB algorithm for all the studies topologies, especially at low loads (< 0.3) where the proposed method required up to the half of wavelengths than the SLB algorithm. For traffic loads higher than 0.6 both algorithms (TLB and SLB) yield the same network cost. For values of $B=10^{-6}$, the proposed algorithm performs better than the reference algorithm for loads lower than 0.4-0.5 in all the studies topologies. As most networks currently operate at low loads (typically<0.3 [24,25]), TLB will ensure the greatest benefits where it is needed most – that is in allowing the network cost to decrease whilst satisfying the required blocking.

From Fig. 3 it can also be seen that the commonly used approach of allocating the same number of wavelengths to all the links (HD) leads to the highest network cost for a wide range of traffic loads. For example, for high loads, the HD algorithm required up to 1.8 times more wavelengths than the reference SLB algorithm. Given that the reference algorithm guarantees zero blocking probability, it is highly inefficient to utilize the HD algorithm whose normalized cost is higher than 1.

The number of wavelengths per link obtained by dimensioning the different networks of Fig. 2, with algorithms TLB and SLB was validated by means of simulation: the target blocking probability per connection was set to 10^{-3} and each link of the networks shown in Fig. 2 was equipped first with the number of wavelengths determined by the TLB dimensioning algorithm and then, equipped with the number of wavelengths defined by the SLB algorithm. The blocking probability of each of the $N(N-1)$ connections was evaluated for both algorithms in the different topologies (results not shown here due to space constraints). In all the cases the dimensioning algorithms proved to be effective since the maximum blocking of all connections was always slightly lower than the target. An additional advantage of TLB was its low execution time – less than one second - making it several orders of magnitude faster than the simulation-based approach of the HD algorithm.

5.   Summary

In this paper a new algorithm, called TLB, for the dimensioning of end-to-end optical burst switched networks under ON-OFF traffic model was proposed and compared to other 2 commonly used approaches. The proposed method was proved to be effective (the target blocking was met) and better than existing approaches in allowing savings of up to half the number of wavelengths required by the static dimensioning. By being effective and fast, the proposed method is very attractive for implementation as a dimensioning tool for future dynamic OBS networks.

Acknowledgements. *Financial support from Fondecyt Projects 1000055 and 1050361 (Chilean Government), USM Projects 23.07.27 and 23.07.28 (Universidad Santa María) and DIPUV project 45/2006 (Universidad de Valparaíso) is gratefully acknowledged.*

References

1. Blumenthal, D. J.: Optical packet switching. In Proc. IEEE Leos Annual Meeting Conference, paper ThU1, Puerto Rico, November 2004
2. El-Bawab, T.S., Shin, J-D.: Optical packet switching in core networks: between vision and reality. IEEE Communications Magazine, Vol.40 (9) (2002) 60-65
3. Qiao, C., Yoo, M.: Optical Burst Switching (OBS) – a new paradigm for an optical internet. Journal of High Speed Networks, Vol.8 (1999) 69-84
4. Turner, J.: Terabit burst switching. Journal of High Speed Networks, Vol.8 (1999) 3-16
5. Düser, M., Bayvel, P.: Analysis of a dynamically wavelength-routed optical burst switched network architecture. J. of Lightwave Technology, Vol.4 (2002) 574-585
6. Arakawa, S., Miyamoto, K., Murata, M., Miyahara, H.: Delay analyses of wavelength reservation methods for high-speed burst transfer in photonic networks. In Proc. International Technical Conference on Circuits and Systems, Computers and Comms. (ITC-CSCC99) July 1999
7. Zapata, A., de Miguel, I., Dueser, M., Spencer, J., Bayvel, P., Breuer, D., Hanik, N., Gladish, A.: Next Generation 100-Gigabit Metro Ethernet (100GbME) using multiwavelength optical rings. IEEE/OSA Journal of Lightwave Technology, special issue on Metro & Access Networks, Vol. 22 (11), (2004), 2420-2434

8. Teng, J.; Rouskas, G.N.: Routing path optimization in optical burst switched networks. Optical Networks Design and Modeling Conference (ONDM), Milan, February 2005
9. Yu, J.; Yamashita, I.; Seikai, S.; Kitayama, K.: Upgrade design of survivable wavelength-routed networks for increase of traffic loads. Optical Networks Design and Modeling Conference (ONDM), Milan, February 2005
10. Kozlovski, E.; Düser, M.; de Miguel, I.; Bayvel, P.: Analysis of burst scheduling for dynamic wavelength assignment in optical burst switched networks. 14th Annual Meeting of the IEEE Lasers & Electro-Optics Society, LEOS 2001. San Diego, California. 11-15 November 2001. paper TuD2
11. Zapata A., Bayvel P.: Dynamic Wavelength-Routed Optical Burst Switched Networks: Scalability Analysis and Comparison with Static Wavelength-Routed Optical Networks. In Proceedings of Optical Fiber Communications Conference 2003 (OFC2003), 23-28 March 2003, Atlanta, USA.
12. Gauger C. M., Kohn M., Zhang J., Mukherjee B.: Network Performance of optical burst/packet switching: the impact of dimensioning, routing and contention resolution. In Proceedings of ITG-Fachtgung Photonic Networks, May 2005, Leipzig, Germany
13. Choi, J.Y., Choi, J.S., Kang M.: Link dimensioning in burst-switching based optical networks: In Proceedings of Optical Fiber Communications Conference, OFC 2006
14. Choi, J. Y. , Vu, H.L., Cameron, C. W., Zukerman, M., Kang, M.: The effect of burst assembly on performance of optical burst switched networks. Lecture Notes on Computer Science, Vol.3090 (2004) 729-739
15. Tancevski, L., Bononi, A., Rusch, L. A.: Output power and SNR swings in cascades of EDFAs for circuit- and packet-switched optical networks. Journal of Lightwave Technology, Vol.17 (5) (1999) 733-742
16. Zukerman, M., Wong, E., Rosberg, Z., Lee, G., Lu, H.V.: On teletraffic applications to OBS. IEEE Communications Letters, Vol.8 (2) (2004) 731-733
17. Elmirghani J., Moutfah H. T.: All-optical wavelength conversion: technologies and applications in DWDM networks. IEEE Communications Magazine, vol. 38, No. 3, pp. 86-92, March 2000
18. Zapata A., Bayvel P.: Optimality of resource allocation algorithms in dynamic WDM networks. In Proceedings of 10th European Conference on Network and Optical Communications, NOC 2005, London, UK, 5-7 July 2005, pp.131-138
19. Zang H., Jue J., Mukherjee B.: Review of routing and wavelength assignment approaches for wavelength-routed optical WDM networks. Optical Networks Magazine, vol.1, pp.47-60, January 2000
20. Adas A: Traffic models in broadband networks, IEEE Communications Magazine, vol. 35, no. 7, (July 1997), pp. 82-89
21. Ross S.: Introduction to probability models. 6th Edition Academic Press, 1997, (Eq. 7.15)
22. Schwartz M.: Telecommunication networks. Protocols, modelling and analysis. Addison Wesley, 1987.
23. Vallejos R., Zapata A., Aravena M.: Fast Blocking Probability Evaluation of End-to-End Optical Burst Switching Networks with non-uniform ON-OFF Input Traffic Model. Photonic Network Communications (D.O.I. *10.1007/s11107-006-0037-y*)
24. Odlyzko, A.: Data networks are lightly utilized, and will stay that way. Review of Network Economics, Vol. 2, (2003), 210-237
25. Bhattacharyya, S.; Diot, C.; Jetcheva, J.; Taft, N.: POP-Level and access-link-level traffic dynamics in a Tier-1 POP. ACM SIGCOMM Internet Measurement Workshop, Vol. 1, 39-53, San Francisco, CA, November 2001

Prudent Creditization Polling (PCP): A novel adaptive polling service for an EPON

Burak Kantarcı, Mehmet Tahir Sandıkkaya, Ayşegül Gençata, and Sema Oktuğ

Istanbul Technical University,
Department of Computer Engineering,
Computer Networks Research Laboratory
34469 Maslak, Istanbul, Turkey
{bkantarci,sandikkaya,
gencata,oktug}@itu.edu.tr
http://www.ce.itu.edu.tr

Abstract. In this paper a novel adaptive bandwidth allocation scheme called Prudent Creditization Polling (PCP) is proposed and adapted with the reservation protocol IPACT. PCP attempts to reduce the average queue length at the ONUs and the average packet delay throughout the network. At each polling period, PCP attempts to creditize the request of the ONUs that suffer from high RTT, high average delay per packet, and high buffer growth. We implemented PCP with a fuzzy functional approach and a linear regression approach. We simulated PCP and several service disciplines in IPACT and compared their performance in terms of average packet delay and queue length. We show that PCP improves the performance of IPACT significantly in terms of the performance metrics aforementioned.

Key words: Optical Access Network, Passive Optical Network, EPON, IPACT, Multipoint Control Protocol.

1   Introduction

Although deployment of optical technology in the backbone has provided sufficient solutions, the bottleneck between the high capacity local area network and the backbone seems to suffer from bandwidth problems. Hence, Passive Optical Networks (PONs) are being thought as an attractive solution to this, so-called *first-mile* problem [1].

A PON is a point-to-multipoint optical network with no active elements in the path between source and destination. The data transmission in a PON is performed between the optical line terminal (OLT) and a number of optical network units (ONUs). OLT is the unit by which the access network is connected to the backbone. End users directly get service from the ONU [2]. EPON seems to be the most attractive PON type for the next generation access networks among several PON standards [3].

Since an OLT does not share the transmission media with any other device, there is no possibility of contention in downstream transmission. The challenging

I. Tomkos et al. (Eds.): ONDM 2007, LNCS 4534, pp. 388–397, 2007.

problem in communication through PON is in the upstream data transmission where a number of ONUs have to share one fiber link to send their frames to the OLT. When more than one ONU attempt to transmit data simultaneously to the OLT, their corresponding frames may collide. The ONUs cannot be aware of each other [3].

Multi-Point Control Protocol (MPCP- IEEE 802.3ah) solves the contention problem in the upstream transmission by the REPORT and GATE messages [3]. Upon registering the network, each ONU generates REPORT messages in which its bandwidth requirement is included. Based on the REPORT messages collected from the ONUs, the OLT grants distinct transmission timeslots to the ONUs by the GATE messages.

Several bandwidth allocation algorithms are proposed to assign non-overlapping timeslots to the ONUs so that each ONU can transmit its frames without a contention although it is not aware of any other ONU in the EPON. A popular scheme is *Interleaved Polling with Adaptive Cycle Time* (IPACT) [4] which offers a number of service disciplines to assure high throughput and low packet delay throughout the EPON. Besides these, there are also several dynamic bandwidth allocation schemes that attempt to enhance QoS by adapting with DiffServ as proposed in [5–7]. A weighted fair queueing model is proposed in [8] that considers the QoS requirements of different traffic types, and schedules the packet traffic in the same ONU based on their urgency. In [9] each ONU employs a database to keep the granted bandwidths to the other ONUs so that the IPACT scheme is implemented in a decentralized manner. In [10], the authors propose a bandwidth allocation scheme where the excess bandwidth of the lightly loaded ONUs is shared among the heavily loaded ONUs, and the modified IPACT leads to a slight decrease in delay. In [11] a dynamic bandwidth allocation scheme based on analytical linear regression of the ONU buffer size is proposed.

In the IPACT scheme, OLT keeps track of $T_{schedule}$, the earliest scheduling time, on the upstream link. Whenever OLT allocates a new timeslot for an ONU, the value of $T_{schedule}$ is updated, and whenever a REPORT message containing an ONU's bandwidth request arrives at the OLT, the OLT provides a guard time just after the previously reserved timeslot, and allocates the start of the ONU's transmission window. Based on the service disciplines employed, the OLT grants a timeslot for the ONU, and using the length of this timeslot (L), it immediately updates $T_{schedule}$ for the following REPORT messages. These service disciplines can be limited service, TDMA (fixed service), constant credit service, gated service, linear credit service, elastic service schemes [4]. The previous research on this subject [4, 9, 10] shows that all of the service schemes coincide in terms of average delay per packet and average queue length under heavy traffic load.

In this paper, we propose a novel adaptive polling algorithm called Prudent Creditization Polling (PCP) that attempts to decrease average packet delay and average queue size. Based on the decrease in these parameters; the overall packet loss rate due to buffer overflow decreases. As we define in the following sections, PCP adapts the polling cycle time based on the creditization of the ONUs. The request of an ONU is creditized based on its running average buffer size (or its differential), average delay it is exposed to, and its RTT value which is computed at the previous transmission period. Hence, the algorithm aims to decrease the average packet delay by supplying longer transmission window to the ONUs which are exposed to higher delays and faster running average buffer growth. The creditization of the ONUs based on these parameters are done by either a merit based fuzzy functional approach (PCP-Merit) or an analytical linear regression approach (PCP-Regression). We adapt PCP to the IPACT and compare its performance with several service disciplines in IPACT by simulation. The simulation results show that PCP brings a significant improvement on the performance of IPACT in terms of average packet delay and average queue size at high loads.

The paper is organized as follows: The two models of the proposed polling scheme and its adaptation to IPACT is described in detail in Section II. The simulation environment and the results obtained are presented in Section III. Finally, Section IV concludes the paper by giving future considerations.

2   Prudent Creditization Polling (PCP)

The first reason of high packet delay and the large amount of queue size is the round-trip times (RTTs) of the ONUs. The signaling between the OLT and the ONU, (including the REPORT and GATE messages) and the packet transmission in the assigned timeslot lead to a buffering delay for the packets that cannot be transmitted at the corresponding timeslot. Another reason for packet delay is the fast growth of the ONU buffer. Hence, the higher running average number of queued packets leads to the higher delay per packet. Besides these, as the packets of an ONU are exposed to a queueing delay, the running average packet delay increases at the ONU. The running average packet delay, in turn, increases the average delay per packet in the EPON. In the former service schemes, the OLT attempts to creditize or weight the ONU requests just based on their instant buffer sizes. However, it does not provision the change of the buffer size in time. Besides this, the OLT considers neither the effect of current request nor buffer growth characteristics, nor the effect of RTT on the weight of the ONU request. Thus, a service scheme that considers the joint effect of these three factors is emergent for the IPACT scheme.

We derive two models to implement the proposed scheme. The first model is based on an analytical linear regression approach (PCP-Regression). The second model is based on a fuzzy functional merit based approach (PCP-Merit).

2.1   PCP-Regression

In the *PCP-Regression* scheme, the difference between the requested and the granted window size is predicted by using an analytical linear regression. We use two previous values of *round trip time*, *buffer size difference* and *running average packet delay* in regression. Modeling buffer size difference (δBS) as shown in (1) it is possible to determine the values of α (3) and β (4) using well-known *least squares method* which produces the equation set in (2). Using the coefficients, difference between the requested and the granted window size can be calculated as in (5).

$$\delta BS_i(t) = \alpha \cdot RAPD_i(t-1) + \beta \cdot RTT_i(t-1) \tag{1}$$

$$\begin{bmatrix} \sum RTT_i^2 & \sum RAPD_i \cdot RTT_i \\ \sum RAPD_i \cdot RTT_i & \sum RAPD_i^2 \end{bmatrix} \times \begin{bmatrix} \alpha \\ \beta \end{bmatrix} = \begin{bmatrix} \sum \delta BS_i \cdot RTT_i \\ \sum \delta BS_i \cdot RAPD_i \end{bmatrix} \tag{2}$$

$$\alpha = \frac{\delta BS_i(t-1) \cdot RAPD_i(t) - \delta BS_i(t) \cdot RAPD_i(t-1)}{RTT_i(t-1) \cdot RAPD_i(t) - RTT_i(t) \cdot RAPD_i(t-1)} \tag{3}$$

$$\beta = \frac{\delta BS_i(t-1) \cdot RTT_i(t) - \delta BS_i(t) \cdot RTT_i(t-1)}{RAPD_i(t-1) \cdot RTT_i(t) - RAPD_i(t) \cdot RTT_i(t-1)} \tag{4}$$

$$\delta BS_i(t+1) = \alpha \cdot RAPD_i(t) + \beta \cdot RTT_i(t) \tag{5}$$

If this result is negative, OLT grants the ONU with the requested window size and if it is greater than zero, OLT grants the ONU by adding this recently calculated value to the requested window size as shown in (6).

$$GRANT_i(t+1) = REQ_i(i) + \delta BS_i(t+1) \tag{6}$$

The above computation expects the growth with respect to time in the buffers of the ONUs by featuring linear regression. Using linear regression OLT is capable of estimating the future tendency of ONUs and reserve a more suitable transmission window than the ONUs requested. A suitable transmission window is created by determining whether the specified ONU needs a longer transmission window than it requested, or not. If the OLT determines that the tendency is at the negative side, it does not reserve additional time for the ONU. If it determines that the tendency is at positive side, it reserves additional time for the ONU. To keep the system stable, the OLT normalizes the total bandwidth grant. Although, this approach needs a bit more computation than the following one, its time complexity is still O(N).

2.2 PCP-Merit

When we construct our model, we consider three factors stated above as the reasons that increase the queuing delay and average delay per packet. Hence, we force OLT to supply more credit to the ONUs which are more likely to be affected by queuing delay. Running average packet delay (RAPD) is computed locally at each ONU just before sending the REPORT message. RAPD is used to calculate the Prudent Creditization (PC) which will later be used to send the state of the ONU to the OLT. This parameter can be computed by (7) where $RAPD_i(t)$ stands for the running average packet delay computed at the ONU_i at each burstification time t.

$$RAPD_i(t) = \frac{AveragePacketDelay}{AverageBufferSize} \tag{7}$$

When calculating the RAPD parameter, the parameters in the numerator and the denominator are measured continuously. For example, if B bytes arrive at time t to the ONU_i, the $AverageBufferSize$ is measured as shown below where $\alpha = 0.8$. $AveragePacketDelay$ is measured in a similar way;

$$AverageBufferSize_{new} = AverageBufferSize_{old} \cdot \alpha + (1 - \alpha) \cdot B$$

Moreover to compute the collective effect of the average packet delay, RTT, and the current buffer size (currently requested bytes) on the throughput of an ONU is also used. We call this parameter prudent creditization (PC) product. For the ONU_i at discrete time t we represent it by $PC_i(t)$, and compute it as shown in (8). Here, $RTT_i(t)$ is the round trip time that is computed by ONU_i at time t, and $BufferSize_i(t)$ is the queue size of the ONU at time t. $PC_i(t)$ is similar to a fuzzy-AND function [12] of these three parameters although it does not take values in $[0, 1]$. Therefore, based on an analogy to a fuzzy-AND function, $PC_i(t+1)$ can be interpreted as "increase the creditization of the ONU whose running average packet delay AND round trip time AND instant buffer size are high enough".

$$PC_i(t+1) = \frac{RAPD_i(t) \cdot RTT_i(t) \cdot BufferSize_i(t)}{RAPD_i(t) + RTT_i(t) + BufferSize_i(t)} \tag{8}$$

The computed $PC_i(t+1)$ value can be transmitted in 4 bytes by padding it into the REPORT message.

The OLT keeps a table for the state of the network where each row of the table represents the metrics computed or collected from the ONUs, namely RTT, requested window size, and PC product. Upon receiving the REPORT message from an ONU, the OLT updates $PC_i(t)$ and derives a normalized value (∇PC) with respect to the sum of the current PC product values of all ONUs as given in (3) where N is the total number of ONUs. Normalization provides fairness among the ONUs by avoiding the possibility of fiber monopolization by an ONU with a large RTT, high packet delay and high buffer growth. Therefore ∇PC takes values in $[0, 1]$.

$$\nabla PC_i(t+1) = \frac{PC_i}{\sum_{k=1}^{N} PC_k} \tag{9}$$

The request of the ith ONU for the $(t+1)^{th}$ cycle is granted by incrementing it proportional to the normalized value of the prudent creditization factor, as shown in (9) where $GRANT_i(t)$ is the transmission window size responsed by OLT for ONU_i and $REQ_i(t)$ stands for the requested transmission windows size by the ith ONU at its transmission in the end of tth cycle.

$$GRANT_i(t+1) = REQ_i(t+1) + REQ_i(t+1) \cdot \nabla PC_i(t+1) \tag{10}$$

The above computation flow can be summarized as follows: Upon updating the $GRANT$, the RTT value for the corresponding ONU is re-calculated by the OLT. When calculating the $GRANT$, the OLT considers the RTT, running average delay, and buffer size. Those considered factors, are combined with the same factors of the other ONUs, and a variable credit is obtained. That variable credit is added to the requested window. This approach is expected to decrease the overall packet delay and queue length in the network. Since in the PON, the packet loss is due to the buffer overflow, as a result of decrease in the average queue length, the packet loss ratio is also expected to decrease. We support this inference in Section IV by the simulation results under different traffic types.

The time complexity of calculating $\nabla\, PC_i(t + 1)$ is O(N). This shows that our scheme does not cause high computational complexity at the OLT. The cost of this technique is an additional $4 \cdot N$ bytes in the state table kept at the OLT to keep the PC product values. In the network traffic point of view, it brings an additional 4 bytes of control information padded to the REPORT message which corresponds to 32 ns at 1 Gbps line rate. These analysis show that PCP-Merit is also cost effective for implementation.

3 Simulation Study

3.1 Simulation Environment

The 16-node tree topology, given in [4], is used in our simulation scenarios. We assume that the line rate between the users and the ONUs (R_D) is 100Mbps while the fiber between the ONUs and the OLT has the capacity of 1 Gbps.

We generate two different traffic types at the ONUs: Self-similar traffic with Hurst parameter 0.8, and Poisson traffic. Self-similar traffic is generated as defined in [13] by aggregating 256 sub-streams on 100Mbps line rate.

The incoming packet size is distributed uniformly in [64, 1500] bytes. We generate 1000000 packet traces at each ONU.

The REPORT and GRANT messages are 84 bytes long. We employ a $5\mu s$ guard time between two adjacent transmission windows. At the beginning of the simulation, in order to construct the initial state table at the OLT, the RTTs of the ONUs are selected uniformly from the interval $[100, 200]\mu s$. At the end of each transmission cycle, the dynamic values of the RTTs are re-calculated. In order to observe the difference between behavior of the services in terms of packet delay and queue length, we set the maximum buffer size to a large value of 1 Gbytes (to introduce delay to all the buffered packets).

We compare the performance of PCP with four different service schemes in IPACT, namely *limited, fixed, constant credit,* and *linear credit* services. In *limited service* scheme we set the W_{max} to be equal to 15000 bytes, where in *fixed service* each transmission window is adjusted to transmit 10000 bytes. In *constant credit* service discipline we add a 1000 bytes credit to the requested window size while we add % 0.1 of the requested window size to the incoming request in order to employ *linear credit service* scheme (since the traffic is bursty, this ratio is observed to be sufficient to be added). Each point in the figures represents the average of ten runs.

3.2   Simulation Results

Our simulation results are two-fold: Average packet delay and Average queue length. First, we take the results under long range dependent traffic ($H = 0.8$).

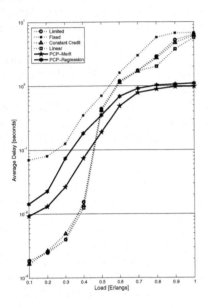
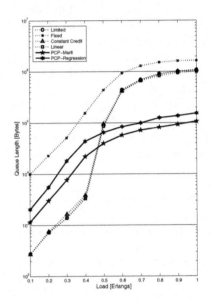

Figure 1. Average delay under long range dependent traffic

Figure 2. Average queue length under long-range dependent traffic.

As it is seen in Figure 1, at light loads, *fixed service* has the highest packet delay due to fixed transmission cycle with idle timeslots. PCP-Merit and PCP-Regression has significantly lower delay since the average delay, RTT and the buffer growth of the ONUs are taken into account both by individually and collaboratively. *Limited, constant credit* and *linear credit* services show performance similar to each other at each load. These triple perform the best at light loads. However, as the traffic offered by the sources increases to moderate and heavy loads, PCP schemes perform significantly better than all of the techniques while the other services lead to the closer delay values to each other. The reason of such a behavior is due to the fact that these services do not provision the network using a dependent scheme among the ONUs while PCP does. Therefore as the offered load gets higher PCP succeeds in delivering a fair polling service to the ONUs by causing an average delay around some 100ms. Besides fairness, the resource utilization is also improved since the average delay per packet is decreased under heavy traffic. It is also seen that PCP-Merit shows better performance than PCP-Regression.

The results showing that the average queue length of each service discipline are closely related to the results obtained for the average packet delay. The delayed packets force the buffer of the ONU to increase in size. Therefore, it is expected that the behavior of the services due to the average queue length would be similar to the behavior of the results due to the average packet delay. This intuitive analysis is supported by the results shown in Figure 2. At heavy loads, all the other schemes have results close to each other. At light loads, fixed service causes larger queue length since as a result of fixed assignment it assigns insufficient timeslots to the ONUs with heavier buffers. Here, PCP schemes result significantly lower queue lengths as the offered load increases since it reduces the packet delay duration as it is seen in the previous figure.

In the simulation results, it can also be seen that at low loads (less than 0.4 Erlang) PCP-Merit and PCP-Regression leads to higher average delay and average queue length in comparison to *limited service, constant credit service,* and *linear credit service* (it still outperforms *fixed service*). The reason is that PCP schemes attempt to creditize each incoming request based on the RTT, average delay and the buffer growth. However, at low loads, the buffer growth level of the ONUs are so low that an additional credit for the incoming requests may lead to some small idle timeslots which may increase the average delay.

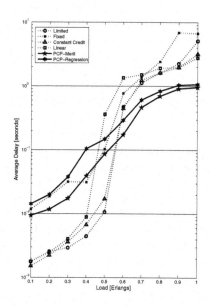

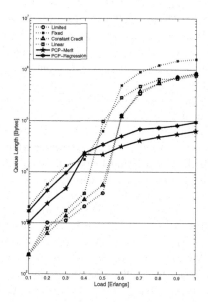

Figure 3. Average delay under Poisson traffic

Figure 4. Average queue length under Poisson traffic.

The second set of results is collected under Poisson traffic. The performance behaviors of the services are similar compared to the behavior under long range dependent traffic. However, all of the services lead to slightly lower packet delays as shown in Figure 3. The performance improvement of PCP schemes is still significant under Poisson traffic. On the other hand, it has to be mentioned that the numeric values of average packet delays of all schemes are slightly less than the results taken under long range dependent traffic. However, that slight decrease cannot be observed explicitly since these figures are in logarithmic scale.

Based on Figure 4, average queue lengths lead by the services can intuitively be expected to show a slight decrease. However, since the queue length is strongly related to the delayed packets, PCP-Merit and PCP-Regression are expected to lead to a shorter average queue length as the load gets higher. Besides these, since the delay characteristics of the remaining techniques show similarity as the load gets higher, the average queue lengths that are caused by the techniques tend to get closer to each other. The results shown in Figure 3 support these inferences.

The experimental results in this section show that as the EPON traffic gets heavier PCP schemes increase the performance of IPACT protocol by improving the average packet delay and average queue length for each traffic pattern.

4   Conclusion

In this paper, we proposed a new bandwidth allocation scheme, Prudent Creditization Polling (PCP) to improve the performance of IPACT in terms of average packet delay and average queue length. PCP is implemented in two different approaches: 1) a fuzzy functional merit based approach (PCP-Merit), and 2) a linear regression based approach (PCP-Regression). PCP adapts the polling cycle time, based on the creditization of the ONUs. The request of an ONU is creditized based on its running average buffer size (or its differential), average delay it is exposed to, and its RTT value which is computed at the previous transmission period. Hence, the algorithm aims to decrease the average packet delay by supplying longer transmission window to the ONUs which are exposed to higher delays and faster running buffer growth. We compare the performance of PCP-Merit and PCP-Regression with the other service schemes (Limited, Fixed, Constant Credit, and Linear Credit) under self-similar and Poisson traffic by simulation. The simulation results show that as the offered load gets higher, PCP leads to a significantly lower average packet delay and average queue length in comparison with the previously proposed schemes.

We are working on the effects of the service schemes on the Hurst parameter value of the traffic going through the system. We also plan to construct an enhanced version of PCP where it pays attention to the QoS requirements of the incoming traffic.

References

1. Effenberger F. J., Ichibangase H., Yamashita H.: Advances in broadband passive optical networking technologies. IEEE Communications Magazine, Vol. 39, Issue 12, (Dec. 2001) 118–124

2. Shin D-B., Lee H-S., Lee H-H., Kim D-Y.: An ONU design for an EPON-based access network. Proc. The 9th Asia-Pacific Conference on Communications, APCC, Vol. 3, (Sept. 2003) 1194 – 1197
3. Zhongjin W., Yeo D., Xiaodan G., Bin Z., Layhong T., Lam J., Cheng T. H., Walla R.: MPCP design in prototyping optical network unit of ethernet in the first mile. Proc. The Ninth International Conference on Communications Systems, ICCS (Sept. 2004) 121 – 125
4. Kramer G., Mukherjee B., Pesavento G.: IPACT a dynamic protocol for an Ethernet PON (EPON). IEEE Communications Magazine, Vol 40, Issue 2, (Feb. 2002) 74 – 80
5. Wei Q., Chen F.: A high efficient dynamic bandwidth scheme for QoS over EPON system. Proc. International Conference on Communications, Circuits and Systems, Vol. 1, (May 2005) 599 – 603
6. Ghani N., Shami N. A., Assi C., A Raja M. Y.: Quality of service in Ethernet passive optical networks. Proc IEEE/Sarnoff Symposium on Advances in Wired and Wireless Communication, (Apr 2004) 161 – 165
7. Assi C. M., Yinghua Y., Sudhir D., Ali M. A.: Dynamic bandwidth allocation for quality-of-service over Ethernet PONs. IEEE Journal on Selected Areas in Communications, Vol 21, Issue 9, (Nov. 2003) 1467 – 1477
8. Zhu Y., Ma M., Cheng T. H.: An Urgency Fair Queueing Scheduling to Support Differentiated Services in EPONs. Proc. IEEE Global Telecommunications Conference (GLOBECOM), Vol. 4, (Dec 2005), 1925–1929
9. Hossain A. D, Dorsinville R., Ali M.: Supporting Private Networking Capability in EPON. Proc. IEEE International Conference on Communications (ICC), Vol. 6, (June 2006) 2655–2660
10. Zheng J., Zheng S.: Dynamic Bandwidth Allocation with High Efficiency for EPONs. Proc. IEEE International Conference on Communications (ICC), Vol. 6, (June 2006) 2699–2703
11. Byun H., Nho J., Lim J., Dynamic bandwidth allocation algorithm in ethernet passive optical networks, Electronics Letters, Vol. 39, (June 2003) 1001–1002
12. Miyamoto S., Fuzzy Sets in Information Retrieval and Cluster Analysis, Kluwer Academic Publisher, (1990), Boston
13. Taqqu M. S., Willinger W., Sherman R.: Proof of a fundamental result in self-similar traffic modeling. ACM/SIGCOMM Computer Communication Review, Vol. 27, (1997) 5–23

Adaptive Moblile Spot Diffusing Transmiter for an Indoor Optical Wireless System

Jamal M. Alattar[+] and Jaafar M. H. Elmirghani[†]

Institute of Advanced Telecommunications
Swansea University
Singleton Park, Swansea SA2 8PP, UK
[+]e-mail: 181853@swan.ac.uk
[†]e-mail:J.M.H.Elmirghani@swan.ac.uk

Abstract: An adaptive approach is used to adjust the transmit power of each beam in a line strip multi-beam optical transmitter based on information about the quality of the received signal due to each spot. The system assumes a feedback link between the optical transceivers. Diversity detection is employed to combat the degrading effect of the background noise in the channel with a 7-detectors angle diversity receiver. Both the optical transmitter and the multi-branch receiver are fully mobile. The performance of the system is evaluated for two weakest links and compared against that of a non-adaptive spot diffusing transmitter. Our results show that the adaptive transmitter produces a considerable gain of 12.9 dB in signal-to-noise-ratio (SNR) when the distance separating the transceivers is largest in the case of a non mobile transmitter. Furthermore, a 6.8 dB increase in SNR is obtained for the mobile transmitter compared to the non-adaptive transmitter's scenario.

Keywords: Spot diffusing, mobile transmitter, multi-beam, optical wireless, indoor channel, adaptive.

1. Introduction

Optical wireless connectivity has recently emerged as a good candidate for indoor local area networks (LANs) and connection of computer peripherals, due to lack of interference as well as governmental regulation and inexpensive components. Optical wireless application growth is extending to outdoor laser links between buildings and campuses using portable terminals for high-data rate (hundreds of Mbps) links for internet backbones, transfer of medical, banking, and computer data, and other applications where high-data rate links are needed. Indoor links are still in need to upgrade and offer wireless communication with high quality signals and meet the increased data rates when portable devices are used.

Non directed line-of-sight (LOS) links, also known as "diffuse links" allow the system to operate even when barriers are placed between the transmitter and receiver; and are therefore becoming increasingly popular. A diffuse transmitter points vertically upwards towards the ceiling, emitting a wide beam of infrared energy and the receiver has a wide field-of-view (FOV), to enable it collect the signal from all reflective surfaces after it has undergone multiple reflections from the ceiling, walls and room objects. The propagation of the transmitted signals follows multiple paths before reaching the receiver's collection area in the indoor environment thus causing temporal

I. Tomkos et al. (Eds.): ONDM 2007, LNCS 4534, pp. 398–407, 2007.
© IFIP International Federation for Information Processing 2007

dispersion on the received pulses. The pulse spread in turn causes the binary transmitted symbols to overlap which introduces Inter Symbol Interference (ISI) that, if not completely corrected, will result in erroneous message detection. Directive noise sources (natural and artificial) further impair the signal and reduce SNR.

A proven technique to mitigate these effects is to replace the fully diffuse optical transmitter by one that produces multiple narrow-beams casting small diffusing spots on the ceiling [1]-[3]. The narrow beams can be practically produced using a holographic optical diffuser mounted on the face of the transmitter [4], [5]. Holograms produced using computer methods can flexibly take advantage of varying the intensity of a particular spot and/or the intensity distribution of the spots. The spots, cast on the ceiling, become secondary Lambertian transmitters (diffusers). Further improvement in the signal quality in the indoor environment has been shown with the employment of angle diversity reception (with multiple photo detectors) to replace the single-detector wide field-of-view receiver traditionally used in conventional systems. This approach was mainly implemented to confine the signal rays to within the field-of-view of the receiver detectors as well as limit the amount of received background noise through the selection of an optimum reception direction. Such designs have detectors pointed to different directions and therefore can reduce the effects of ambient light noise. Various designs of angle diversity receivers have been considered including the three branch square-based fly-eye angle diversity receiver [6], [7], the hexagonal based 7-detector diversity receiver [7] - [9] and the multi-layered narrow FOV detectors composite angle diversity receiver [10]. Since noise is directional in this environment, circuitry in the optical receiver can implement signal combining techniques which enable the diversity receiver to combine its branch detectors' signals or select the one with best SNR. An efficient angle diversity receiver design has to produce a high and uniform SNR distribution within the room. Increasing the number of branches of the diversity receiver with an optical photo-detector on each has been shown to improve the gain achieved with spot diffusing configurations [7]. Combining the detected signals from each photo detector using maximum ratio combining (MRC) yields better output than either selecting the detector with the highest signal-to-noise-ratio (SNR) (selection combining) or equally combining all detectors' signals (equal gain combining, EGC).

In this work, we present a study of a novel approach to optimise signal reception with mobile transceivers. The method focuses on varying the power allocated to each beam in a single line strip multi-beam transmitter so as to optimise the received SNR for a given receiver location. A new adaptive method is proposed and used to set the amount of power transmitted on each beam based on information about the signal quality fed back to the transmitter by the receiver. Effectively, the proposed method varies the optical power in each beam or spot (secondary transmitter) in a manner that optimises the SNR at the receiver. As a result of the transmitter's mobility, the diffusing spots will change their locations [11],[12]. Spots that are closer to the receiver are to be given higher transmit powers than those far away from the receiver especially when the distance separating the transceivers becomes large (e.g., transmitter near one room corner and the receiver near an opposite room corner). The following proposed protocol is used to obtain the power distribution ratios among the line strip beams: periodically the transmitter switches on one of the beams in turn and the receiver measures the SNR due to that beam (diffusing spot). The process is repeated with the transmitter switching on a beam at a time. The receiver then transmits back to the transmitter at a low data rate (for reliability) the measured SNR. The transmitter then sets its beams power tap weights in proportion to the SNR. The tap weights are adjusted periodically depending on the degree of mobility anticipated in the environment. Our approach has achieved an SNR gain of as much as 12.9 dB without transmitter mobility and as high as 9.7dB with transmitter mobility. Furthermore, an improvement of 6.8 dB is obtained at the weakest links.

2.   System Setup

The optical wireless link is established on a communication plane (CP) 1m above the floor in an empty mid-sized room (size: 4m×8m×3m Width, Length, Height). The room has no doors or windows and the walls and ceiling plastering results in 80% reflectivity of incident light whereas the floor produces reflectivity of 30%. A 1 W upright optical transmitter (elevation = 90°) is placed in the centre of the communication plane (at x = 2m and y = 4m). Multiple narrow beams are produced with a holographic device mounted on the face of the optical transmitter resulting in a cluster of 80 equally-separated narrow beams casting a spot on the ceiling as depicted in Fig. 1. The power associated with each spot can be adjusted using a liquid crystal device for example.

The 7-detectors angle diversity receiver used has a hexagonal base as shown in Fig. 2. One detector is placed on its top and one on each of the six elevated branches with 20° elevation angle. The side detectors have symmetric azimuth angles with the x and y axes. Hence the azimuth angles: 45°, 90°, 135°, 225°, 270° and 315° and 0° respectively were used. The detectors' FOVs were optimised to achieve highest SNR from the line of diffusing spots. The detectors that face two walls (detectors 1, 3, 4 and 6) were given larger FOVs than those facing only one wall (detectors 2 and 5) so as to collect most of the signals reflected off both walls. Hence, the FOVs of the six detectors on the side branches were 60°, 50°, 60°, 60°, 50° and 60°. The FOV of the top detector was chosen smaller (24°) in order to avoid the noise direct power component as the noise sources are directive.

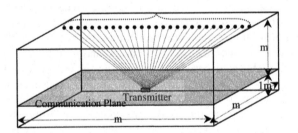
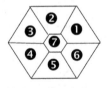
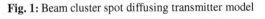

Fig. 1: Beam cluster spot diffusing transmitter model   **Fig. 2:** 7-detectors angle diversity receiver (top view)

The background noise was evaluated using eight incandescent light sources placed equidistantly on the ceiling, 2m apart (along the lines x = 1m and x = 3m) starting at x = 1m and y = 1m; thus producing a well-illuminated environment. Each lamp is of type

Philips PAR 38 which emits an optical power of 65W where the mode number n = 33.1.

Ray-tracing simulation was used to compute the received optical signal as well as the background noise following the algorithm introduced by Barry et al. [13]; the simulation was developed in C++. Simulation was carried out along two lines on the communication plane: one closer to a wall, x = 1m (where the signal reception is expected to be influenced by reflections) and the other along the middle line of the CP, x = 2m, with 1m receiver position intervals. Results were produced for 14 different receiver locations (starting 1m away from the wall). Due to symmetry in the room, results for the line x = 3m will be the same as for the line x = 1m.

3.   System Analysis

3.1 Mobile transmitter analysis

The evenly spaced beams are produced with equal emitted angles in one direction. Based on the fact that as the transmitter moves around on the communication plane, the beams' angles remain unchanged, the new coordinates of the spots can be computed with respect to the original transmitter location (at the centre of the communication plane). As the transmitter approaches a wall, some spots will start appearing on the wall as their beams get intercepted by that wall, Fig. 3. A transmitter's motion in either the y or x direction causes spots to fall on the wall that the transmitter is approaching. When the transmitter moves in the y direction, for example, (across room length) the diffusing spots will move away from one wall and become closer to the other and vice versa. Further transmitter movement will cause some spots to move from the ceiling and appear on the wall.

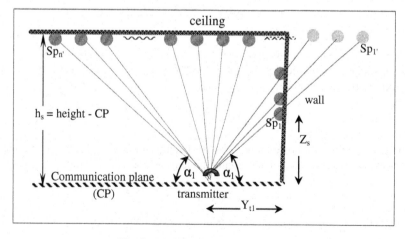

Fig. 3: Mobile transmitter scenario

Therefore, as shown in Fig. 3, the new vertical distance Z_s (from the CP) of the spot that appears on a wall is found from

$$Z_s = Y_{t1} \cdot \tan(\alpha_i) \tag{1}$$

where Y_{t1} is the new distance of the transmitter - on the y axis - from the wall it is moving to, and α_i is the angle of beam i with respect to the communication plane given by

$$\tan(\alpha_i) = h_s/d_{yi} \tag{2}$$

where h_s = height – CP = 2m and the distance d_{yi} is found from

$$d_{yi} = \frac{length}{2} - \left(\frac{length}{2N_y} + \left(\frac{length}{N_y} \right) \cdot i \right), \qquad 1 \le i \le \frac{N_y}{2} \tag{3}$$

where length is the room length (dimensions along the y-axis) and the first spot is assumed to be $length/2N_y$ from the wall, and for the other half of the spots to the right of the transmitter

$$d_{yi} = \left(\left(\frac{length}{N_y} \right) \cdot i - \frac{length}{2N_y} \right) - \frac{length}{2}, \qquad \frac{N_y}{2} + 1 \le i \le N_y \tag{4}$$

3.2 Signal-to- noise ratio calculation

The simplest modulation technique for OW systems is On-Off keying (OOK) which employs a rectangular pulse with duration equal to the bit period for each binary bit. The bit rate for our OW system is 50 Mbits/s giving a bit period of 20 ns. The SNR associated with the received signal is given by

$$SNR = \left(\frac{R \times (P_{s1} - P_{s0})}{\sigma_t} \right)^2 \tag{5}$$

where R is the detector responsivity (R = 0.5 A/W in this study), P_{s1} and P_{s0} are the powers associated with a logic "1" and logic "0" received signals respectively and σ_t^2 is the total noise variance which can be classified into three categories as

$$\sigma_t^2 = \sigma_{bn}^2 + \sigma_{pr}^2 + \sigma_s^2 \tag{6}$$

The first, σ_{bn}, is the shot noise induced by the background light which can be computed from its respective associated background noise power level P_{bn} using

$$\sigma_{bn} = \sqrt{2 \times q \times P_{bn} \times R \times BW} \tag{7}$$

where q is the electron charge and BW is the receiver bandwidth.

The second noise component is the receiver noise generated in the preamplifier components. The preamplifier used in this work is the positive-intrinsic-negative bipolar-junction-transistor (PIN-BJT) design proposed by Elmirghani et al [14]. This preamplifier structure has a noise current density of $\sigma_{pr} = 2.7$ pA/√Hz and a bandwidth of 70 MHz. Therefore, the preamplifier shot noise σ_{pr} is

$$\sigma_{pr} = 2.7 \times 10^{-12} \sqrt{70 \times 10^6} = 0.023 \, \mu A \tag{8}$$

Finally, the noise induced by the received signal power σ_s, consists of two parts depending on the logical level of the received signal. The shot noise current is σ_{s1} when a signal logic "1" is received and a different shot noise current σ_{s0} when a signal logic "0" is received. This signal dependant noise is very small in this work and therefore can be neglected.

Hence, the SNR can be calculated using equations (5) through (8) as

$$SNR = \frac{\left(P_{s1} - P_{s0}\right)^2}{8 \times \left(\left(0.023 \times 10^{-6}\right)^2 + 70 \times 10^6 \times q \times P_{bn}\right)} \tag{9}$$

3.3   Maximum ratio combining

In an optical receiver with multiple branches, the signal collected from each branch detector is processed separately to produce the resulting output electrical signal Circuitry integrated within the optical receiver has the purpose of either the selection of one branch detector or the combination (with some predefined criteria) of the detected optical signal from some or all branches. The select best scheme also known as selection combining (SC) chooses the branch with best SNR value. Two widely known combining techniques in diversity reception are the equal gain combining (EGC) and the maximum ratio combining (MRC).

While the EGC method adds the detected signals from all branches together, the MRC combines these signals according to weights proportional to their collected noises. It turns out that the EGC technique is a special case of the MRC with the combining weights set to unity (i.e., 1). For the MRC receiver, a signal multiplier circuit is added before the combiner circuit which takes its weight factor from the SNR estimator of a branch detector to produce the proportional gain of that branch. The maximal-ratio combiner circuit requires a variable gain amplifier per sector and a summing circuit. Clearly the differences between the combining methods are most noticeable as the unbalance in the distribution of the SNR among the sectors increases. Under the assumption of independent noise, the optimum output SNR is achieved by the maximal-ratio combining receiver [15]. The SNR using the Maximum Ratio Combining method is given by

$$SNR_{MRC} = \frac{\left(\sum_{i=1}^{J} \left(w_i \cdot I_i\right)\right)^2}{\sum_{i=1}^{J} \left(w_i \cdot \sigma_i\right)^2}, \qquad 1 = i = J \tag{10}$$

where σ_i is the standard deviation of the total noise in branch i.

Setting the weight w to $\dfrac{I}{\sigma^2}$, it can be shown that the output SNR is

$$\text{SNR}_{\text{MRC}} = \sum_{i=1}^{J} \left(\frac{I_i}{\sigma_i} \right)^2, \qquad 1 = i = J \tag{11}$$

4.   Results

Figure 4 shows the SNR distribution along two lines: the room centre line x = 2m and its edge line, x = 1m. Results are given at two positions in addition to the reference position at the centre of the communication plane (at 2m, 4m, 1m). The results are quoted for an adaptive transmitter versus those obtained with a non-adaptive transmitter. The positions selected are at the two weakest links: near the corner at (1m, 1m, 1m) and near a side wall at (2m, 7m, 1m). It is seen that the use of an adaptive transmitter produced enhancement in the performance of the spot diffusing OW system. This increase in SNR with an adaptive transmitter is seen to be more along the centre line than along the edge line. On the edge line, the line of spots lies directly underneath the noise sources whose influence on the system's performance is clear with lower SNR levels than along the room centre line. In spite of the strong influence and the directive nature of the noise sources, the employment of a line strip spot diffusing transmitter combined with a multi-detector angle diversity receiver produced a uniform SNR distribution across the length of the room.

The introduction of transmitter mobility alters this uniform SNR distribution. This can be seen in the plots where the highest SNR levels are where the distance between the transceivers is shortest and vice versa. The SNR levels with the adaptive transmitter at the farthest receiver positions from the transmitter have increased by as much as 6.8 dB as can be clearly seen from second plot of Fig. 4b. Although with the adaptive transmitter a considerable gain in SNR has been obtained even with transmitter mobility, some receiver positions did not benefit from this technique. When the transmitter is near the room corner; at (1m, 1m, 1m) and the receiver at y = 5m (x = 1m and x = 2m) for example, no gain in SNR is observed when the adaptive method is used. This is attributed to the small optical power collected from the diffusing spots as a result of the transmitter moving to the corner and the weak signal reflected off the wall as the receiver is now farthest from the wall (distance = 3m). The same can be said to explain the equal SNR levels for the case of transmitter at (2m, 7m, 1m) and receiver at x = 1m and y = 3m when both adaptive and non adaptive transmitters are used. The effect of noise along the room centre line is minimal which produced higher SNR than along the room edge line.

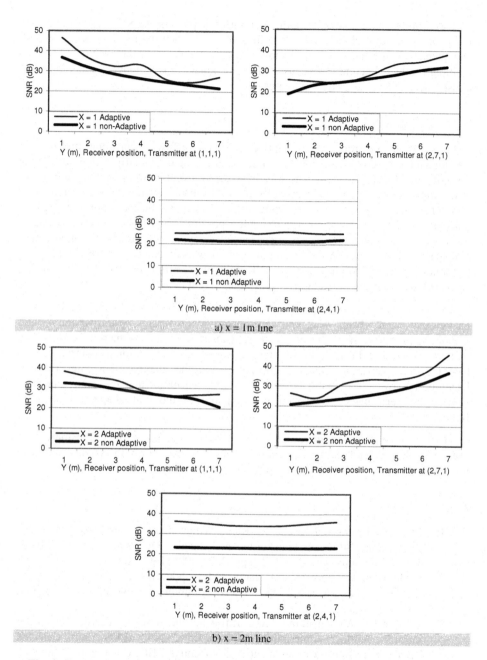

Fig. 4: Signal-to-noise ratio (SNR) comparisons of mobile spot diffusing transmitter across room length (adaptive and a non-adaptive)

Multipath reflection is seen to improve the received signal quality when comparing the SNR levels as the receiver is closer to a wall (at y = 1m and 7m) with that when it is away from the same wall (at y = 2m and 6m). The plots of the SNR distributions due to the adaptive transmitter show a rise in SNR level when the receiver is near a wall and then a sloped decrease as it moves away towards the room centre where the multipath effect becomes less as reflected signals travel longer paths and the transmitter-receiver separation increases.

5.   Conclusions

In this paper, we proposed an adaptive multi-spot diffusing transmitter and showed the advantage of employing the technique with a single line strip along with angle diversity reception. The aim was to adjust the transmit powers of the multi-beam transmitter relative to the receiver's location. Performance evaluation was reported for a system employing a 7-detectors angle diversity receiver. Furthermore, the adaptive transmitter approach was applied to a mobile spot diffusing transmitter and the results show a considerable performance improvement compared to the system employing a non adaptive transmitter. An SNR gain of 12.9 dB was obtained over the non adaptive transmitter without mobility and an SNR gain of over 6.8 dB was obtained at the weakest links when the non adaptive transmitter is mobile.

The use of angle diversity reception is very effective in reducing the effect of background noise. Employing a 7-detectors angle diversity with maximum ratio combining produced the highest uniform SNR distribution across the room length. The mobility of the spot diffusing transmitter influenced this distribution as the transceivers have maximum separation distance. Moreover, with the adaptive transmitter, receiver locations far away from both the spot diffusing transmitter and the room side wall showed no performance improvement compared to the non adaptive transmitter figures.

The multipath reflection increased the SNR gain when the adaptive transmitter was used. The delay spread and the optical path loss distribution of both systems were also investigated but not included in this study for brevity.

References

1.   Pouyan Djahani and Joseph M. Kahn, "Analysis of Infrared Wireless Links Employing Multibeam Transmitters and Imaging Diversity Receivers," IEEE Trans. Commun., vol. 48, no. 12, pp. 2077-2088, Dec. 2000.
2.   S. T. Jivokova and M. Kavehard: "Multispot diffusing configuration for wireless infrared access," IEEE Trans. Commun., vol. 48, pp. 970-978, Jun. 2000.
3.   G. Yun and M. Kavehrad, "Spot diffusing and fly-eye receivers for indoor infrared wireless communications," in Proc. 1992 IEEE Conf. Selected Topics in Wireless Communications, Vancouver, BC, Canada, pp. 286–292, Jun. 1992.
4.   M. R. Pakravan, E. Simova, and M. Kavehard, "Holographic diffusers for indoor infrared communication systems," Int. J. Wireless Inform. Networks, vol. 4, no. 4, pp. 259-274, 1997.

5. P. L. Eardley, D. R. Wiely, D. Wood, and P. McKee, "Holograms for optical wireless LANs," IEE Proc. Optoelectron., vol. 143, no. 6, pp. 365-369, Dec. 1996.
6. A. G. Al-Ghamdi and J. M. H., Elmirghani, "Line Strip-Diffusing Transmitter Configuration for Optical Wireless Systems Influenced by Background Noise and Multipath Dispersion," IEEE Trans. Commun., vol. 52, no. 1, pp. 37-45, Jan. 2004.
7. A. G. Al-Ghamdi and J. M. H., Elmirghani, "Performance Evaluation of a Triangular Pyramidal Fly-Eye Diversity Detection for Optical Wireless Communications," IEEE Commun. Mag., pp. 80-86, Mar. 2003.
8. A. G. Al-Ghamdi and J. M. H., Elmirghani, "Performance evaluation of a pyramidal fly-eye diversity antenna in an indoor optical wireless multipath propagation environment under very directive noise sources," IEE Proc.-Opto., vol. 150, no. 5, pp. 482-489, Oct. 2003.
9. A. G. Al-Ghamdi and J. M. H., Elmirghani, "Spot diffusing technique and angle diversity performance for high speed indoor diffuse infra-red wireless transmission," IEE Proc.-Optoelectron., Vol. 151, No. 1, pp. 46-52, Feb. 2004.
10. J. M. Alattar, J. M.H. Elmirghani, "Multi-line Multi-Spot Diffusing Indoor OW Channel with A 7-Detectors Diversity Receiver," Proc. of the London Communications Symposium, University College London, pp. 37-40, Sept. 2006.
11. A. G. Al-Ghamdi and J. M. H., Elmirghani, "Characterization of mobile spot diffusing optical wireless system with diversity receiver," IEEE Commun. Society, pp. 133-138, Jan. 2004.
12. J. M. Alattar, J. M.H. Elmirghani, "Evaluation of An Indoor OW Channel Employing A Mobile Multi-line Multi-Spot Diffusing Transmitter and A Seven Detectors Angle Diversity Receiver," Proc. of the London Communications Symposium, University College London, pp. 29-32, Sept. 2006.
13. John R. Barry, et al. "Simulation of Multipath Impulse Response for Indoor Wireless Optical Channels," IEEE Journal on selected areas in communication, Vol. 11, no. 3, pp. 367–379, Apr. 1993.
14. J. M. H. Elmirghani, H. H. Chan, and R. A. Cryan, "Sensitivity evaluation of optical wireless PPM systems utilising PIN-BJT receivers," IEE Proc.-Optoelectron., vol. 143, no. 6, pp. 355-359, Dec. 1996.
15. Rui Tomaz Valadas and A. M. de Oliveira Duarte, "Sectored Receivers for Indoor Wireless Optical Communication Systems," 5th IEEE International Symposium on Personal Indoor and Mobile Radio Communications (PIMRC'94),The Netherlands, pp. 1090-1095, Sept. 1994.

Extra Window Scheme for Dynamic Bandwidth Allocation in EPON *

Sang-Hun Cho, Tae-Jin Lee, Min Young Chung, and Hyunseung Choo

School of Information and Communication Engineering
Sungkyunkwan University 440-746, Suwon, Korea
{shcho,tjlee,mychung,choo}@ece.skku.ac.kr

Abstract. To ensure efficient data transmission for multimedia services in Ethernet passive optical networks (EPON) which are considered as a promising solution to the last-mile problem in the broadband access network, they employ the media access control (MAC) mechanism by sharing efficiently the bandwidth of all optical network units (ONUs) and by avoiding data collisions in the upstream channel. The representative dynamic bandwidth allocation scheme, Interleaved Polling with Adaptive Cycle Time (IPACT), is considered as a standard approach in services for requests of ONUs. It reduces the performance of the entire network in terms of mean packet delay and packet loss ratio, due to congestion for the case that an ONU has burst traffic or highly loaded traffic. To handle this, the proposed scheme varies the cycle length in the basic period center and guarantees a maximum window size per ONU. In this paper, the proposed scheme demonstrates enhanced performance in terms of mean packet delay and packet loss ratio, of up to 58% and 10%, respectively.

1 Introduction

EPON is a next generation broadband access network selected by the IEEE 802.3ah Task Force [1], as the solution to the last-mile problem, and keeps the advantages of wide Ethernet deployment, while reducing the cost of fiber infrastructure. EPON is composed of an optical line termination (OLT) and several ONUs, such as asymptotic structure, which is a point-to-multipoint network in the downstream direction and a multipoint-to-point network in the upstream direction [2, 7]. To avoid data collision in the upstream channel, it uses the multipoint control protocol (MPCP) for sharing efficiently the upstream bandwidth as exchanging of REPORT and GATE massages. That is, an ONU reports its bandwidth requests to the OLT and then transmits only bandwidth granted by

---

* This research was supported by the MIC(Ministry of Information and Communication), Korea, under the ITRC(Information Technology Research Center) support program supervised by the IITA(Institute of Information Technology Assessment), IITA-2006-(C1090-0603-0046), and by the Korea Research Foundation Grant funded by the Korean Government(MOEHRD) (KRF-2005-042-D00248). Corresponding author: H. Choo.

I. Tomkos et al. (Eds.): ONDM 2007, LNCS 4534, pp. 408–417, 2007.

the OLT [1]. The bandwidth allocation problem is an important issue for the passive optical network. EPON systems need to use an efficient bandwidth allocation algorithm for providing users guaranteed network services and enhancing the network performance.

In general, bandwidth allocation schemes are classified into static ones and dynamic ones. Static bandwidth allocation (SBA) schemes allocate a fixed time slot regardless of the variable requests for ONUs. DBA schemes allocate a time slot of appropriate size for variable bandwidth requests of each ONU. SBA is more easily implemented than DBA, nevertheless, there is considerable DBA research being conducted [3–11], because SBA is not adaptable to the burst nature of network traffic. DBA uses an interleaved polling mechanism, overlapping upstream and downstream at the same time for using efficiently optical channel and reducing packet delay by exchanging MPCP messages. These polling mechanisms can be classified as interleaved polling and interleaved polling with a stop for upstream transmission. Typical interleaved polling schemes are IPACT [3] and Sliding Cycle Time (SLICT) [4]. In addition, DBA for Quality-of-Service (QoS) [5] and Two-Layer Bandwidth Allocation (TLBA) [6] exists in interleaved polling with a stop.

IPACT is an adaptive cycle scheme with a changable cycle time according to each ONU's bandwidth request. The paper introduces the *gated* service that allocates unlimitedly about requests of each ONU, and variable services such as *limited, constant credit, linear credit* and *elastic*, which prevent monopolization of the entire bandwidth and reducing bandwidth to waste [3]. SLICT is an improved DBA algorithm based on *limited* and *elastic* services in IPACT [4]. In DBA for QoS [5], the OLT allocates optimized bandwidth via total computation after receiving REPORT messages from all ONUs in order to assure QoS. TLBA [6] first divides the entire bandwidth of a cycle into three priority class (Class-layer allocation), then it computes allocation bandwidth of each ONU after dividing again a class into the number of ONUs (ONU-layer allocation). In this paper, based on interleaved polling having comparatively high network throughput, we propose an efficient algorithm with better performance than existing algorithms.

In this paper, to support the best service in EPON systems, without considering a service level agreement (SLA), we make up for the weak points of *elastic* service in IPACT. The first of them is that some ONUs are allocated instable bandwidth every cycle, the other is that the *elastic* service whose average cycle time is approximately 1.887 ms has more overheads than *limited* service because it uses more cycles for transmission of the same bandwidth. The former is solved into assurance of the maximum window size, the latter is solved by changing the sum of N windows into the sum of $N + 1$ windows when the entire bandwidth is computed. According to the proposed scheme, mean packet delay decreases by up to 54.25 % than that of *elastic* service.

In section 2 of this paper, we describe typical DBA algorithms headed by IPACT and discuss problems with the existing schemes. In section 3, we propose an algorithm that solves above problems. In section 4, we compare and evaluate the performance of the proposed scheme with *limited* and *elastic* services in IPACT. We conclude this paper in section 5.

2   Related Work

In EPON architectures, according to the size of round trip time (RTT) caused by propagation delay that arises from the distance among the OLT and ONUs, network throughput decreases. To avoid this decrease, it is an interleaved polling mechanism that overlaps messages and data without interference. According to this mechanism, bandwidth which excepts for the overheads for messages, can be used very efficiently. Fig. 1 represents an example that works using an interleaved polling algorithm. The OLT sends a GATE message that makes ONU1 transmit 5,000 bytes which it has requested, and the ONU1 reports a request of 8,000 bytes, after it transmits the upstream message during a time slot granted by the OLT. The OLT received a REPORT message and renews the request information in its polling table, transmitting a GATE message to arrive at ONU1 before ONU1 starts transmission in the next cycle. Where a cycle is a period from when the OLT sends a GATE message into ONU1 to when the OLT sends the next GATE message into ONU1.

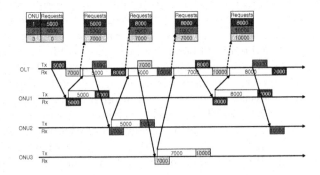

Fig. 1. Interleaved polling algorithm's operation.

IPACT suggests five services according to bandwidth allocation policies. Table 1 summarizes these five services, where $W^{[i]}$ is the size of bandwidth that OLT allocates to the i-th ONU, and $V^{[i]}$ is the size of bandwidth that the $i-th$ ONU requests. Δ is a constant value, and W_{Max} is the maximum window size. The *gated* service that does not limit the size of bandwidth has better performance than others, however, it is inappropriate for high quality services because cycle time becomes unlimitedly longer, according to queue size of ONUs. *Limited* service assures the same maximum window size for all ONUs. If the request bandwidth is smaller than the maximum window size, then the OLT grants only request bandwidth. *Constant credit* service, a modification of *limited* service, allocates additional bandwidth as much as a constant size, when the request bandwidth is smaller than the maximum window size; however, it has an opposite effect of additional overhead because amount of arrival traffic during the waiting time is not regular.

Linear credit service allocates also additional bandwidth as much as a proportion of request bandwidth if this is smaller than the maximum window size. This service improves *limited* service, like *constant credit* service; however, its performances decrease rather than the *limited* service in terms of mean packet delay, average queue size, and mean cycle times according to results in [3]. At last, *elastic* service is designed to allow other ONUs to use the remaining bandwidth after consumed by the previous ONUs. A limit factor is not maximum window size but the entire bandwidth (the number of ONUs times the maximum window size) unlike *limited* service. If the number of ONUs is N, OLT grants the smaller one between the request bandwidth and the remaining bandwidth excluding the size granted of the previous N ONUs. In these five services, *limited* service shows steadily better performance than the others [3].

Table 1. Several services of IPACT.

Service type	DBA Computation Formula
gated	$W^{[i]} = V^{[i]}$
limited	$W^{[i]} = MIN\{V^{[i]}, W_{Max}\}$
constant credit	$W^{[i]} = MIN\{V^{[i]} + \Delta, W_{Max}\}$
linear credit	$W^{[i]} = MIN\{V^{[i]} \cdot \Delta, W_{Max}\}$
elastic	$W^{[i]} = MIN\{V^{[i]}, N \cdot W_{Max} - \sum_{j \equiv (i-N) mod N}^{i-1} W^{[j]}\}$

The existing *elastic* service of IPACT is free from constraints of maximum window size unlike *limited* service. The only limitation is the maximum cycle time. The maximum bandwidth possible during the maximum cycle time is $N \cdot W_{Max}$. It allocates the smaller one between the last N accumulative allocation bandwidth and request bandwidth. Thus, the allocation bandwidth for the i-th ONU is

$$W^{[i]} = MIN\{V^{[i]}, N \cdot W_{Max} - \sum_{j \equiv (i-N) mod N}^{i-1} W^{[j]}\}. \qquad (1)$$

Shortcoming of this service is that the present available bandwidth can be limited into zero resluting from allocation size in the previous cycle. On the other hand, when an ONU fully uses available bandwidth, any of the next ONUs cannot be allocated a bandwidth because its available bandwidth can be zero if its previous request is zero and the summation is larger than $N \cdot W_{Max}$ in equation (1). In addition, average cycle time is shortened due to the same effect that $N + 1$ ONUs share the entire bandwidth of N times maximum window size. In other words, it needs more cycles to transmit the same bandwidth, and increases overheads of basic requirements for each cycle, such as guard time, frame gaps, messages, and so on.

Fig. 2 presents the problems occurring when *elastic* service is used. We assume that the number of ONUs is three and the maximum window size is 5,000 bytes. Thus, the available entire bandwidth during a cycle time is 15,000 bytes. Before the OLT allocates bandwidth for ONU1, it knows the previous requests in the table T_1 and grants in the table T_2 information from its polling table. From the table T_2, the sum of the previous grants is 15,000 bytes, and allocation bandwidth of ONU1 becomes zero byte since its request is zero byte. While the OLT is granting zero byte into ONU1, it updates the grant infomation in the grants table T_3. In table T_3, available bandwidth excluding 10,000 bytes, sum of the previous grants from the entire bandwidth is 5,000 bytes; and a request of ONU2 is 7,000 bytes, and then allocation bandwidth is 5,000 bytes. In the same way the OLT grants 5,000 bytes into ONU2, and updates table T_4. Because the available bandwidth in table T_4 is 5,000 bytes and its request is 8,000 bytes, 5,000 bytes is allocated. In table T_5 5,000 bytes is allocated. Although a request in table T_6 is 9,000 bytes, zero byte are allocated because the available bandwidth is zero byte. That is, if *elastic* service is used, any ONU may not be allocated nevertheless it has requests.

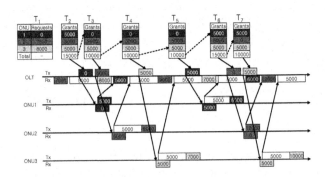

Fig. 2. Non-grant case in *elastic* service.

3   The Proposed Algorithm

In EPON architecture, many users are accessed on an ONU. It is quite probable that the characteristic of users accessed on each ONU is similar. The differences of ONUs' offered traffic loads are clearly heavy and light. So, we treat an EPON system with unbalanced load in each ONU. In the mean time, SLA assures the bandwidth decided according to service contract with users, and it is considered by many researches; however, it sometimes becomes a factor that decreases the efficiency of the access network in terms of resources utilization because it is allowed to transmit bandwidth predetermined by SLA; nevertheless, the network is idle. Therefore, we assume that EPON systems don't use SLA on purpose to maximize utilization of network resource.

In the considered environment, if the system uses the *limited* service in IPACT, it cannot respond to the purpose for requests of ONUs with unbalanced load, because all ONUs have the same maximum window size. For example, in case that an ONU' traffic load is high and the other is light, packet transmission delay of the ONU is much increased since its maximum window size is limited and overheads by light loaded ONUs increase. *Elastic* service is more adaptable than *limited* service. However, according to bandwidth that is allocated to the last $N - 1$ ONUs, the present ONU cannot be guaranteed in bandwidth allocation. When the number of ONUs is 16 and the cycle time is $2ms$, in case of using this scheme, average cycle time is not more than $1.887ms$. Thus, the *elastic* service cannot be a counterplan of *limited* service. To solve this problem, we improve *elastic* service so that it can assure the maximum window size for each ONU and its average cycle time can increase by increasing the entire bandwidth.

We allow the OLT to guarantee bandwidth up to the maximum window size for the ONU that requests more bandwidth than the maximum window size. For instance, in the case that an ONU requests less bandwidth than the maximum window size, the OLT grants bandwidth on demand; otherwise, the OLT grants bandwidth up to the maximum window size. It inherits the strength of the *limited* service. And we inherit the strength of the *elastic* service that uses efficiently entire bandwidth. Unlikely the *elastic* service, by adding an extra window to an entire bandwidth in a cycle, it allows $N + 1$ ONUs to share $N + 1$ windows. The extra window is only used in DBA computation and is excluded from composing a cycle. Consequently, allocation bandwidth for the i-th ONU computed by this scheme is

$$W^{[i]} = MIN[V^{[i]}, MAX\{W_{Max}, (N + 1)W_{Max} - \sum_{j \equiv (i-N)modN}^{i-1} W^{[j]}\}]. \quad (2)$$

The pseudo code in Fig. 3(a) is that the OLT computes bandwidth to grant to the $i - th$ ONU after receiving the REPORT message from the ONU. Where W_{Max} is the maximum window size for an ONU and W_A is available window size for the current ONU. The number of ONUs is N, G_L^i is the last grant size for $i-th$ ONU, and G_S^N is a summation of the last N grant sizes. If a request is more than maximum window size and is more than the available window size, then the OLT grants the larger window size between two window sizes; otherwise, it grants as much as the request bandwidth. Then G_S^N, the sum of previous N grants, is renewed into the sum of the recent N grants, the G_S^N minuses a grant for the $i - th$ ONU in the previous cycle and pluses a grant for the ONU in this cycle. For similar computation in the next cycle, an array G_L^i saves the grant information in this cycle.

To help understanding, we represent the grant situation in Fig. 3(b), where the OLT decides grant bandwidth by a relation about a request Ri, an available window size W_A and maximum window size W_{Max}. (b)-1 shows that the OLT computes the available window size, which is $N + 1$ times the maximum window size minus the recent N grants, where G_L^i is the last grant size for $i - th$ ONU

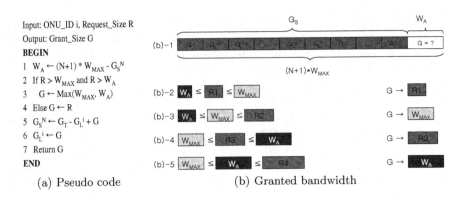

```
Input: ONU_ID i, Request_Size R
Output: Grant_Size G
BEGIN
1  W_A ← (N+1) * W_MAX - G_S^N
2  If R > W_MAX and R > W_A
3     G ← Max(W_MAX, W_A)
4  Else G ← R
5  G_S^N ← G_T - G_L^i + G
6  G_L^i ← G
7  Return G
END
```

(a) Pseudo code   (b) Granted bandwidth

Fig. 3. Pseudo code and bandwidth granting in the proposed scheme.

and G_S^N is the sum of the last N grant sizes. When the OLT receives a request R1 from an ONU in (b)-2, it grants all of R1 since R1 is more than W_A and less than W_{Max}. In (b)-3 since R2 is more than W_A and W_{Max}, the OLT grants bandwidth as much as W_{Max}. In (b)-4 and (b)-5, if the request is more than W_{Max}, the OLT grants bandwidth not more than W_A.

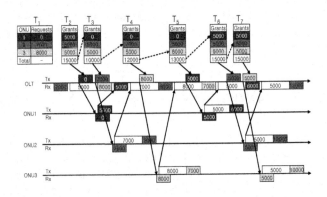

Fig. 4. Operation in Extra Window scheme.

Fig. 4 shows an example which the Extra Window scheme operates. Since the number of ONUs is 3, the available entire bandwidth is the sum of 4 windows and each ONU is guaranteed up to bandwidth of 5,000 bytes. The previous requests T_1 and grants T_2 are similar to that shown in Fig. 2. In T_2, since the sum of previous grants is 15,000 bytes and a request of ONU1 is zero byte, the grant bandwidth is zero byte. While the OLT grants zero byte into ONU1, it updates the last grant information in table T_3. From table T_3, available bandwidth of entire bandwidth minus 10,000 bytes, a sum of the previous grants, is 10,000 bytes. Since a request of ONU2 is 7,000 bytes, grant bandwidth is 7,000 bytes.

In the same way, while the OLT is granting 7,000 bytes to ONU2, it updates the last grant information in table T_4. In table T_4, available bandwidth is 8,000 bytes, and a request of ONU3 is 8,000 bytes. Thus, all of 8,000 bytes are granted to ONU3. In table T_5, 5,000 bytes is granted for ONU1. In table T_6, 5,000 bytes are granted for ONU2 while zero byte is granted in Fig. 2. In the table T_7 5,000 bytes are granted for ONU3. Therefore, the Extra Window scheme can guarantee bandwidth as much as the maximum window size, in contrast to *elastic* service, in case that a request exists.

4   Performance Evaluations

In this paper, we consider an EPON system consisting of an OLT and 16 ONUs, and each ONU contains 32 users. The data rate of access link from a user to an ONU is 100 Mbps, and the rate of the upstream link from an ONU to the OLT is 1 Gbps. The propagation delay between the OLT and each ONU is 5 ns/m, and distance between the OLT and ONUs ranges from 0.5 to 20 km. The length of the average cycle time is $2ms$, and a guard time between ONUs is 5 μs. We assume that the queue size for each ONU is 10 Mbytes. For generating self-similar traffic, after gathering user data generated by ON/OFF periods according to the Pareto distribution, an ONU receives the aggregated user traffic streams. The average offered load for ONUs is varied from 0.05 to 0.9. We consider two different ways of setting the load. The first is the case that the offered load of all ONUs is equal, the other is the case that the offered load of ONUs is mutually different. In both cases, we compare the proposed scheme with the *limited* and *elastic* service of IPACT.

We compare the results of mean packet delay in Fig. 5(a). The term '-different' in the legend represents a simulation result for the different load case. When the same load is offered to all ONUs, *limited* service has the longest mean packet delay, and the performance of *elastic* and Extra Window are almost equal. When loads are 0.6 and 0.65, improvement of performance is remarkable. When the same load is offered, since the queue of each ONU is almost full in the case of load more than 0.6, all packets through the queue have long queuing delay. When loads are mutually different, the packets of the ONUs with light load are not accumulated and have low delay. Thus delay with relatively different load for ONUs is lower than that of the case with the same load. In the simulation of different load, e.g., at 0.5, the proposed Extra Window scheme shows lower delay up to 58.1% than the *limited* service, and up to 54.25% than the *elastic* service.

Fig. 5(b) shows a change of the average queue length for all ONUs. We know that the average queue length and the mean packet delay are proportional, and the former shows the similar trend to the later. When offered load is more than 0.4, since the *limited* service on average has heavy use of the queue, the mean packet delay becomes high. At the loads of 0.6 and 0.65, it shows the similar results due to the same reason with the mean packet delay. In the proposed Extra Window scheme, at the load of 0.5 reduces queue occupancy is reduced

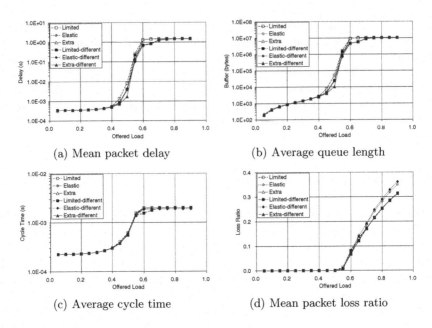

(a) Mean packet delay   (b) Average queue length

(c) Average cycle time   (d) Mean packet loss ratio

Fig. 5. Performance of proposed scheme and IPACT.

up to 58%, compared to that of the *limited* service, and up to 55.6% compared to that of the *elastic* service.

Fig. 5(c) shows the comparison of the average cycle time. When the offered load is less than 0.6, the average cycle time is long in order of *limited* > *elastic* > Extra Window; however, in the other load, in order of Extra Window > *elastic* > *limited*. In addition, *limited* service shows that two simulations have different results at a load of approximately 0.6. Thus in case of the same load, since all ONUs have similar requests, granted slots increase constantly; however, in case of the different load since ONUs' slots with relatively low packet load is small, total cycle time is reduced. In this case, the total time of a cycle cannot be used all because light loaded ONUs don't occupy the guaranteed maximum window according to *limited* service. Therefore, it has more overheads due to such factors as guard time since it needs more cycles than the other services for equal bandwidth. The Extra Window scheme demonstrates better performance in the mean packet delay and the average queue length, when its offered load is light, and in heavy load, it shows better performance with maximal use of cycle time.

In Fig. 5(d), according to the increase in offered load, it shows the packet loss ratio. The packet loss ratio of *elastic* service is highest, then one of *limited* service is middle. When the former variable results are compared totally, *elastic* service has better performance than *limited* service; however, it has higher packet loss. Finally, Extra Window scheme has the lowest packet loss ratio, moreover it has the most valuable performances among three services.

5   Conclusion

In this paper, we consider that EPON is a system that variable users access with similar pattern in an ONU. In this environment, the existing IPACT reduced the total network performance, such as packet loss, since an ONU in burst traffic situation results in relatively lower throughput than the others. In this paper, the proposed Extra Window scheme solved the shortcomings of existing *elastic* service in IPACT, and shows advanced performance up to 58% in terms of mean packet delay. In addition, it has good performance in terms of packet loss ratio and variation in mean packet delay among ONUs.

References

1. S. Jiang and J. Xie, "A Frame Division Method for Prioritized DBA in EPON," IEEE Journal on Selected Areas in Communications, vol. 24, no. 4, pp. 83-94, April 2006.
2. J. Zheng, "Efficient bandwidth allocation algorithm for Ethernet passive optical networks," IEE Proc.-Commun., vol. 153, no. 3, pp. 464-468, June 2006.
3. G. Kramer, B. Mukherjee and G. Pesavento, "IPACT: A Dynamic Protocol for an Ethernet PON (EPON)," IEEE Communications Magazine, pp. 74-80, February 2002.
4. H. Kim, H. Park, D. K. Kang, C. Kim and G. I. Yoo, "Sliding Cycle Time-based MAC Protocol for Service Level Agreeable Ethernet Passive Optical Networks," in Proc. of IEEE International Conference on Communications 2005, vol. 3, pp. 1848-1852, 2005.
5. C. M. Assi, Y. Ye, S. Dixit and M. A. Ali "Dynamic Bandwidth Allocation for Quality-of-Service over Ethernet PONS," IEEE JSAC, vol. 21. no. 9, pp. 1467-1477, November 2003.
6. J. Xie, S. Jiang and Y. Jiang, "A Dynamic Bandwidth Allocation Scheme for Differentiated Services in EPONS," IEEE Communications Magazine, vol 42, issue 8, pp. S32-S39, August 2004.
7. G. Kramer and G. Pesavento, "Ethernet Passive Optical Network (EPON): Building a Next-Generation Optical Access Network," IEEE Communications Magazine, pp. 66-73, February 2002.
8. S. Choi and J. Huh, "Dynamic Bandwidth Allocation Algorithm for Multimedia Services over Ethernet PONs," ETRI Journal, vol. 24, no. 6, pp. 465-468, December 2002.
9. N. Ghani, A. Shami, C. Assi, and M. Y. A. Raja, "Quality of Service in Ethernet Passive Optical Networks," 2004 IEEE/Sarnoff Symposium on Advances in Wired and Wireless Communication, pp. 161-165, Apr 2004 .
10. Y. Zhu, M. Ma, and T. H. Cheng, "A Novel Multiple Access Scheme for Ethernet Passive Optical Networks," IEEE GLOBECOM 2003, vol. 5, pp. 2649-2653, 2003.
11. J. Zheng and H. T. Mouftah, "An Adaptive MAC Polling Protocol for Ethernet Passive Optical Networks," IEEE ICC 2005, vol. 3, pp. 1874-1878, May 2005.

Cost versus flexibility of different capacity leasing approaches on the optical network layer

Sofie Verbrugge, Didier Colle, Mario Pickavet, Piet Demeester

Dept. of Information Technology, Ghent University - IBBT,
Sint-Pietersnieuwstraat 41, B-9000 Gent, Belgium,
tel. +32 9 33 14900, fax +32 9 33 14899
{sofie.verbrugge, didier.colle, mario.pickavet, piet.demeester}intec.ugent.be

Abstract. This paper discusses different capacity leasing scenarios for the optical network. It aims at providing systematic understanding of cost structure of optical networks, without focusing on technical details. It reviews different kinds of capacity leasing approaches discussing the relative costs and the flexibility associated to them to adapt to future traffic needs. Then it applies the Real Options principles to evaluate the most convenient planning solution. Real Option valuation is shown to be the formalization of the natural way of valuating different alternatives under uncertainty, taking into account that information becoming available during the course of the planning horizon might influence the strategy followed. The goal of the paper is to give insight in the practical applicability of the technique for the network operator.

Keywords: optical layer capacity, dark fiber, IRU, flexibility, Real Options

1 Introduction

The presence of multiple telecommunication providers and their past investments in optical layer capacity, have led to a wide availability of dark fiber throughout Europe. Therefore, for a network operator planning to deploy a new network, there is no need to acquire all physical capacity himself. Leasing capacity gives more flexibility and is therefore considered an attractive solution. A distinction can be made between traditional leasing and longer term Indefeasible Right of Use. Moreover, either dark fiber of wavelengths can be the subject of the lease.

This paper describes several capacity leasing scenarios, indicating typical costs and contract terms. We indicate useful approaches to evaluate the associated investment decisions from the perspective of the user of the lease and focus on the flexibility to switch between network scenarios and the relation between future traffic uncertainty and network flexibility, using Real Options thinking.

I. Tomkos et al. (Eds.): ONDM 2007, LNCS 4534, pp. 418–427, 2007.

2   Capacity leasing approaches

2.1   IRU versus lease

Leasing is a concept where the *grantor* grants the use of the asset to the *grantee (user)* for the duration of the lease, a well-known example is leasing a car where the asset considered is the car. In telecom, the lease of capacity on someone else's network can for example take the form of dark fiber or wavelength lease. A lease usually applies for a relatively short term, e.g. one to five years.

An Indefeasible Right of Use (IRU) is similar to a lease, but usually applies to a longer term, e.g. 15 or 20 years. For instance, assume operator X aggressively builds a worldwide fiber-optic network. If another operator Y is building a network but not in the same places as operator X, to expand its reach, operator Y might buy an IRU for two fibers in operator X's network for 20 years [1].

2.2   Dark fiber versus wavelength

After the liberalization of the telecom market and the strong past investment in optical layer capacity, a lot of dark fiber is available. This means that, apart from the traditional deployment or acquisition of an optical layer network, a network operator nowadays also has the possibility to lease the optical layer capacity. He can deploy the higher layer network infrastructure based on this.

The form of the optical layer capacity the grantor of the lease offers to the grantee, can either be dark fiber or wavelengths. Both a dark fiber and a wavelength network are based on wavelength infrastructure in the network nodes. In a dark fiber network, the transmission infrastructure like multiplexers and transponders is owned by the grantee, whereas in a leased wavelength network this equipment is owned by the grantor of the lease. Several candidate node configurations can be distinguished, depending on the physical topology at hand and the requirement to have grooming capabilities (packing low capacity traffic in high capacity streams). More information on possible node configurations can be found in [2]. If we consider an optical ring network based on optical add-drop multiplexers (OADMs), on top of which DXCs and higher layer equipment is deployed, the costs for the optical network layer for both situations include:

– In case of a leased wavelength network: costs of the lease contracts
– In case of a dark fiber network: costs of the lease contracts, costs of the optical add-drop multiplexers (OADMs), costs of the transponders and costs of the WDM multiplexers

The costs of the DXCs and the other higher layer equipment will be similar in both cases.

2.3   Cost and flexibility

As indicated above, IRUs usually have longer contract terms than ordinary leasing contracts. Furthermore, leasing of dark fiber also usually applies for longer terms than leasing of wavelengths. Fig. 1 compares IRU on dark fiber (IRU on DF) with a cost of 0,4 euro/meter/year, dark fiber lease (DF lease) with a cost of 1,2 euro/meter/year and wavelength lease (lambda lease) with a cost of 1,5 euro/meter/year. The mentioned link prices are in line with values found in the literature [3].

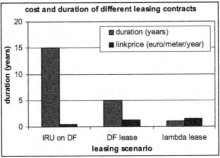

Fig. 1. Comparison of leasing contracts in term of costs

In order to make a full comparison of the optical layer costs of the different network deployment scenarios, also equipment costs[1] for the transmission equipment are to be taken into account. Some reference costs can be found in Tab. 1.

Tab. 1. Reference costs for optical layer equipment

equipment	OADM	WDM mux (40 lambdas)	transponder 10G
cost (euro)	17000	37500	23000

The duration of the contract has some relation with the flexibility: shorter contract terms lead to additional flexibility because the leased capacity can easily be increased or decreased at the end of the lease contract, e.g. by taking a lease for a higher number of wavelengths or by releasing a wavelength after one year in case of the lambda lease scenario. This type of flexibility is illustrated in **Fig. 2**, where the horizontal arrows represent the lease contracts (full lines for the IRU on DF, dashed lines for the DF lease and dotted lines for the lambda lease scenarios) and the vertical arrows indicate the possibility to increase or decrease the capacity at the end of the contract term by extending or releasing the leased capacity for the considered contract. Apart from the limited flexibility to increase or decrease capacity within a certain scenario (at the end of a contract term) indicated above, there is also the flexibility to switch between the considered scenarios at the end of the contract term, this means after 15 years for the

---

[1] The cost to set up the network and install this equipment is mainly labour costs and constitutes produced fixed assets, which are to be counted as CapEx together with the equipment itself. The labour costs for first-time network installation could be treated using an activity-based process description model [11], however, they are neglected here.

IRU on DF, after 5 years for the DF lease and yearly for the leased wavelengths. Note that, in real life, there is also the possibility to prematurely end the contract, e.g. end the IRU before the end of the 15 year period by paying a kind of penalty fee, but this is not considered here.

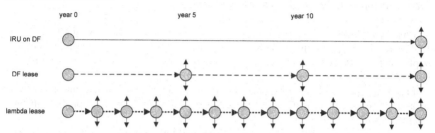

Fig. 2. Flexibility based on leasing contract duration

Given the fact the longer leasing contracts tend to be cheaper on a yearly basis than shorter ones, it may be beneficial to use IRU on DF in cases where the required capacity for the future years is expected to be non-decreasing. However, in cases where capacity decrease can be expected, shorter term contracts can be worthwhile, e.g. because of a competitor taking over some part of the market share in a particular area of the network where the considered operator is no longer interested in). Moreover, in an uncertain situation, shorter leasing contracts allow to follow a wait-and-see approach. If the situation becomes clear after some time, it is possible to quickly react to it by releasing capacity or by moving to a leasing contract with a longer duration.

3   Traffic uncertainty and how to cope with it

3.1   Uncertain traffic evolutions

Unfortunately, there is no single, unbiased source of information on IP traffic growth. The National Science Foundation in the US stopped measuring the growth of the IP traffic on the Internet backbone in 1995. Nevertheless, some numbers can be found in literature. Growth rates till 150%, 200% or even more were often cited before the Internet bubble exploded [4]. Although overall IP traffic can be expected to continue growing, the actual growth rate is unclear. Moreover, when considering a particular network offering particular services the situation becomes even more cumbersome. Several forecasting methods are distinguished in the literature [5][6]. Based on accurate observations and the use of a suitable planning technique, several cases can be distinguished where part of the initial uncertainty disappears in the beginning of the planning horizon, so that more accurate planning becomes possible after that time, the so-called learning time. This can for example be the case when a competitor is entering the market but is unclear beforehand what market segment he will actually focus on. We will focus on these cases in the remainder of this paper.

3.2   Flexibility to react to uncertain evolutions

Based on additional information becoming available during the learning time in the first part of the planning horizon, the comparison of different possible network deployment scenarios might be very different from the initial expected situation. Consider the example of

Fig. 3. where we observe two possible network deployment paths, one consisting of a single path (no subpaths) and the other one containing two subpaths with two alternatives for the last subpath.

–   When evaluating the single deployment path under traffic uncertainty, the only solution is to calculate its average cost over the different uncertain future evolutions. In case of a 60% chance of a traffic growth and a 40% chance of a status quo in the traffic, the expected cost for this deployment path would be

$$0.6 * cost_{traffic\ growth}(a) + 0.4 * cost_{consta\ nt\ traffic}(a) = E_{uncertain\ traffic\ evolutions}[cost(a)] \quad (1)$$

–   When evaluating the path with the alternative subpaths, the expected cost for the first subpath (calculated in the same way as the path without flexibility above) will be augmented with the cost of the cheapest alternative for the second subpath. If the choice between alternative c and d is to be made at the beginning of the deployment path, there is no flexibility to react to changes happening during the first subpath. In that case the expected cost of the last subpath is to be calculated as the minimum value of the average expected costs for both alternative subpaths

$$\min\left[E_{uncertain\ traffic\ evolutions}[cost(c), cost(d)]\right] \quad (2)^2$$

–   However, when the decision between c and d is only made at the end of subpath b, there is true flexibility, so that at that point in time the cheapest alternative will be chosen, taking into account the actual situation at the end of subpath b (i.e. knowing possible evolutions that have taken place during the course of b). In that case the value of the second subpath is to be calculated as the average value over all expected traffic scenarios of this minimum cost

$$E_{uncertain\ traffic\ evolutions}\left[\min[cost(c), cost(d)]\right] \quad (3)$$

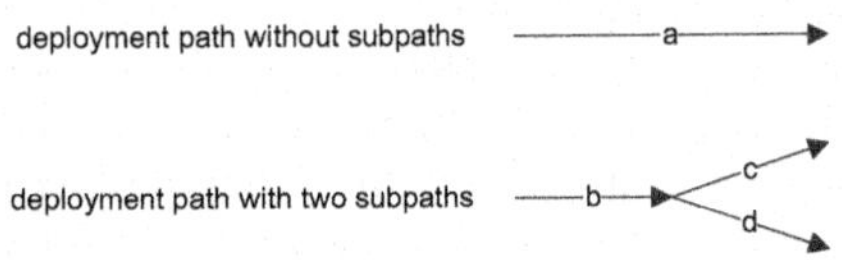

deployment path without subpaths   ——a——→

deployment path with two subpaths   ——b——→ < c
                                                          d

Fig. 3. Different deployment paths

---

[2] Remark that we use E[x,y] as a short notation for E[x],E[y].

The above observations indicate that the solution with the lowest cost based on the expected situation before the start of the deployment path (choice without flexibility) may be different from the solution with the lowest cost based on the knowledge after the first subpath (choice taking into account flexibility). When comparing the deployment paths a (without flexibility) and the paths b-c-d with the real flexibility to choose between c and d after the first subpath, the costs to be compared for the overall paths are

$$E_{uncertain\,traffic\,evolutions}\left[cost(a)\right] \qquad (4)$$

versus

$$E_{uncertain\,traffic\,evolutions}\left[b + \min\left[cost(c),cost(d)\right]\right] \qquad (5)$$

3.3   Real Options thinking

Real Options thinking originates from the financial world. An option can be defined as the right for a limited time, to buy or sell the underlying value for a predetermined exercise price. Exercising the option (i.e. buying or selling the underlying value) is always optional; it is a right, not an obligation. This right holds for a predetermined time, till the so-called exercise date. The underlying value is the asset which the option concerns; this may be assets, real estate, precious metals, etc. The exercise price is the price for which the option can be exercised by its holder.
The term Real Options was introduced in 1977 [7]. It referred to the application of option pricing theory to the valuation of investments in real assets where a large part of the value is attributable to flexibility and learning over time. After some academic attention in the 1980s, interest in real options from industry rose considerably since mid 1990s. Real Options thinking has been successfully applied to determine the value of flexibility when deploying mines, manufacturing plants, etc. Application in the telecom business started recently [8][9], however, those papers do not aim at practical applicability of the methodology for the network operator. [10] provides a comprehensive introduction to Real Options theory, with a lot of practical examples.
Real Options valuation can be seen as the formalization of the natural valuation for a deployment path with flexibility as described above. Referring back to
Fig. 3, the deployment path with flexibility gives us the options to choose between subpaths c and d later on, when choosing subpath b in the beginning of the planning horizon. On the other hand, when taking the single deployment path a immediately, there is no option later on, we need to stick to this path anyway. The path with the two alternative subpaths can be seen as holding an option to choose between them later on, when more information is available. The option price (price to obtain the option) in this case is the price difference between the subpath b (providing the flexibility) and the first part of the path a (without flexibility).
As illustrated for the example above, Real Options valuation is the natural valuation technique for two-phased investment decisions, with an optional second phase (e.g. only performed if market situation is favourable). By the time of the second phase of

the investment, the market situation is already more clear, so that a well-advised decision can be taken[3].

4   Case study: comparison of capacity leasing approaches for Belgian scale network

We consider the three network deployment scenarios (IRU on DF, DF lease, lambda lease) discussed in Fig. 2, but assume now that the first five years of the planning horizon can be seen as a kind of learning time, over which some information considering the unknown evolutions in the rest of the planning horizon (the last 10 years in this example) becomes clear. We consider the case of Fig. 4, where one of the considered scenarios is chosen in year 0 and this scenario is followed during 5 years, after which we can choose to switch to another scenario. This means that year 5 is the only point of flexibility in this example (which is a simplification of the reality).

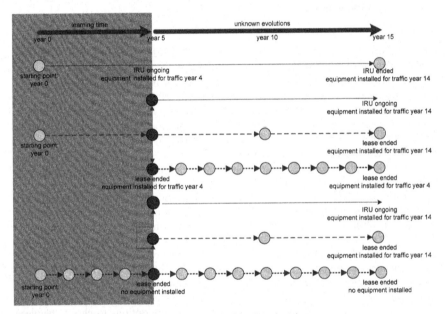

Fig. 4. Flexibility to react to changes at the end of the learning time

During the learning time of the first 5 years, a competitor might have entered the market, so that we can expect a drop in the market share in the coming years. Performing some customer surveys, it might even be possible to get an idea of the actual expected decrease of the market share and other implications. On the other hand, if a complementary service is becoming available, an increased market share

---

[3] Note that several option valuation techniques are distinguished in the literature. In this paper we only consider valuation through simulation, which is the most intuitive technique.

can be estimated in a similar way. If we have started our network roll-out in year 0 using DF lease or lambda lease, the contract is ended and we can enroll in any new leasing contract of our choice.

We consider a network on the Belgian scale (32,545 km2) consisting of 3 interconnected rings with a total link length of 1488 km. We consider a planning horizon of 15 years, with an expected annual traffic growth of 60%. The reference traffic described in [2] was assumed to be obtained in the fourth year of the period studied. We assume OADMs in the nodes, with the costs described in Tab. 1. Equipment cost is expected to undergo a cost erosion of 10% a year. The costs and durations for the different considered network deployment scenarios are those of Fig. 1. The costs for dark fiber and wavelength lease are considered to be constant over time. This assumption originates from the observation that labour costs increase over time (contrary to equipment costs) and that in the future dark fiber might become scarcer.

We have calculated the cash-out flows for the three considered scenarios (IRU on DF, DF lease and lambda lease), i.e. the expenses for the leasing contracts as well as the required optical layer equipment. All network scenarios are dimensioned in order to cope with the expected traffic demand (so that the network operator's revenues or cash-in flows are equal for all scenarios). For lease contracts with durations of over one year (granularity of the traffic predictions), i.e. the dark fiber lease for 5 years and the IRU on dark fiber for 15 years in our study, the expected need of capacity is calculated at the beginning of the contract term in order to determine the required contract size. In case the traffic demand is rather static or well predictable, this approach leads to good results, as better rates can be negotiated in case the contract is taken for multiple fibers at the same time. When the actual demand exceeds the forecasts, additional wavelengths/fibers are leased or equipment is bought according to necessity.

We have considered an uncertain future traffic scenario, in order to simulate the fact that the actual traffic will be equal to the predicted traffic. In case the actual required line capacity is bigger than expected, the lease contract is immediately expanded and the additional required node capacity is bought and installed. In case of a traffic stagnation or decrease, leasing contracts are changed at the end of the term according to necessity, no new equipment is installed.

We consider the following uncertain traffic scenarios for the last 10 years of the planning horizon. The traffic is supposed to follow the forecasted traffic growth of 60% a year, except from an abrupt change in that trend in year 6. We assume the information about this change to be known at the end of year 5 (e.g. competitor has entered the market). The abrupt change is modelled as a traffic increase or decrease somewhere between 0.01 times its current size and 10 times is current size and simulated using Monte Carlo simulation, performed by Crystal Ball [12]. A continuous variable with uniform distribution is assumed in the simulation.

The cost of the second phase of our planning horizon (last ten years of Fig. 4), calculating both with and without flexibility (using the formulae (3) and (2 respectively) is depicted in Fig. 5 for the cases where either DF lease or lambda lease is used for the first phase[4]. The difference between the bars without and with

---

[4] In case of IRU on DF in phase 1, there is no flexibility anyhow.

flexibility indicates the additional cost for the second phase in the absence of flexibility. As we assume the revenues for all scenarios to be equal, the difference in costs equals the difference in value (different sign). The cost difference therefore represents the option value. This illustrates that Real Options valuation (taking into account the value of the inherent options of the migration paths) is the natural way to evaluate the network deployment cost in this example.

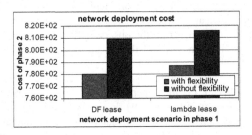

Fig. 5. Cost of second phase of network deployment

The real-life planning question to be answered in the considered case, is what network deployment scenario to start using in the beginning of the planning horizon (year 0 of our study). Traditionally, the evaluation will be made based on the expected traffic and cost evolutions, leading to the costs of the left bars in
Fig. 6. This evaluation shows that, in case everything evolves as expected, the dark fiber scenarios outperform lambda leasing in terms of costs.
However, when taking into account the uncertain evolutions in the second part of the planning horizon (abrupt change in year 6), the evaluation should consider the flexibility of some scenarios to easily react to those changes, by using formula (5). This means using the left bars of Fig. 5 (with flexibility) as the cost of the second phase, leading to the right bars in
Fig. 6 as the costs for the total planning horizon, which is exactly what Real Options evaluation suggests. This analysis shows the highest cost for the IRU on DF scenario, indicating that this is scenario is least flexible. The DF and lambda lease scenarios are more flexible. DF lease still comes out as the best solution.

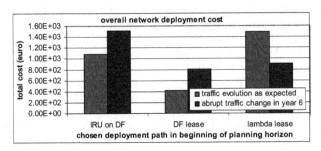

Fig. 6. Network deployment costs taken into account in the beginning of the planning horizon

5   Conclusions

In this paper, we have compared different optical layer capacity leasing scenarios based on costs versus flexibility. Whereas an Indefeasible Right of Use on dark fiber is the cheapest per capacity unit per year and allows to cope with growing traffic without the need to update the contract immediately, the DF lease scenario was shown to offer the best trade-off between cost and flexibility as it allows to respond faster to decreasing traffic and it allows to switch sooner to another network scenario (IRU on DF or lambda lease). The Real Option valuation technique was shown to be the formalization of the natural way of evaluating the flexibility to switch between the scenarios, taking into account that information becoming available during the course of the planning horizon might influence the strategy followed.

Acknowledgments. We were supported by the EC through IST-Nobel2 and e1+, by IWT-Flanders through a grant for S.Verbrugge and by UGent through BOF-RODEO.

References

1. K. Maney, " Indefeasible rights of use is capturing nation", USA Today – Cyberspeak, February 27th 2002.
2. S. Verbrugge et al.: "Planning of transmission infrastructure to support next generation BELNET network," IST BroadBand Europe conference, Bordeaux, France, December 12-14, 2005, papers online at www.bbeurope.org, p. W03B.02.
3. W. Van Dijk: "Acquisition and deployment of dark fiber within SURFnet", presented on CEF workshop, Prague, Czech Republic, May 25th, 2004.
4. A. Odlyzko: "Internet growth: myth and reality, use and abuse," Information Impacts Magazine, November 2000.
5. L. Lee: "An Introduction to Telecommunications Network Traffic Engineering", Alta Telecom International Ltd., Edmonton, Alberta, Canada, 1986.
6. E. Jantsch: "Technological Forecasting in Perspective: A Framework for Technological Forecasting, its Techniques and Organisation", Organisation for Economic Co-operation and Development, available online..
7. S. Myers: "Determinants of Corporate Borrowing", Journal of Financial Economics, 1997, vol. 5, no. 2, pp.147-175.
8. Y. d'Halluin, P.A. Forsyth, K.R. Vetzal: "Managing Capacity for Telecommunications Networks under Uncertainty", IEEE/ACM Transaction on Networking, Aug. 2002, vol, 10, pp. 579-588.
9. C. Kenyon, G. Cheliotis: "Dark fiber valuation", The Engineering Economist, 2002, vol. 47, no 3, pp. 264 – 308.
10. W. De Maeseneire, "The real options approach to strategic capital budgeting and company valuation", Financiële Cahiers, LARCIER, 2006, ISBN: 2-8044-2318-2.
11. S. Verbrugge et al.: "Methodology and input availability parameters for calculating OpEx and CapEx costs for realistic network scenarios," Journal of Optical Networking, Feature Issue: Optical Network Availability, June 2006, vol. 5, no. 6, pp. 509-520.
12. Crystal Ball, http://www.decisioneering.com/

A Bayesian decision theory approach for the techno-economic analysis of an all-optical router

Víctor López[1], José Alberto Hernández[1], Javier Aracil[1],
Juan P. Fernández Palacios[2] and Óscar González de Dios[2]

[1] Universidad Autónoma de Madrid, Spain
[victor.lopez, jose.hernandez, javier.aracil]@uam.es,
WWW home page: http://www.ii.uam.es/~networking
[2] Telefónica I+D, Spain
[jpfpg, ogondio]@tid.es

Abstract. Typically, core networks are provided with both optical and electronic physical layers. However, the interaction between the two layers is at present limited, since most of the traditional transport functionalities, such as traffic engineering, switching and restoration, are carried in the IP/MPLS layer.

In this light, the research community has paid little attention to the potential benefits of the interaction between layers, multilayer capabilities, on attempts to improve the Quality of Service control.

This work shows when to move incoming Label Switched Paths (LSPs) between layers based on a multilayer mechanism that trades off a QoS metric, such as end-to-end delay, and techno-economic aspects. Such mechanism follows the Bayesian decision theory, and is tested with a set of representative case scenarios.

1 Introduction

Core networks are often provided with both electronic and optical routing capabilities. Essentially, electronic routing has the well-known advantages of statistical multiplexing and granularity, but is a hard-computational process which introduces queuing delay to packets. On the other hand, data packets switched in the optical domain only experience propagation delay. However, optical resources provide a granularity which is too coarse for typical Internet streams, even if they come from the multiplex of many users.

As noted in previous work [1–3], it is highly desirable to efficiently combine the benefits of both optical and electronic domains, according to some policy. In a typical scenario, incoming Label Switched Paths (LSPs) arrive to a multilayer-capable router, which has to decide whether to perform optical or electronic switching (fig. 1). If an incoming LSP is routed in the electronic domain, it

---

* The authors would like to thank the support from the European Union VI Framework Programme e-Photon/ONe+ Network of Excellence (FP6-IST-027497). This work has also been partially funded by the IST Project NOBEL II (FP6-IST-027305).

I. Tomkos et al. (Eds.): ONDM 2007, LNCS 4534, pp. 428–437, 2007.

suffers hop-by-hop opto-electronic conversion (with subsequent delay), but if it is routed optically a lightpath is reserved end-to-end. Some LSPs share the end-to-end lightpath. The choice of electronic or optical switching is based upon a set of previously-defined rules in the multilayer-capable router.

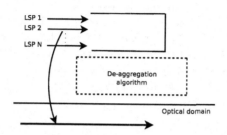

Fig. 1. Multilayer-capable router scenario

In this paper, we propose a techno-economic model to help routers to take the decision of optical or electronic switching of their LSPs. Such approach makes use of Bayesian decision theory, and takes into account several aspects concerning the Quality of Service perceived by packets, by means of queuing delay, and also techno-economic aspects such as the relative cost associated to switching LSPs in either the optical or the electronic domain. The algorithm computation is low cost, because operations are not recursive and only has to be computed when a new LSP crosses the router.

In this light, the remainder of this work is organised as follows: Section 2 covers the mathematical foundations for such techno-economic analysis with a Bayesian decisor. Section 3 provides a set of experiments and numerical examples to show how to reach to an optimal decision. Finally, section 4 outlines a summary of the results obtained and further lines of investigation.

2 Analysis

2.1 Problem statement

As previously stated, the aim is to define a mathematically rigorous set of rules to help such multilayer-capable core routers decide whether to switch a given LSP in the optical domain or the electronic domain.

At a given time, a multilayer router handles a set of LSPs. The router can switch each LSP either on the electronic domain or the optical domain. Typically, due to QoS constraints, it is preferred optical switching due to the lack of queueing delay. However, optical switching resources are limited (for example, due to wavelength conversion capabilities) and their cost is larger than electronic switching. Thus, the router must trade-off these two parameters: queuing delay versus the cost associated to optical switching, and needs to have a set of rules predefined to make a decision on how many LSPs should be switched in the optical domain and how many in the electronic domain.

To do so, let N refer to the number of LSPs handled at a given random time by the multilayer router, and let $L(d_i, x)$ refer to the loss function. The loss function $L(d_i, x)$ denotes the cost or loss of switching i LSPs in the electronic domain (thus, $N - i$ LSPs in the optical domain) with subsequent queueing delay experienced by the packets of the electronically switched LSPs, which is denoted x (for simplicity, the optically switched LSPs have been assumed to experience zero delay). The term d_i denotes the "decision" of routing i LSPs out of a total of N in the electronic domain, and is defined for some decision space $\Omega = \{d_1, \ldots, d_N\}$. In this light, $L(d_i, x)$ is given by:

$$L(d_i, x) = (C_e(i) + C_o(N - i)) - U(x), \qquad i = 1, \ldots, N, \quad x > 0 \qquad (1)$$

where $C_e(i)$ and $C_o(N - i)$ refer to the cost associated to routing i LSPs in the electronic domain and $U(x)$ refers to the utility associated to a queuing delay of x units of time, experienced by the electronically switched LSPs.

Following [4], the Bayes risk, which is essentially the expectation of the loss function with respect to x, equals:

$$R(d_i) = \mathbb{E}_x L(d_i, x) = (C_e(i) + C_o(N - i)) - \mathbb{E}_x U(x), \qquad i = 1, \ldots, N \qquad (2)$$

The goal is to obtain the optimal decision d^* such that the Bayes risk $R(d^*)$ is minimum. In other words:

$$\text{find } d^* \text{such that } R(d^*) = \min_{d_i, i=1,\ldots,N} R(d_i)$$

The next section proposes a set of utility functions, $U(x)$, that measures the QoS experienced (in terms of queuing delay) by the electronically-switched packets; and also introduces a metric for quantifying the relative cost of optical switching with respect to electronic switching.

2.2   The utility function $U(x)$

As previously stated, the utility function $U(x)$ is defined over the random variable x, which represents the queuing delay experienced by the packets of electronically switched LSPs. The queuing delay shall be assumed to be Weibull distributed, since this has been shown to accurately capture the queueing delay behaviour of a router with self-similar input traffic [5–7]. In this light, the delay probability density function is given by [5]:

$$p(x) = (2-2H)C\frac{(C-m)^{2H}}{2K(H)^2 am}(Cx)^{1-2H} \exp\left(-\frac{(C-m)^{2H}}{2K(H)^2 am}(Cx)^{2-2H}\right), \qquad x > 0$$

$$(3)$$

where C is the lightpath capacity, m is the average input traffic and a is a variance coefficient such that $am = \sigma^2$ (with σ^2 being the input traffic variance) and H is the Hurst parameter.

Once $p(x)$ has been defined, the next step is to define a measure of the "utility" associated to routing LSPs in the electronic domain.

Delay based utility In its simplest way, we can easily evaluate the utility based on the observed delay, that is, $U_{delay}(x) = -x$. The utility function is thus opposite to the queuing delay x, since the more utility occurs for smaller delays. Thus, computing the Bayes risk defined in eq. 2 yields:

$$E_x[U_{delay}(x)] = E_x[-x] = -\int_0^\infty xp(x)dx \qquad (4)$$

which equals the average queuing delay experienced by the electronically-switched packets. Such value takes the following analytical expression:

$$E_x[U_{delay}(x)] = -\frac{1}{C}\left(\frac{(C-mi)^{2H}}{2K(H)^2 ami}\right)^{1/(2-2H)}\Gamma\left(\frac{3-2H}{2-2H}\right) \qquad (5)$$

However, the average delay is not always a useful (or at least, representative) metric in the evaluation of the Quality of Service experienced by certain applications, especially when quantifying the relative QoS experienced by real-time applications. The following considers two other utility functions used in the literature for hard-real time and elastic applications [8, 9].

Hard real-time utility Hard real-time applications are those which tolerate a delay of up to a certain value, say T_{max}, but their performance degrades very significantly when the delay exceeds such value. Examples are: online gaming, back-up services and grid applications. The parameter T_{max} denotes the tolerated delay threshold for each particular application. The ITU-T recommendation Y.1541 [10] and the 3GPP recommendation S.R0035 [11] defined service classes based on thresholds.

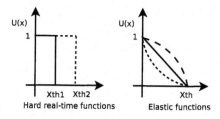

Fig. 2. Utility functions: hard real-time (left) and elastic (right)

Hard real-time utility can thus be modelled by a step function as shown in figure 2 left, and takes the expression:

$$U_{\text{step}}(x) = \begin{cases} 1 & \text{if } x < x_p \\ 0 & \text{otherwise} \end{cases} \tag{6}$$

To compute the Bayes risk requires the average utility:

$$E_x[U_{\text{step}}(x)] = \int_0^{T_{\max}} p(x)dx = 1 - \int_{T_{\max}}^{\infty} p(x)dx = 1 - \mathbb{P}(x > T_{\max}) \tag{7}$$

which, according to 3, leads to:

$$E_x[U_{\text{step}}(x)] \sim 1 - exp\left(-\frac{(C-mi)^{2H}}{2K(H)^2 ami}(Cx)^{2-2H}\right), \quad x > 0 \tag{8}$$

Elastic utility Other services consider a more flexible QoS function, since the service is degraded little by little (fig. 2 right).These services considers zero delay as the maximum possible utility, but the utility reduces with increasing delay. The ITU-T recommendation G.107 defines the "E model" [12], which explains in detail the degradation of voice service in humans. In other utility function studios, the exponential function has been used to describe the degradation of elastic services [9].

Thus, the elastic utility function is modelled as:

$$U_{\text{exp}}(x) = \lambda e^{-\lambda x}, \quad x > 0 \tag{9}$$

where λ refers to decay ratio of the exponential function. Following the definition of $T_{\max}$ above, the value of λ has been chosen such that 90% of the utility lies before $T_{\max}$. That is:

$$\lambda = \frac{1}{T_{\max} \log(1 - 0.9)} \tag{10}$$

Finally, the average elastic utility follows:

$$E_x[U_{\text{exp}}(x)] = \int_0^{\infty} \lambda e^{-\lambda x} p(x)dx \tag{11}$$

which has no analytical form. However, we can use the Taylor expansion to approximate it, since:

$$E[f(x)] \approx \int_0^{\infty} p(x)\left(f(E[x]) + f'(E[x])(x - E[x]) + \frac{1}{2}f''(E[x])(E[x] - x)^2\right)dx$$
$$= f(E[x]) + \frac{1}{2}f''(E[x])\sigma_x^2 \tag{12}$$

Thus:

$$E_x[U_{\text{exp}}(x)] \approx U_{\text{exp}}(E_x[x]) + \frac{1}{2}U_{\text{exp}}''(E_x[x])\sigma_x^2 \tag{13}$$

where the variance σ_x^2 can be easily derived from eq. 3:

$$\sigma_x^2 = \frac{1}{C^2} \left(\frac{2K(H)^2 a m i}{(C - mi)^{2H}} \right)^{1/(1-H)} \left(\Gamma \left(\frac{2 - H}{1 - H} \right) + \Gamma^2 \left(\frac{3 - 2H}{2 - 2H} \right) \right) \qquad (14)$$

2.3   The economic cost of electronic and optical switching

As previously stated, the values of $C_e(i)$ and $C_o(N - i)$ in eq. 1 represent the cost associated to switching i LSPs in the electronic domain and $N - i$ in the optical domain. Typically, the optical resources are more precious than the electronic resources, hence we will penalise the optical switching more than electronic switching.

For simplicity purposes, we have considered a *linear* cost approach, at which electronic switching is penalised as $C_e(i) = Ki$ for some $K > 0$, and the cost of optical switching is $C_o(N - i) = R_{\text{cost}} K(N - i)$. The value of R_{cost} denotes the relative optical-electronic cost, that is, the ratio at which the optical cost increases with respect to the electronic cost.

3   Experiments and results

This section provides a few numerical examples applied to real case scenarios. The aim is to show a few practical cases at which the implemented algorithm at a given core multilayer switch decides the number of optically-switched LSPs that should be transmitted according to three sets of parameters: (1) QoS parameters, essentially the $T_{\max}$ value introduced above; (2) the relative cost R_{cost} which provides a measure of the economic cost of switching LSPs in the optical domain with respect to the electronic switching; and, (3) the self-similar characteristics of the incoming flows, represented by the Hurst parameter H.

The simulation scenario assumes a 2.5 Gbps core network, which carries a number of $N = 72$ standard VC-3 LSPs (typically 34.358 Mbps each). The values of m, σ and H, which represent the characteristics of the traffic flows, i.e. average traffic load, variability and Hurst parameter, have been chosen as $H = 0.6$ (according to [13]) and m and σ such that $\frac{\sigma}{m} = 0.3$.

Finally, the value of K has been chosen as $K = \frac{1}{N}$, in order to get the electrical cost normalised, i.e. within the range $[0, 1]$.

3.1   Study of threshold $T_{\max}$

This experiment shows the influence of the choice of $T_{\max}$ in the decision to be made by the multi-layer router. Fig. 3 shows this case for several values of $T_{\max}$ assuming the step utility function (left) and the exponential utility function (right). The values of $T_{\max}$ have been chosen to cover a wide range from 0.1 ms to 100 ms. Clearly, the number of optically-switched LSPs should increase with increasing values of $T_{\max}$, since high QoS constrains require small delays in the

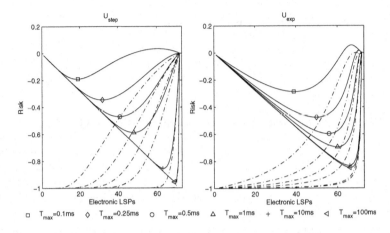

Fig. 3. Optimal decisions for several T_{max} values assuming hard real-time utility function (left) and elastic utility function (right). Dashed line = Utility.

packet transmission (thus larger number of optically-switched LSPs to reduce latency).

Typically, most of the end-to-end delay suffered by applications occur in the access network, and it is widely accepted that the core network should be designed to introduce delay of no more than $1-10\%$ of the total end-to-end delay. For hard real-time applications, which may demand a maximum end-to-end of 100 ms, the core delay is thus in the range of $1-10$ ms, which implies a total number of electronically-switched LSPs of $d^* = d_{48}$ and $d^* = d_{63}$ respectively of a total of $N = 72$ LSPs. For the same delay constrains, elastic applications impose a number of electronically-switched LSPs of $d^* = d_{60}$ and $d^* = d_{66}$ respectively.

Obviously, the delay requirements for hard real-time applications are tighter than those for elastic applications, thus demanding a larger number of optically-switched LSPs, as shown in fig. 3.

3.2 Analysis with different R_{cost} values

This experiment shows the impact of the relative cost R_{cost}, which refers to the relative cost of optical switching with respect to electronic switching, in the final decision d^*, to be taken by the multilayer router. Fig. 4 left shows where the optimal decision lies (minimum cost) for different R_{cost} values considering the case of mean utility function ($T_{max} = 10$ ms). As shown, the more expensive optical switching is (large values of R_{cost}), the less number of LSPs are switched optically.

Fig. 4 right shows where the optimal decision lies considering the mean (dashed), exponential (dotted) and step (solid) utility functions, for different values of R_{cost}. Again, as the R_{cost} value increases, the number of optically-switched LSPs decreases (hence, larger number of LSPs switched in the electronic

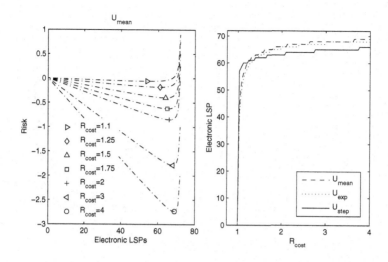

Fig. 4. Study of relative cost (R_{cost}) variation

domain). As shown, the step utility function, which represents the highest QoS constraints, demands more optically-switched LSPs than the other two, for large values of R_{cost}.

Finally, it is worth noticing that, for values $R_{\text{cost}} \geq 1.5$, the optimal decision lies in the range d_{65} to d_{72}. The reason for this is that, when optical switching becomes too expensive, the R_{cost} is critical in the optimal decision, thus cancelling any influence of the QoS parameter $T_{\max}$. In this light, the network operator has a means to decide where the optimal decision lies, trading off the R_{cost} parameter and the QoS values.

3.3 Influence of the Hurst parameter H

The previous two numerical examples have assumed a value of $H = 0.6$, as observed in real backbone traces [13]. However, other scenarios may show different values of H and it is interesting to study its impact on the bayesian decisor. In this light, fig. 5 shows the influence (left) or no-influence (right) of such parameter H in the optimal decision. In spite that long-range dependence degrades queuing performance generally, at high-delay values, the delay variability is smaller for high values of H (see [5], fig. 5).

Thus, the characteristics of the incoming traffic have a more or less impact on the bayesian decisor, depending on the QoS parameters. When $T_{\max} \geq 10$ ms, there is little influence of H (fig 5 right), but for $T_{\max} = 1$ ms and smaller, the value of H is key since it moves the decision in a wide range of optimal values: from d_{31} in the case of $H = 0.5$ to d_{64} for $H = 0.9$ (fig 5 left).

The level curves shown in fig. 6 shows such behaviour for the three utility functions ($T_{max} = 1ms$). Each level curve corresponds to a different utility. Fig. 6 left (case of mean utility function) should read as no influence with the

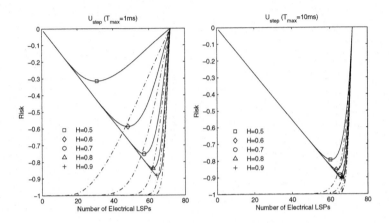

Fig. 5. Hurst parameter variation

Hurst parameter (i.e. parallel level curves = optimal decision independent of H value). On the contrary, fig. 6 middle (case of exponential utility function) and right (case of step utility function) shows more influence with the H value (i.e. less parallelness in the level curves).

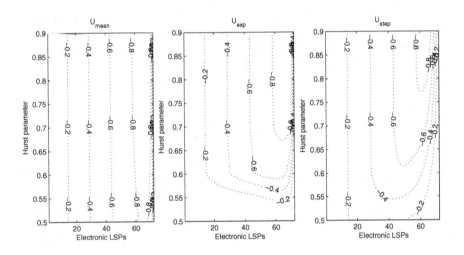

Fig. 6. Hurst parameter variation

4   Summary and conclusions

This work's main contribution is two-fold: First, it presents a novel methodology, based on the Bayesian decision theory, that helps multilayer-capable routers to take the decision of either optical or electronic switching of incoming LSPs. Such decision is made based on technical aspects such as QoS constrains and long-range dependence characteristics of the incoming traffic, but also considers the economic differences of optical and electrical switching. This way permits high flexibility to the network operator to trade off both economic and technical aspects.

The algorithm proposed is of low complexity, and can easily adapt to changing conditions: QoS guarantees, traffic profiles, economic conditions and network operator preferences.

This algorithm can be implemented in a per node basis by using local and independent parameters (e.g delay thresholds) in each node. However, in further extensions of this mechanism, the local parameters used in each node will be based on information about end-to-end delay through the whole network.

References

1. Sato, K., Yamanaka, N., Takigawa, Y., Koga, M., Okamoto, S., Shiomoto, K., Oki, E., Imajuku, W.: GMPLS-based photonic multilayer router (Hikari router) architecture: an overview of traffic engineering and signaling technology. IEEE Communications Magazine **40**(3) (2002) 96–101
2. Puype, B., Yan, Q., De Maesschalck, S., Colle, D., Pickavet, M., Demeester, P.: Optical cost metrics in multi-layer traffic engineering for IP-over-optical networks. In: Transparent Optical Networks, 2004. Proceedings of 2004 6th International Conference on. Volume 1. (2004) 75–80
3. Vigoureux, M., Berde, B., Andersson, L., Cinkler, T., Levrau, L., Ondata, M., Colle, D., Fernandez-Palacios, J., Jager, M.: Multilayer traffic engineering for GMPLS-enabled networks. IEEE Communications Magazine **43**(7) (2005) 44–50
4. French, S., Ríos Insúa, D.: Statistical decision theory. Oxford University Press Inc. (2000)
5. Norros, I.: On the use of fractional Brownian motion in the theory of connectionless networks. IEEE J. Selected Areas in Communications **13**(6) (1995) 953–962
6. Papagiannaki, K., Moon, S., Fraleigh, C., Thiran, P., Diot, C.: Measurement and analysis of single-hop delay on an ip backbone network. IEEE J. Selected Areas in Communications **21**(6) (2003) 908–921
7. Hernández, J.A., Phillips, I.W.: Weibull mixture model to characterise end-to-end Internet delay at coarse time-scales. IEE Proc. Communications **153**(2) (2005)
8. Shenker, S.: Fundamental design issues for the future internet. IEEE Journal on Selected Areas in Communications **13**(7) (1995) 1176–1188
9. Choi, Y.J., Bahk, S.: Qos scheduling for multimedia traffic in packet data cellular networks. In: IEEE International Conference on Communications, 2003. Volume 1. (2003) 358–362
10. ITU-T: ITU-T Recommendation Y.1541 - Network Performance Objectives for IP-Based Services (2003)
11. 3GPP: 3GPP Recommendation s.r0035-0 v1.0. - Quality of Service (2002)
12. ITU-T: ITU-T Recommendation G.107 : The E-model, a computational model for use in transmission plannings (2005)
13. Clegg, R.G.: Markov-modulated on/off processes for long-range dependent internet traffic. ArXiv Computer Science e-prints (2006)

Regenerator Placement with Guaranteed Connectivity in Optical Networks

Marcio S. Savasini[1], Paolo Monti[2], Marco Tacca[2],
Andrea Fumagalli[2], and Helio Waldman[1]

[1] Optical Networking Lab
Faculdade de Engenharia Eletrica e de Computacao
State University of Campinas, Campinas, SP, Brazil
{savasini, waldman}@decom.fee.unicamp.br
[2] Open Networking Advanced Research (OpNeAR) Lab
Erik Jonsson School of Engineering and Computer Science
The University of Texas at Dallas, Richardson, TX, USA
{paolo, mtacca, andreaf}@utdallas.edu

Abstract. The problem of minimizing the number of optical nodes with signal regeneration capability can be constrained to guarantee a desired degree of end-to-end connectivity in the all-optical transport network. The problem can be formulated using a k-connected, k-dominating node set, which is a known approach in mobile ad hoc wireless networks. This paper presents a preliminary study aimed at establishing whether efficient centralized solutions to this problem in optical networking ought to be investigated to improve the decentralized solutions already available for wireless networks.

Index Terms– Regenerator Placement, Connected Dominating Set, k-Connectivity, Greedy Algorithm, Optical Networks.

1 Introduction

Optical networks can provide transport capabilities to routers (and other electronic nodes) via the combination of optical circuits, or lightpaths [1], and optical signal regenerators. The former reduce the required number of total transponders in the network, while the latter regenerate the optical signal when it is weakened by transmission impairments [2]. If well designed, this approach may lead to a substantial cost reduction of the transport network [3, 4].

The transmission impairments of the optical medium limit the number of network elements that can be traversed by the optical signal without making use of regenerators [2], i.e., the maximum reach of the lightpath. In practice, one can define for each (optical) node the *transparency island* (TI) [5, 6] as the subset of nodes that can be reached all-optically — at a given minimum transmission rate — without a significant degradation of the signal quality, e.g., bit error rate. If the two end nodes that must be connected are not within each other's

I. Tomkos et al. (Eds.): ONDM 2007, LNCS 4534, pp. 438–447, 2007.

TI, the optical signal must be regenerated at some intermediate node(s). One type of regeneration is reamplification, reshaping and retiming (3R) [7]. This type of regeneration may take place in both the electronic (OEO conversion) or the optical domain [8]. Either solution adds to the overall network cost.

Assume that both the optical topology and nodes are given. Then a cost function to be minimized is the number of regenerators, which are required in the given optical topology to provide end-to-end connectivity. Some solutions addressing this problem can be found in the literature. For example, minimal cost placement algorithms [9] compute which subset of nodes must be provided with regeneration capability to guarantee connectivity against any single link failure in the network. Blocking probability (when attempting to create end-to-end connections) due to the lack of available regenerators can be minimized by special algorithms for placing regenerators [10–12]. The regenerator placement problem has some similarities with the wavelength converter placement problem and solutions devised to deal with the latter problem may be used to handle the former [13, 14]. None of these solutions, however, looks into the regenerator placement problem while guarantying an arbitrary degree of end-to-end connectivity in the optical network.

The preliminary study in this paper tackles the problem of minimizing the number of (optical) nodes that must be enabled with 3R capability, while guarantying a desired minimum degree of end-to-end (optical) connectivity. More formally, the 3R nodes must be placed to obtain a k-connected, k-dominating 3R-node set (k-CD3S), while minimizing the number of 3R-nodes in the set. With a guaranteed end-to-end k connectivity, the resulting transport network offers both known resilience benefits against 3R-node failures and known load balancing benefits when routing and setting up connections.

The problem of minimizing the size of k-CD3S is found to be the same problem already defined in mobile ad hoc wireless networks [15], whereby a subset of nodes is selected to form a routing backbone with a guaranteed degree of end-to-end connectivity. In the wireless network the k-connected, k-dominating set is found using a decentralized algorithm to cope with the mobility of the nodes. The k-CD3S minimization problem in optical networks, on the contrary, is solved in this paper by using a centralized algorithm that is expected to be more efficient, when compared to the decentralized one. A lower bound on the size of k-CD3S is provided, to assess the efficiency of both algorithms.

As discussed in the paper, it appears that centralized solutions to address the problem of minimizing k-CD3S are worth exploring as they yield results that are substantially closer to the lower bound, when compared to the decentralized solutions devised to solve the same problem in mobile ad hoc wireless networks.

2   Network Model and Problem Definition

This section contains the assumptions and properties of the (optical) networks under investigation, along with the problem definition.

A network with arbitrary physical topology is considered. Every node in the network functions as OXC. Only a subset of the nodes has regeneration capabilities, i.e., the 3R-nodes.

The physical topology is modeled as a graph $G(\mathcal{N}, \mathcal{A})$, where $\mathcal{N}$ is the set of nodes in the network and $\mathcal{A}$ is the set of directed (fiber) links connecting the nodes. Each node is uniquely identified, e.g., node i is denoted as N_i. The link connecting node N_i to N_v is denoted as $l_{(i,v)}$. Let $C_{(i,j)}^{(r)}$ be the set of all links and nodes that can be used to establish a lightpath[3] connecting node N_i to N_j with transmission rate r, without requiring 3R. Set $C_{(i,j)}^{(r)}$ defines a subgraph of $G(\mathcal{N}, \mathcal{A})$. Let $TI_i^{(r)}$ be the transparency island of node N_i, for lightpaths operating at transmission rate r. $TI_i^{(r)}$ is a subgraph of G, defined as

$$TI_i^{(r)} = \cup_{j \in \mathcal{N}} \left(C_{(i,j)}^{(r)} \right). \tag{1}$$

In summary, lightpaths originating at N_i, operating at transmission rate r, and not requiring 3R can only be simple paths in $TI_i^{(r)}$. Note that incoming links $l_{(j,i)}$ are not in $TI_i^{(r)} \forall r, j$, as lightpaths containing loops are not allowed.

A second graph, i.e., $G'^{(r)}(\mathcal{N}, \mathcal{A}'^{(r)})$, is used to represent the nodes' connectivity based on their own TI's. Let set $\mathcal{A}'^{(r)}$ be

$$\mathcal{A}'^{(r)} = \bigcup_{i \in \mathcal{N}} \bar{l}_{(i,j)}^{(r)}, \ \forall j \in TI_i^{(r)}. \tag{2}$$

In simple terms, if $l_{(i,j)}^{(r)} \in \mathcal{A}'^{(r)}$ it is possible to set up a lightpath from node N_i to N_j without requiring 3R. Notice that $\mathcal{A}'^{(r)}$ is a function of the employed transmission rate r. To simplify the notation in the remainder of the paper, it is assumed that all lightpaths operate at the same rate r, and index r is dropped, e.g., $G'^{(r)}(\mathcal{N}, \mathcal{A}'^{(r)})$ is simply denoted as $G'(\mathcal{N}, \mathcal{A}')$.

1-connectivity from node N_i to N_j in $G'(\mathcal{N}, \mathcal{A}')$ is defined as follows. Let $\mathcal{S}_{3R} \subseteq \mathcal{N}$ be the set of 3R-nodes. N_i is 1-connected to N_j if there exists a simple path [16] $p =< N_i, N_{a_1}, N_{a_2}, \ldots, N_{a_{m-1}}, N_j >$ in $G'(\mathcal{N}, \mathcal{A}')$, such that $(N_{a_1}, N_{a_2}, \ldots, N_{a_{m-1}}) \in \mathcal{S}_{3R}$. Note that intra TI ($m = 1$) connectivity does not require 3R-nodes. Inter TI connectivity requires that the lightpath signal be regenerated at selected 3R-nodes, which form a backbone for the optical signal to propagate and reach the end node. $G'(\mathcal{N}, \mathcal{A}')$ is 1-connected if all the pairs $(N_i, N_j), i \neq j \in \mathcal{N}$ are 1-connected.

N_i is k-connected to N_j if there exist k node-disjoint simple paths $p^{(v)} =< N_i, N_{a_1^{(v)}}, N_{a_2^{(v)}}, \ldots, N_{a_{m^{(v)}-1}^{(v)}}, N_j >, v = 1, 2, \ldots, k$ in $G'(\mathcal{N}, \mathcal{A}')$, such that $(N_{a_1^{(v)}}, N_{a_2^{(v)}}, \ldots, N_{a_{m^{(v)}-1}^{(v)}}) \in \mathcal{S}_{3R}, \forall v = 1, 2, \ldots, k$. $G'(\mathcal{N}, \mathcal{A}')$ is k-connected if all node pairs $(N_i, N_j), i \neq j \in \mathcal{N}$ are k-connected.

Set $\mathcal{S}_{3R}$ is a k-connected, k-dominating 3R-node set (k-CD3S) of $G'(\mathcal{N}, \mathcal{A}')$, if and only if the following constraints are met:

---

[3] It is assumed that if a lightpath can be established, then its performance in terms of bit error rate is satisfactory.

(a) k-dominating constraint: each node that is not in the 3R-node set must be intra TI connected to at least k 3R-nodes, i.e.,

$$\sum_{j \in \mathcal{S}_{3R}} \mathcal{L}_{(i,j)} \geq k, \ \forall i \in (\mathcal{N} \backslash \mathcal{S}_{3R}) \tag{3}$$

where $\mathcal{L}_{(i,j)}$ is a binary variable defined as:

$$\mathcal{L}_{(i,j)} = \begin{cases} 1 & \text{if } \bar{l}_{(i,j)} \in \mathcal{A}' \\ 0 & \text{otherwise} \end{cases} \tag{4}$$

(b) k-node connectivity constraint: subgraph $\tilde{G}(\mathcal{S}_{3R}, \tilde{\mathcal{A}})$, where

$$\tilde{\mathcal{A}} = \{\bar{l}_{i,j} \in \mathcal{A}' \ : \ i, j \in \mathcal{S}_{3R}\} \tag{5}$$

must be k-node connected. A graph is defined to be k-node connected if and only if the removal of any of its $k - 1$ nodes does not cause a partition [16].

A centralized algorithm able to find a 3R-node placement such that $\mathcal{S}_{3R}$ satisfies both constraints (a) and (b), while minimizing the number of 3R-nodes, is presented next.

3   A Centralized Algorithm to Compute k-CD3S

In this section a two-step approach, called Select-&-Prune (S-&-P), is presented to find a sub-optimal solution to the k-CD3S problem with minimum number of 3R-nodes. In step 1, an initial solution for the k-CD3S problem is found, by selecting a number of nodes that are potential candidates for the 3R-node set. In step 2, a greedy algorithm is applied to prune from the initial set of 3R-nodes as many nodes as possible without violating the k-connected, k-dominating constraints on k-CD3S. Table 1 contains a pseudo code description of the two sequential steps.

3.1   Step 1: Select

The objective of this step is to find an initial set of 3R-nodes that may be included in k-CD3S of $G'(\mathcal{N}, \mathcal{A}')$. This initial selection consists of two substeps. In the first substep, a set of 3R-nodes ($\mathcal{S}_{3R}$) if found solving a variation of the vertex cover problem. In the second substep, the just computed $\mathcal{S}_{3R}$ is checked to verify the k-node connectivity constraint defined in Section 2. If $\mathcal{S}_{3R}$ is k-connected, no further action is required in step 1. Conversely, if $\mathcal{S}_{3R}$ is not k-connected additional nodes are sequentially selected and added to $\mathcal{S}_{3R}$, until the k-node connectivity constraint is met. The two substeps are described next.

Substep 1.1: Vertex Covering : the objective of this step is to find a set $(S_{3R} \subseteq \mathcal{N})$, with minimum number of 3R-nodes, that is a solution for the following problem:

$$\min \sum_{i \in \mathcal{N}} r_i \qquad (6)$$

subject to:

$$\sum_{j, \mathcal{L}_{(i,j)} \neq 0} r_j \geq k, \quad \forall i \in \mathcal{N}; \qquad (7)$$

where r_i is a binary variable defined as:

$$r_i = \begin{cases} 1 & \text{if } N_i \in S_{3R} \\ 0 & \text{otherwise.} \end{cases} \qquad (8)$$

Note that constraint (7) is more stringent than the k-dominating constraint in (3). Constraint (7) requires all 3R-nodes $\in S_{3R}$ to have at least k 3R-node neighbors themselves. Since the objective of the S-&-P algorithm is to find k-CD3S, this additional requirement may increase the probability that the resulting S_{3R} is k-connected.

The size of S_{3R} computed so far represents a lower bound (LB) on the size of k-CD3S. This claim is based on the following simple observation. S_{3R} is by definition an optimal solution under constraint (7), i.e., (a) defined in Section 2. Then, the optimal solution under both constraints (a) and (b) defined in Section 2 must require the same number of 3R-nodes or more.

Substep 1.2: Connectivity Check : the objective of this step is to check the connectivity degree of S_{3R} and, if necessary, to select additional 3R-nodes for S_{3R} so that both constraints (a) and (b) defined in Section 2 are satisfied.

This step, first checks if set S_{3R} generated by the vertex covering set is k-connected. If S_{3R} is k-connected the algorithm stops and set S_{3R} is returned as the optimal solution of the minimum k-CD3S problem.

If set S_{3R} is not k-connected, a node is selected from $(\mathcal{N} \backslash S_{3R})$ and added to S_{3R}) as follows. For each source destination pair in $G'(\mathcal{N}, \mathcal{A}')$ a k node-disjoint shortest path [16], with weights 0 ($\forall N_i \in S_{3R}$), weight $2 \cdot |\mathcal{N}|$ ($\forall N_i \in (\mathcal{N} \backslash S_{3R})$) and weight 1 ($\forall l_{(i,j)} \in \mathcal{A}'$), is computed. This choice of weights forces the shortest path algorithm to compute paths that make use of already selected 3R-nodes as much as possible. All nodes $N_i \in (\mathcal{N} \backslash S_{3R})$ are then scored based on the number of times they are chosen for a shortest path. The node with the highest score — i.e., the node most used in connecting source destination pairs — is added to S_{3R}. This process is repeated until S_{3R} is k-connected.

3.2 Step 2: Prune

In this final step, the algorithm attempts to remove as many nodes from S_{3R} as possible without violating both constraints (a) and (b). Nodes in S_{3R} are first sorted randomly. In that order, an attempt to prune each node from $\mathcal{S}_{3R}$ is made as follows. If removing the node from $\mathcal{S}_{3R}$ does not violate both constraints (a) and (b) defined in Section 2, the node is removed permanently. Otherwise, the node is labeled permanently to remain in $\mathcal{S}_{3R}$. Once the pruning is complete, k-CD3S=$\mathcal{S}_{3R}$.

Table 1. Pseudo code of the Select-&-Prune algorithm

```
begin algorithm
 if (G'(N, A') k-connected) {
   S₃ᵣ = Vertex Covering()
   LB = |S₃ᵣ|
   if (S₃ᵣ k-connected) {              \\ -> Connectivity Check()
     return S₃ᵣ
   }
   else {
     while (S₃ᵣ not k-connected) {
       Run kNodeDisjointSP(Nᵢ, Nⱼ), ∀(i, j) ∈ N
       Score(Nᵢ) = number of SP's traversing i, ∀i ∈ N
       Nᵤ : [Score(Nᵤ)] ≥ [Score(Nᵢ)], ∀Nᵢ ∈ (N\S₃ᵣ)
       S₃ᵣ = S₃ᵣ ∪ Nᵤ
     }
   }
   while(all Nᵢ ∈ S₃ᵣ have been checked) {  \\ -> Pruning()
     Pick randomly Nᵢ
     S₃ᵣ = S₃ᵣ\Nᵢ
     if(S₃ᵣ not k-CD3S){
       S₃ᵣ = S₃ᵣ ∪ Nᵢ
     }
   }
   return S₃ᵣ
 }
 S₃ᵣ = ∅
 return S₃ᵣ
end algorithm
```

4   Simulation Results

This section presents a collection of simulation results obtained for the Select-&-Prune algorithm presented in Section 3. The results are compared against both the lower bound (LB) given in Section 3.1, and the solution found using the decentralized algorithm k-*coverage* proposed for mobile and ad hoc wireless networks [15]. Results for the k-coverage algorithm are obtained by first selecting

$\mathcal{S}_{3R} = \mathcal{N}$, and then pruning $\mathcal{S}_{3R}$ as described in Section 3.2. The comparison is carried out to assess whether there is a substantial advantage in devising centralized solutions for the problem at hand. The ILP formulation, presented in Section 3.1 is solved using LP Solve 5.5.0.10 [18].

A number of simulation experiments is performed using network topologies that have the same minimum and average nodal degree (2, 3 respectively) and varying number of nodes. Each network topology is randomly generated using the Doer and Leslie's formula [19]. The total number of experiments is chosen so that the presented average values have a confidence interval of 12% or better at 90% confidence level. The TI of each node is chosen based on the following simplistic assumption: any lightpath without regeneration can span at most $TI = 1, 2, 3$, and 4 physical links. The value of TI is varied to analyze its effect on the algorithms' efficiency. The algorithms' performance is assessed in terms of the total number of 3R-nodes that are necessary to obtain a k-connected and k-dominating 3R-backbone. This value is normalized to the total number of nodes, i.e., the 3R-percentage is defined as $\%3R = |\mathcal{S}_{3R}|/|\mathcal{N}|$.

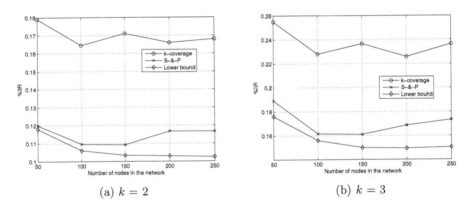

(a) $k = 2$ (b) $k = 3$

Fig. 1. Percentage of 3R-nodes (%3R) vs. the network size ($|\mathcal{N}|$), $TI = 3$.

Fig. 1 shows the value of $\%3R$ as a function of the number of nodes in the network, when $TI = 3$. Two degrees of connectivity are considered: $k = 2$ (Fig.1(a)) and $k = 3$ (Fig. 1(b)). In both of the cases, the centralized S-&-P algorithm yields better results, when compared to the decentralized k-coverage algorithm. In Fig. 1(a) the difference between the lower bound and the S-&-P algorithm is around 1%, while the k-coverage algorithm percentage averages almost 7% higher. Similar results are found when the required degree of connectivity increases to $k = 3$ (Fig. 1(b)). In general, the two figures confirm that the number of 3R nodes selected by the decentralized k-coverage algorithm is at least 40% higher when compared to the centralized S-&-P algorithm. This result suggests that a careful selection of the initial solution for k-CD3S may lead to a substantial cost reduction in terms of number of 3R-nodes required in the 3R-backbone.

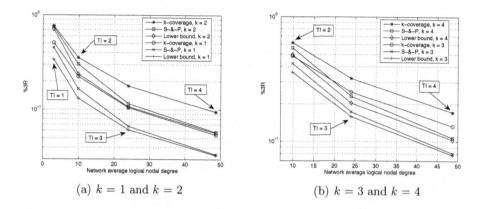

(a) $k = 1$ and $k = 2$ (b) $k = 3$ and $k = 4$

Fig. 2. Total number of 3R-nodes (%3R) vs. the average nodal degree in G'.

Fig. 2 shows the value of %$3R$ as a function of the average logical nodal degree in G'. The nodal degree is a function of the TI size, which is reported on the plots. Four degrees of connectivity are presented: $k = 1$ and $k = 2$ in Fig. 2(a), $k = 3$ and $k = 4$ in Fig. 2(b). Note that when $TI = 1$, $G' = G$. Then the average nodal degree for the two graphs is 3, making it impossible[4] to provide a guaranteed degree of connectivity $k=4$ in G'. As intuition suggests, the higher the value of the nodal degree the smaller is the number of required 3R-nodes. The centralized algorithm is consistently more efficient than the decentralized one. The figures also show that with approximately the same number of 3R-nodes the centralized algorithm guarantees a connectivity that is one degree higher than the one provided by the decentralized algorithm.

5   Conclusion

Based on the preliminary study presented in this paper, it appears that centralized algorithms may offer a substantial efficiency improvement (up to 40% in topologies with up to 250 nodes) over the already existing decentralized algorithms, when it comes to finding a sub-optimal solution to the problem of minimizing the k-connected, k-dominating node set. This result suggests that it may be worth investigating centralized algorithms in more details, as their improved efficiency translates directly into a reduction of the number of nodes with signal regeneration capability that are required in the all-optical transport network to guarantee a desired degree of end-to-end connectivity.

---

[4] All generated networks had at least one node with nodal degree = 2, making it impossible to provide also a guaranteed degree of connectivity $k=3$.

Acknowledgments This research was supported in part by NSF Grant No. CNS-043593, CAPES Process No. BEX4403/05-3, and the Italian Ministry of University (MIUR) (contract # RBNEO1KNFP).

References

1. I. Chlamtac, A. Ganz, and G. Karmi, "Lightpath communications: an approach to high-bandwidth optical WAN's," *IEEE Trans. Commun.* **40**, pp. 1171–1182, July 1992.
2. B. Ramamurthy, H. Feng, D. Datta, J. Heritage, and B. Mukerjee, "Transparent vs. opaque vs. translucent wavelength-routed optical networks," in *Optical Fiber Communication (OFC '99) Technical Digest, San Diego, CA, Feb. 1999,*
3. R. Ramaswami, "Optical networking technologies: What worked and what didn't," *IEEE Commun. Mag.* **44**, pp. 132–139, September 2006.
4. I. Cerutti and A. Fumagalli, "Traffic grooming in static wavelength division multiplexing networks," *IEEE Commun. Mag.* **43**, pp. 101–107, January 2005.
5. A. A. M. Saleh, "Islands of transparency - an emerging reality in multiwave optical networking," in *11th Annual Meeting IEEE Lasers and Electro-Optics Society,* 1998.
6. J. Strand and A. Chiu, "Impairments and other constraints on optical layer routing," RFC 4054, Internet Engineering Task Force, 2005.
7. R. Ramaswami and K. N. Sivarajan, *Optical Networks: a Practical Perspective,* Morgan Kaufmann Publishers, 2002.
8. H.-P. Nolting, "All-optical 3R-regeneration for photonic networks," in *ONDM 2003: Proceedings of the 7th IFIP Working Conference on Optical Network Design & Modelling,* 2003.
9. E. Yetginer and E. Karasan, "Regenerator placement and traffic engineering with restoration in GMPLS networks," *Photonic Network Communications* **2**, pp. 139–149, September 2003.
10. S.-W. Kim, S.-W. Seo, and S. C. Kim, "Regenerator placement algorithms for connection establishment in all-optical networks," in *IEEE Global Telecommunications Conference,* **2**, pp. 1205–1209, Nov. 2000.
11. X. Yang and B. Ramamurthy, "Sparse regeneration in a translucent WDM optical network," in *Proceedings of APOC 2001,* pp. 61–70, 2001.
12. G. Shen, W. D. Grover, T. Hiang Cheng, and S. K. Bose, "Sparse placement of electronic switching nodes for low blocking in translucent optical networks," *Journal of Optical Networking, vol. 1, Issue 12, p.424* **1**, Dec. 2002.
13. S. Subramaniam, M. Azizoglu, and A. Somani, "On optimal converter placement in wavelength-routed networks," *INFOCOM* **1**, pp. 500–507, Apr. 1997.
14. X. Chu, B. Li, and I. Chlamtac, "Wavelength converter placement under different RWA algorithms in wavelength-routed all-optical networks," *IEEE Trans. Commun.* **51**, pp. 607–617, Apr. 2003.
15. F. Dai and J. Wu, "On constructing k-connected k-dominating set in wireless ad hoc and sensor networks," *Journal of Parallel and Distributed Computing* **66**, pp. 947–958, July 2006.
16. T. H. Cormen, C. E. Leiserson, R. L. Rivest, and C. Stein, *Introduction to Algorithms (Second Edition),* MIT Press, Cambridge, MA, 2001.
17. M. R. Garey and D. S. Johnson, *Computers and Intractability: A Guide to the Theory of NP-Completeness,* W. H. Freeman & Co., New York, NY, USA, 1979.

18. "Lp solve 5.5.5.10." http://lpsolve.sourceforge.net/5.5/.
19. M. Doar and I. M. Leslie, "How bad is naive multicast routing?," in *INFOCOM*, **1**, pp. 82–89, March/April 1993.

Optimal routing for minimum wavelength requirements in end-to-end optical burst switching rings

Reinaldo Vallejos[1], Alejandra Zapata[1], Víctor Albornoz[2]

[1] Telematics Group, Electronic Engineering Department, Universidad Técnica Federico
Santa María, Av. España 1680, Valparaíso, Chile
[2] Industrial Engineering Department, Universidad Técnica Federico Santa María, Av. Santa
María 6400, Santiago, Chile
{reinaldo.vallejos, alejandra.zapata, victor.albornoz}@usm.cl

Abstract. A novel routing and link dimensioning optimisation method which minimises the total wavelength requirements of dynamic optical WDM rings - the most popular topology in metropolitan networks- is proposed. The method finds the solution (set of routes) of minimum cost by solving an integer linear optimisation problem. Contrary to the common belief, results show that the optimal routes found by the proposed method are not necessarily balanced and that significant wavelength savings are achieved compared to the usual balanced-load routing approach in rings. This makes the proposed method the best choice for implementation in future dynamic WDM ring networks.

Keywords: optical burst switching networks, rings, routing, dimensioning, wavelength requirements.

1 Introduction

One promising dynamic WDM network architecture, given the current unavailability of optical buffers and processors (which hampers the implementation of the already successful packet-switching technique very much used in Internet), is Optical Burst Switching (OBS) [1-4]. The operation principle of OBS combines the dynamic reservation/release of optical network resources with low speed requirements for the electronic layer (in charge of buffering and processing). To do so, OBS networks electronically aggregate packets at the edge of the network into a burst. Once a burst is ready to be transmitted, corresponding optical network resources are requested. The allocation of optical network resources can be implemented in a hop-by-hop [1,2] or in an end-to-end [3,4] basis. This paper focuses in the latter, as it has been shown to significantly reduce the burst loss rate with respect to a hop-by-hop OBS network of equivalent complexity [5]. Thus, only after all the resources from source to destination have been reserved the data burst is released into the optical core.

In an end-to-end OBS network each node consists of an optical switch locally connected to an edge router (see Figure 1). Every edge router, where burst aggregation is carried out, is equipped with one buffer per destination. Incoming

I. Tomkos et al. (Eds.): ONDM 2007, LNCS 4534, pp. 448–457, 2007.

packets are classified according to destination and stored in the corresponding buffer. During the burst aggregation process the resource allocation request state is triggered once a pre-defined condition is met (e.g., latency or burst size), and a request is sent through the network to reserve resources for the burst. Resource reservation can be done in a centralised or distributed way. In each case, requests are processed according to the implemented resource allocation algorithm. If such algorithm finds an available route, the network is configured to establish the end-to-end route and an acknowledgement is sent to the corresponding source node. Otherwise, the request is dropped with a NACK message sent to the source node which blocks the burst from entering the network. In this paper an end-to-end OBS network with full wavelength conversion capability is considered, as such feature is key for a dynamic network to achieve significant wavelength savings with respect to the static approach [6].

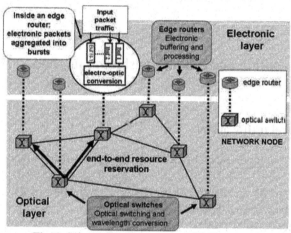

Fig. 1. Schematic of an end-to-end OBS network

Given that wavelengths are costly resources, the efficient dimensioning of an end-to-end OBS network to achieve a target blocking probability is of fundamental importance. The dimensioning of a network is strongly dependant on the routing algorithm, whose aim is to determine the routes to be used by each node pair to transmit information such that minimum resources are required. A common near-optimal routing heuristic in static WDM networks (where a route for each node pair is permanently established) has been to utilise the shortest paths whilst minimising the number of wavelengths in the most loaded link, see for example [7]. This approach (used in rings, for example, in [5, 8]) leads to routes that balance the traffic load throughout the network. Later, in [9], it was shown that using these routes in the dynamic case results in significant lower blocking probability than arbitrarily selecting any shortest path for each node pair. Consequently, this balanced-load routing approach has been commonly used in dynamic networks, see for example [5, 10,11]. However, by maintaining the links uniformly loaded, the statistical multiplexing of connections per link is not fully exploited.

In this paper we propose a novel method for optimal routing and dimensioning of optical network rings, the most popular topology in metropolitan networks [12]. The proposed method determines - by solving an integer linear optimisation problem in an off-line manner- the set of routes (one per node pair) which minimises the total wavelength requirements whilst satisfying a given blocking probability. The resulting routes are such that (contrary to the common belief) the traffic load is not necessarily balanced among the network links and routes might be longer than the shortest paths (as long as the wavelength savings due to the statistical multiplexing, because of the traffic load concentration, are higher than the additional wavelengths required by the extra hops of possibly longer routes).

For dynamic ring operation, the obtained routes are stored in a routing table and looked for when connections must be established on demand (that is, a fixed routing dynamic algorithm is used).

The remaining of this paper is as follows: section 2 describes the optimisation method; section 3 presents the numerical results whilst section 4 summarises the paper.

2   Optimal routing and link dimensioning method

2.1. Network and traffic model

The network is represented by a directed graph $G=(N,L)$ where N is the set of network nodes and L the set of unidirectional links (adjacent nodes are assumed connected by two fibres, one per direction). The cardinality of the sets N and L are denoted by N and L, respectively.

The burst aggregation process transforms the original packet traffic into a burst traffic which can be described by a source which switches its level of activity between two states: ON (burst transmission) and OFF (time between the end of the transmission of a burst and the beginning of the following one). During the ON period, the source transmits at the maximum bit rate (corresponding to the transmission rate of one wavelength). During the OFF period, the source refrains from transmitting data. This behaviour has been modelled by an ON-OFF alternating renewal process, for example in [13-15], and it is also used in this paper to characterise the burst traffic generated at the edge nodes.

In the network there are $N(N-1)$ different connections (source-destination pairs). In this paper, each connection is mapped to a number –denoted by c- between 1 and $N(N-1)$. Each connection c is associated to a route, r_c, determined by a fixed routing algorithm and established only during the ON period of connection c.

According to the usual characterization of an ON-OFF traffic source [16], the demand of connection c is defined by two parameters: ***mean OFF period duration*** (denoted by t_{OFF}) and ***mean ON period duration*** (denoted by t_{ON}), equal for all connections (homogeneous traffic). The mean ON period includes burst transmission as well as the time required to reserve/release resources for the burst. The latter depends on the reservation mechanism used, which is out of the scope of this paper. According to the alternating renewal process theory [17], the traffic load offered by one connection, ρ, is given by $t_{ON}/(t_{ON}+t_{OFF})$. The traffic load can be also though as the stationary state probability that a given connection is in ON state.

2.2. Routing and link dimensioning optimal method

In a network ring a routing algorithm must determine, for each node pair, whether the clockwise or the counter-clockwise route is used. Given a specific routing algorithm (i.e. the route for each node pair has been determined), the network cost C (in terms of number of wavelengths) is given by the sum of the wavelength requirements of all the network links. That is:

$$C = \sum_{l \in L} W_l \qquad (1)$$

where W_l denotes the number of wavelengths required in each link $l \in L$ such that a target blocking probability is guaranteed (in an end-to-end OBS network bursts are blocked from entering the network due to unavailability of network resources as opposed to the conventional hop-by-hop OBS networks where bursts are lost at any point along the path due to wavelength contention).

2.2.1. Evaluation of W_l

The evaluation of W_l is divided in two steps: firstly, the target value of the blocking of each network link is determined (Eqs. (2)-(4) in the following); then, this target value (obtained in Eq. (4)) is used to obtain the value of W_l from an analytical expression that relates W_l with the link blocking probability (Eqs. (5)-(6) in the following).

Target value for the link blocking, B_l
The number of wavelengths that the link l must be equipped with, W_l, depends on the level of acceptable blocking probability for such link, B_l (for example, in the extreme case of $B_l=0$, W_l must be equal to the number of connections using the link l, N_l). B_l can be easily determined from the value of the blocking specified for the network connections according to the following expression (assuming link blocking independence):

$$B_c = 1 - \prod_{\forall l \in r_c} (1 - B_l) \qquad (2)$$

where B_c is the blocking probability of connection c and l denotes a link in the route r_c associated to connection c.
Let H_c be the length (in number of hops) of the route r_c. There are many values for each factor (of the form $(1-B_l)$, $\forall l \in L$, in the right side of Eq. (2)) which satisfy the equality of Eq. (2), including the one where all the H_c factors of Eq. (2) have the same value. To facilitate analytical treatment of Eq. (2) without modifying the target of the dimensioning process (to guarantee that the blocking B_c, $1 \le c \le N(N-1)$, does not exceed a given value B) the same value of B_l for all H_c links in the route r_c is used. This leads to the following expression:

$$1-(1-B_l)^{H_c} \leq B ; \qquad 1 \leq c \leq N(N\text{-}1) \qquad (3)$$

where H_c is the length (in number of hops) of the route r_c.

Because the different connections using a given link l might have different route lengths, to guarantee that all connections (even the longest) achieve a blocking probability lower than B, the value of H_c corresponding to the longest route using link l is used in Eq. (3). Thus, the link l must be dimensioned so the following condition is met:

$$B_l = 1 - \sqrt[\hat{H}_l]{(1-B)} ; \qquad \forall\, l \in L \qquad (4)$$

where $\hat{H}_l$ corresponds to the length (in number of hops) of the longest route using link l.

Evaluation of W_l

Eq.(4) provides a target value for the link blocking probability, which can be expressed as the probability of having more than W_l (W_l: capacity of link l) connections in ON state. Given that the probability of a connection being in ON state corresponds to a Bernoulli random variable with parameter ρ, the probability of having n connections in ON state corresponds to a Binomial random variable with parameters N_l (number of connections whose routes use link l) and ρ (probability of a connection being in ON state). Thus, the link blocking probability can be estimated from:

$$B_l = 1 - \sum_{n=0}^{W_l} \binom{N_l}{n} \rho^n (1-\rho)^{N_l - n} \qquad (5)$$

By combining Eqs. (4) and (5), W_l is determined as the minimum number that satisfies:

$$1 - \sum_{n=0}^{W_l} \binom{N_l}{n} \rho^n (1-\rho)^{N_l - n} \leq 1 - \sqrt[\hat{H}_l]{(1-B)} \qquad (6)$$

where the left side corresponds to the estimation of the actual link blocking as a function of W_l whilst the right side corresponds to the target value of the link blocking.

2.2.2. Integer linear optimisation model

The proposed optimisation method (for minimum wavelength requirements) operates as follows:

1. Determine the set S of all possible different routes, considering all node pairs. For the ring topology there are only two different routes per node pair; thus, the cardinality of S is equal to $2N(N\text{-}1)$.
2. Given the set S, an integer programming model is used to determine an optimal route (in the sense that the sum of the number of wavelengths required at each link

in the network is minimised) for each node pair. The integer programming model utilises the following notation:

Indexes and Parameters. Let

- i and j be the indexes for nodes in $\mathcal{N}$,
- l be the index for a link in $\mathcal{L}$,
- $R_{i,j,k}$ be the k-th route connecting the source node i to the destination node j . If $k=0$, the counter-clockwise route is used; if $k=1$ the clockwise route is selected.
- $\delta_{i,j,k,l}$ be the input parameter that takes the value 1 if the k-th route ($0 \le k \le 1$) connecting the source node i to the destination node j ($i \ne j$) uses link l; and 0 otherwise.

Decision variables. Let

- $\alpha_{i,j,k}$ be a binary decision variable that takes the value 1 if the k-th route ($0 \le k \le 1$) connecting the node i to node j ($i \ne j$) is used by the routing algorithm; and 0 otherwise
- N_l be an integer decision variable that denotes the number of routes that use link l, given by Eq. (10) below.

The integer programming model can be described as follows:

$$\text{Min} \quad C - \sum_{l \in L} W_l \tag{7}$$

subject to:

$$\alpha_{i,j,k} \in \{0,1\} \qquad ; \text{ for all } i \in \mathcal{N}, j \in \mathcal{N}, k \in \{0,1\} \tag{8}$$

$$\sum_{k=0}^{1} \alpha_{i,j,k} = 1 \qquad ; \text{ for all } i \in \mathcal{N} \text{ and } j \in \mathcal{N} \ (i \ne j) \tag{9}$$

$$N_l = \sum_{\forall i \ne j} \sum_{k=0}^{1} \delta_{i,j,k,l} \alpha_{i,j,k} \qquad ; \text{ for all } l \in L \tag{10}$$

Eq. (7) is evaluated using the values of W_l obtained from the routing algorithm and Eq. (6). Constraint (8) imposes the 0-1 value of variables $\alpha_{i,j,k}$ (as stated in the definition of this variable); constraint (9) imposes that only one route connecting nodes i and j must be used and constraint (10) determines the value of the integer variable N_l.

3   Numerical Results

The integer programming model (7)-(10) was solved on a PC Pentium IV (3192 MHz. and 1GB RAM). The algebraic modelling language AMPL was used, with CPLEX 10.0 as solver [18].

The optimal routing and dimensioning method described in section 2 was applied to rings of different sizes under different values of traffic load (ρ) with a target B_l of 10^{-6}. The wavelength requirements were compared to those obtained with Eq. (6) when applying the balanced-load routing algorithm proposed in [8]. The algorithm in [8] determines the routes which minimise the wavelength requirements of WDM networks whilst balancing the traffic load (which has been the common approach for network dimensioning).

Table 1 shows the total wavelength requirements obtained with the method proposed in this paper and the balancing load method of [8] for rings from 6 to 16 nodes (in the table, P.M. and B.L. stand for "proposed method" and "balanced load", respectively) for different values of traffic load. Fig. 1 shows graphically these results for rings of 6, 8, 12 and 16 nodes. The cases where the proposed method requires a number of wavelengths lower than the balanced-load approach have been highlighted in bold in Table 1. In the remaining entries in Table 1 the proposed optimisation method achieves the same cost as the balanced-load approach.

Table 1. Wavelength requirements for rings of 6-16 nodes for the proposed method and the balanced-load approach

ρ	N=6 PM/BL	N=7 PM/BL	N=8 PM/BL	N=9 PM/BL	N=10 PM/BL	N=11 PM/BL	N=12 PM/BL	N=13 PM/BL	N=14 PM/BL	N=15 PM/BL	N=16 PM/BL
0.1	**48/54**	**69/70**	**92/96**	**117/126**	**156/160**	**187/198**	**228/234**	**273/286**	**321/322**	360/360	416/416
0.2	54/54	**82/84**	**116/128**	**153/162**	**200/210**	**250/264**	**300/306**	364/364	434/434	510/510	**592/608**
0.3	54/54	84/84	**127/128**	**171/180**	**227/230**	286/286	**358/366**	**430/442**	518/518	**627/630**	**722/736**
0.4	54/54	84/84	128/128	180/180	**244/250**	**314/330**	**396/402**	**482/494**	588/588	690/690	**816/832**
0.5	54/54	84/84	128/128	180/180	250/250	**329/330**	**420/432**	520/520	**634/644**	**768/780**	**912/928**
0.6	54/54	84/84	128/128	180/180	250/250	330/330	432/432	546/546	**672/686**	810/810	**976/992**
0.7	54/54	84/84	128/128	180/180	250/250	330/330	432/432	546/546	686/686	840/840	**1020/1024**
0.8	54/54	84/84	128/128	180/180	250/250	330/330	432/432	546/546	686/686	840/840	1024/1024
0.9	54/54	84/84	128/128	180/180	250/250	330/330	432/432	546/546	686/686	840/840	1024/1024

It can be seen that, depending on the value of the traffic load, the optimisation method proposed in this paper requires a lower or equal number of wavelengths than the usual approach of determining the routes by means of load balancing. Results show that the range of traffic load values (in steps of 0.1) at which the routing and dimensioning method proposed here achieves wavelength savings with respect to the balanced approach increases with the ring size (in number of nodes): for the 6-node ring a saving of 6 wavelengths was achieved for only one value of the traffic load (0.1) whilst for the 16-node ring savings of up to 16 wavelengths were observed in the range [0.1- 0.7]. As most networks currently operate at low loads (typically<0.3 [19, 20]), the proposed method will ensure benefits where it is needed most.

It can also be seen that the load-balanced approach is optimal for small-size rings (<8 nodes) in a wide range of traffic loads as well as for high loads (> 0.7) in all the studied rings.

As a way of example, Fig. 2 shows the routing matrices of the proposed method (upper left) and the balanced-load approach (upper right) for the 8-node ring with a traffic load of 0.1. In each matrix the element (i,j) corresponds to the parameter k defined in section 2.2. The elements in the matrix of the proposed method which are different from the balanced-load approach are highlighted in bold.

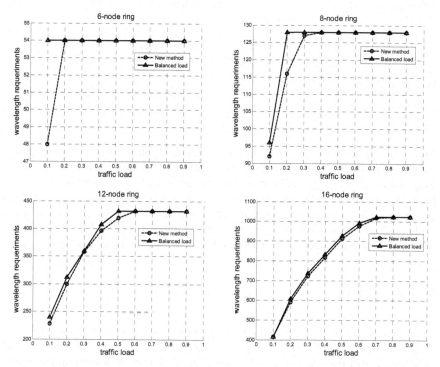

Fig. 1. Wavelength requirements for rings of 6 (upper left), 8 (upper right), 12 (lower left) and 16 (lower right) nodes using balanced-load routing and the method proposed in this paper for ρ=0.1

	1	2	3	4	5	6	7	8
1	-	1	1	1	1	1	1	0
2	0	-	1	1	1	1	1	0
3	1	0	-	1	1	1	1	1
4	1	0	0	-	1	1	1	1
5	1	1	0	0	-	1	1	1
6	1	1	1	1	0	-	1	1
7	1	1	1	1	0	0	-	1
8	1	1	1	1	1	0	0	-

	1	2	3	4	5	6	7	8
1	-	1	1	1	0	0	0	0
2	0	-	1	1	1	0	0	0
3	0	0	-	1	1	1	0	0
4	0	0	0	-	1	1	1	0
5	1	0	0	0	-	1	1	1
6	1	1	0	0	0	-	1	1
7	1	1	1	0	0	0	-	1
8	1	1	1	1	0	0	0	-

Fig.2. Routing matrix (1 denotes the route in clockwise direction; 0 in counter-clockwise direction) for the proposed method (left) and the balanced-load approach (right) for a 8-node ring with a traffic load equal to 0.1

Table 2 shows, for the same case (8-node ring, ρ=0.1), the number of connections per link (N_l) obtained by applying the optimisation method (whereas in the load-balanced approach, N_l =8 for each link).

Table 2. Value of N_l in a 8-node ring. Routing determined by the optimisation method for $\rho=0.1$

	Clockwise links *("i->j" denotes the link from node i to node j)*							
	1->2	2->3	3->4	4->5	5->6	6->7	7->8	8->1
N_l	17	17	18	17	17	18	17	17
	Counter-clockwise links *("i->j" denotes the link from node i to node j)*							
	2->1	3->2	4->3	5->4	6->5	7->6	8->7	1->8
N_l	2	2	3	2	2	3	2	2

From Fig. 2 and Table 2 it can be observed that the proposed method concentrated as many connections as possible in the same link (by using longer routes if necessary) as long as the statistical gain of doing so was higher than the extra wavelengths required by the longer routes. See, for example, the route connecting the node 1 to the node 6 (from Fig.2): whilst under a balanced-load approach the shortest route is used (i.e. the route made of the nodes 1-8-7-6) in the optimal method proposed here a longer route (made of the nodes 1-2-3-4-5-6) is preferable as the statistical gain is higher.

4 Conclusions

In this paper a new routing and link dimensioning optimisation method was proposed. The method aims to minimise the wavelength requirements of dynamic end-to-end OBS ring networks by solving a linear integer optimisation model for the routing problem. Unlike previous approaches, the method does not necessarily balance the traffic load. In fact, obtained results show that in some cases the method does exactly the opposite: it concentrates the load (by using routes longer than the shortest paths if necessary) as long as the additional number of wavelengths due to the longer paths is lower than the number of wavelengths saved due to the higher statistical gain.

The method can be implemented in dynamic optical rings by storing the optimal set of routes (obtained in an off-line manner) in a routing table from which the route information for a specific node pair can be retrieved on demand.

The proposed method was applied to ring networks from 6 to 16 nodes. It was found that, compared to the balanced approach, the proposed method could achieve wavelength savings of up to 16 wavelengths and that the wavelength savings increased with the ring size (in number of nodes). It was also observed that the balanced-load approach is optimal for small-size rings (<8 nodes) in a wide range of traffic loads as well as for high loads (>0.7) in all the studied rings.

These results show that the usual approach of balancing the load in rings does not always yield the optimal results in terms of wavelength requirements and that by applying a selective unbalancing of load wavelength savings can be achieved.

Acknowledgements. Financial support from Fondecyt Project #1050361 (Chilean Government) and USM Projects 23.07.27, 23.07.28 and 28.06.21 is gratefully acknowledged. We would also like to thank Marco Tarifeño for collecting and formatting the results of the optimisation program.

References

1. Qiao, C., Yoo, M.: Optical Burst Switching (OBS) – a new paradigm for an optical internet. Journal of High Speed Networks, Vol.8 (1999) 69-84
2. Turner, J.: Terabit burst switching. Journal of High Speed Networks, Vol.8 (1999) 3-16
3. Düser, M., Bayvel, P.: Analysis of a dynamically wavelength-routed optical burst switched network architecture. J. of Lightwave Technology, Vol.4 (2002) 574-585
4. Arakawa, S., Miyamoto, K., Murata, M., Miyahara, H.: Delay analyses of wavelength reservation methods for high-speed burst transfer in photonic networks. In Proc. International Technical Conference on Circuits and Systems, Computers and Comms. (ITC-CSCC99) July 1999
5. Zapata, A., de Miguel, I., Dueser, M., Spencer, J., Bayvel, P., Breuer, D., Hanik, N., Gladish, A.: Next Generation 100-Gigabit Metro Ethernet (100GbME) using multiwavelength optical rings. IEEE/OSA Journal of Lightwave Technology, special issue on Metro & Access Networks, Vol. 22 (11), (2004), 2420-2434
6. Zapata A., Bayvel P.: Optimality of resource allocation algorithms in dynamic WDM networks. In Proceedings of 10th European Conference on Network and Optical Communications, NOC 2005, London, UK, 5-7 July 2005, pp.131-138
7. Baroni, S.; Bayvel, P.: Wavelength requirements in arbitrarily connected wavelength-routed optical networks. IEEE/OSA Journal of Lightwave Technology, Vol. 5 (2), (1997), 242-251
8. Hunter, D.; Marcenac, D.: Optimal mesh routing in four-fibre WDM rings. Electronic Letters, Vol. 34 (8), (1998), 796-797
9. Van Parys, W.; Van Caenegem, B.; Demeester, P.: Reduction of blocking in arbitrary meshed WDM networks through a biased routing approach. Proceedings Optical Fiber Communicactions Conference, OFC'98, San José, CA, USA, page 94
10. Zapata, A.; Bayvel, P.: Improving the scalability of lightpath assignment algorithms in dynamic networks. Proceedings 31st European Conference on Optical Communications, ECOC 2005, Glasgow, Scotland, September 2005
11. Späth, J.: Dynamic routing and resource allocation in WDM transport networks. Computer Networks 32 (2000), 519-538
12. Ramaswami, R.: Optical networking technologies: what worked and what didn't. IEEE Communications Magazine, v.44 (9), 132-139, September 2006
13. Choi, J. Y. , Vu, H.L., Cameron, C. W., Zukerman, M., Kang, M.: The effect of burst assembly on performance of optical burst switched networks. Lecture Notes on Computer Science, Vol.3090 (2004) 729-739
14. Tancevski, L., Bononi, A., Rusch, L. A.: Output power and SNR swings in cascades of EDFAs for circuit- and packet-switched optical networks. Journal of Lightwave Technology, Vol.17 (5) (1999) 733-742
15. Zukerman, M., Wong, E., Rosberg, Z., Lee, G., Lu, H.V.: On teletraffic applications to OBS. IEEE Communications Letters, Vol.8 (2) (2004) 731-733
16. Adas A: Traffic models in broadband networks, IEEE Communications Magazine, vol. 35, no. 7, (July 1997), pp. 82-89
17. Ross S.: Introduction to probability models. 6th Edition Academic Press, 1997, (Eq. 7.15)
18. Fourer, R.; Gay, D.M.; Kernigham, B.W.: AMPL: a modeling language for mathematical programming. 2nd ed. The Scientific Press, USA, 2000.
19. Odlyzko, A.: Data networks are lightly utilized, and will stay that way. Review of Network Economics, Vol. 2, (2003), 210-237
20. Bhattacharyya, S.; Diot, C.; Jetcheva, J.; Taft, N.: POP-Level and access-link-level traffic dynamics in a Tier-1 POP. ACM SIGCOMM Internet Measurement Workshop, Vol. 1, 39-53, San Francisco, CA, November 2001

Author Index

Printing: Mercedes-Druck, Berlin
Binding: Stein+Lehmann, Berlin

Lecture Notes in Computer Science

For information about Vols. 1–4397

please contact your bookseller or Springer

Vol. 4451: T.S. Huang, A. Nijholt, M. Pantic, A. Pentland (Eds.), Artifical Intelligence for Human Computing. XVI, 359 pages. 2007. (Sublibrary LNAI).

Vol. 4450: T. Okamoto, X. Wang (Eds.), Public Key Cryptography – PKC 2007. XIII, 491 pages. 2007.

Vol. 4448: M. Giacobini et al. (Ed.), Applications of Evolutionary Computing. XXIII, 755 pages. 2007.

Vol. 4447: E. Marchiori, J.H. Moore, J.C. Rajapakse (Eds.), Evolutionary Computation,Machine Learning and Data Mining in Bioinformatics. XI, 302 pages. 2007.

Vol. 4446: C. Cotta, J. van Hemert (Eds.), Evolutionary Computation in Combinatorial Optimization. XII, 241 pages. 2007.

Vol. 4445: M. Ebner, M. O'Neill, A. Ekárt, L. Vanneschi, A.I. Esparcia-Alcázar (Eds.), Genetic Programming. XI, 382 pages. 2007.

Vol. 4444: T. Reps, M. Sagiv, J. Bauer (Eds.), Program Analysis and Compilation, Theory and Practice. X, 361 pages. 2007.

Vol. 4443: R. Kotagiri, P.R. Krishna, M. Mohania, E. Nantajeewarawat (Eds.), Advances in Databases: Concepts, Systems and Applications. XXI, 1126 pages. 2007.

Vol. 4440: B. Liblit, Cooperative Bug Isolation. XV, 101 pages. 2007.

Vol. 4439: W. Abramowicz (Ed.), Business Information Systems. XV, 654 pages. 2007.

Vol. 4438: L. Maicher, A. Sigel, L.M. Garshol (Eds.), Leveraging the Semantics of Topic Maps. X, 257 pages. 2007. (Sublibrary LNAI).

Vol. 4433: E. Şahin, W.M. Spears, A.F.T. Winfield (Eds.), Swarm Robotics. XII, 221 pages. 2007.

Vol. 4432: B. Beliczynski, A. Dzielinski, M. Iwanowski, B. Ribeiro (Eds.), Adaptive and Natural Computing Algorithms, Part II. XXVI, 761 pages. 2007.

Vol. 4431: B. Beliczynski, A. Dzielinski, M. Iwanowski, B. Ribeiro (Eds.), Adaptive and Natural Computing Algorithms, Part I. XXV, 851 pages. 2007.

Vol. 4430: C.C. Yang, D. Zeng, M. Chau, K. Chang, Q. Yang, X. Cheng, J. Wang, F.-Y. Wang, H. Chen (Eds.), Intelligence and Security Informatics. XII, 330 pages. 2007.

Vol. 4429: R. Lu, J.H. Siekmann, C. Ullrich (Eds.), Cognitive Systems. X, 161 pages. 2007. (Sublibrary LNAI).

Vol. 4427: S. Uhlig, K. Papagiannaki, O. Bonaventure (Eds.), Passive and Active Network Measurement. XI, 274 pages. 2007.

Vol. 4426: Z.-H. Zhou, H. Li, Q. Yang (Eds.), Advances in Knowledge Discovery and Data Mining. XXV, 1161 pages. 2007. (Sublibrary LNAI).

Vol. 4425: G. Amati, C. Carpineto, G. Romano (Eds.), Advances in Information Retrieval. XIX, 759 pages. 2007.

Vol. 4424: O. Grumberg, M. Huth (Eds.), Tools and Algorithms for the Construction and Analysis of Systems. XX, 738 pages. 2007.

Vol. 4423: H. Seidl (Ed.), Foundations of Software Science and Computational Structures. XVI, 379 pages. 2007.

Vol. 4422: M.B. Dwyer, A. Lopes (Eds.), Fundamental Approaches to Software Engineering. XV, 440 pages. 2007.

Vol. 4421: R. De Nicola (Ed.), Programming Languages and Systems. XVII, 538 pages. 2007.

Vol. 4420: S. Krishnamurthi, M. Odersky (Eds.), Compiler Construction. XIV, 233 pages. 2007.

Vol. 4419: P.C. Diniz, E. Marques, K. Bertels, M.M. Fernandes, J.M.P. Cardoso (Eds.), Reconfigurable Computing: Architectures, Tools and Applications. XIV, 391 pages. 2007.

Vol. 4418: A. Gagalowicz, W. Philips (Eds.), Computer Vision/Computer Graphics Collaboration Techniques. XV, 620 pages. 2007.

Vol. 4416: A. Bemporad, A. Bicchi, G. Buttazzo (Eds.), Hybrid Systems: Computation and Control. XVII, 797 pages. 2007.

Vol. 4415: P. Lukowicz, L. Thiele, G. Tröster (Eds.), Architecture of Computing Systems - ARCS 2007. X, 297 pages. 2007.

Vol. 4414: S. Hochreiter, R. Wagner (Eds.), Bioinformatics Research and Development. XVI, 482 pages. 2007. (Sublibrary LNBI).

Vol. 4412: F. Stajano, H.J. Kim, J.-S. Chae, S.-D. Kim (Eds.), Ubiquitous Convergence Technology. XI, 302 pages. 2007.

Vol. 4411: R.H. Bordini, M. Dastani, J. Dix, A.E.F. Seghrouchni (Eds.), Programming Multi-Agent Systems. XIV, 249 pages. 2007. (Sublibrary LNAI).

Vol. 4410: A. Branco (Ed.), Anaphora: Analysis, Algorithms and Applications. X, 191 pages. 2007. (Sublibrary LNAI).

Vol. 4409: J.L. Fiadeiro, P.-Y. Schobbens (Eds.), Recent Trends in Algebraic Development Techniques. VII, 171 pages. 2007.

Vol. 4407: G. Puebla (Ed.), Logic-Based Program Synthesis and Transformation. VIII, 237 pages. 2007.

Vol. 4406: W. De Meuter (Ed.), Advances in Smalltalk. VII, 157 pages. 2007.

Vol. 4405: L. Padgham, F. Zambonelli (Eds.), Agent-Oriented Software Engineering VII. XII, 225 pages. 2007.

Vol. 4403: S. Obayashi, K. Deb, C. Poloni, T. Hiroyasu, T. Murata (Eds.), Evolutionary Multi-Criterion Optimization. XIX, 954 pages. 2007.

Vol. 4401: N. Guelfi, D. Buchs (Eds.), Rapid Integration of Software Engineering Techniques. IX, 177 pages. 2007.

Vol. 4400: J.F. Peters, A. Skowron, V.W. Marek, E. Orłowska, R. Słowiński, W. Ziarko (Eds.), Transactions on Rough Sets VII, Part II. X, 381 pages. 2007.

Vol. 4399: T. Kovacs, X. Llorà, K. Takadama, P.L. Lanzi, W. Stolzmann, S.W. Wilson (Eds.), Learning Classifier Systems. XII, 345 pages. 2007. (Sublibrary LNAI).

Vol. 4398: S. Marchand-Maillet, E. Bruno, A. Nürnberger, M. Detyniecki (Eds.), Adaptive Multimedia Retrieval: User, Context, and Feedback. XI, 269 pages. 2007.